Functional Ingredients from Food Waste and By-Products: Processing Technologies, Functional Characteristics and Value-Added Applications

Functional Ingredients from Food Waste and By-Products: Processing Technologies, Functional Characteristics and Value-Added Applications

Guest Editors

Lucía Seguí
Cristina Barrera

MDPI

Basel • Beijing • Wuhan • Barcelona • Belgrade • Novi Sad • Cluj • Manchester

Guest Editors
Lucía Seguí
Institute of Food
Engineering-FoodUPV
Universitat Politècnica de
València
Valencia
Spain

Cristina Barrera
Institute of Food
Engineering-FoodUPV
Universitat Politècnica de
València
Valencia
Spain

Editorial Office
MDPI AG
Grosspeteranlage 5
4052 Basel, Switzerland

This is a reprint of the Special Issue, published open access by the journal *Foods* (ISSN 2304-8158), freely accessible at: https://www.mdpi.com/journal/foods/special_issues/3P2221Q0S8.

For citation purposes, cite each article independently as indicated on the article page online and as indicated below:

Lastname, A.A.; Lastname, B.B. Article Title. *Journal Name* **Year**, *Volume Number*, Page Range.

ISBN 978-3-7258-5055-6 (Hbk)
ISBN 978-3-7258-5056-3 (PDF)
https://doi.org/10.3390/books978-3-7258-5056-3

Cover image courtesy of Lucía Seguí Gil and Cristina Barrera

Contents

About the Editors

Lucía Seguí

Lucía Seguí is an Agri-Food Engineer with a PhD in Food Science and Technology. She is an Associate Professor at the Universitat Politècnica de València, where she teaches Food and Biotechnological Processes courses. She engages in research activity at the Institute of Food Engineering–FoodUPV, focused on sustainable processing technologies for the valorization of fruit and vegetable residues. She has participated in more than 25 research projects, published 45 research papers in indexed journals and 7 book chapters, and is responsible for more than 90 contributions in national and international conferences. She started her research career in Food Process Engineering, particularly mass transfer operations and structure–property relationships. Later, she was involved in developing integral valorization processes of fruit industrial residues through membrane separation and biological transformations, as well as in projects aimed at evaluating functional food ingredients and products design through matrix engineering techniques. At present, she combines her experience in food processing and functional food development to explore biological and thermophysical treatments (microwaves, ultrasounds, fermentation, hydrothermal, and their combinations), along with drying technologies (air drying, freeze drying), to produce ingredients with enhanced biological activity; and evaluates the impact of processing on the properties of the ingredients and products obtained, including functional, technological, and nutritional characteristics. Her research also explores the applications of upcycled products in the agri-food sector.

Cristina Barrera

Cristina Barrera graduated with a degree in Agricultural Engineering from the Universitat Politècnica de València in 2001 and earned a Ph.D. in Food Science and Technology in 2007. She is an Associate Professor with tenure at the Food Technology Department of the Universitat Politècnica de València, where she teaches courses related to Food Process Engineering. Cristina is the Secretary of the Institute of Food Engineering—FoodUPV, and her research focuses on developing functional foods and studying how their structure and processing methods affect their health benefits. Her work includes fortifying plant-based beverages and snacks with probiotics and improving their stability during processing and digestion. She also investigates how thermal treatments and dehydration impact the antioxidant content of plant-based foods. To date, Cristina has led two competitive R&D projects and collaborated on twenty-three others, including one under the KA2 Erasmus+ program. She has co-supervised three doctoral theses, published 41 peer-reviewed articles in international journals, and contributed seven chapters to books from international publishers. She has also presented nearly 90 notable contributions at national and international conferences. In recent years, her research has focused on promoting sustainability in the food chain by valorizing food waste. She works on transforming vegetable discards and by-products from cooperatives into functional powdered ingredients. By applying thermophysical and biological pretreatments, such as fermentation with GRAS microorganisms, along with optimized dehydration techniques, she enhances the bioaccessibility and stability of key nutrients, contributing to the development of more sustainable and nutritious food products.

Preface

To meet current food-industry sustainability and resource-efficiency demands, food waste and by-product management and valorization have become both a must and a challenge. The reduction of food waste is one of the biggest challenges that agricultural and food systems are presently facing. It is estimated that around 13.2% of food produced is lost between harvest and retail, while 19% of total global food production is wasted in the whole food chain. Every year, about 1.3 billion tons of edible food are wasted worldwide, of which 60 million tons are produced in the European Union. Food loss and waste consumes resources, thus contributing to environmental degradation and climate change, and implies a loss of nutrients. According to FAO estimates, the food that is lost and wasted could feed 1.26 billion hungry people every year.

Lucía Seguí and Cristina Barrera
Guest Editors

foods

MDPI

Editorial

Functional Ingredients from Food Waste and By-Products: Processing Technologies, Functional Characteristics and Value-Added Application s

Lucía Seguí * and Cristina Barrera

Instituto Universitario de Ingeniería de Alimentos—FoodUPV, Universitat Politècnica de València, Camino de Vera, s/n, 46022 Valencia, Spain; mcbarpu@tal.upv.es
* Correspondence: lusegil@upvnet.upv.es

1. Introduction

In the current scenario, food waste stands out as a pressing issue, accounting for a significant portion of the waste generated worldwide [1]. It arises throughout the food chain, from production to overharvesting, processing, transportation, storage, and consumption [2]. Every year, about 1.3 billion tons of edible food are wasted worldwide, of which 60 million tons are produced in the European Union [2,3]. Despite over half of food waste being generated at the consumption stage, postharvest and processing losses also account for an important share. Reducing food loss and waste is a key challenge that must be addressed to achieve the Sustainable Development Goals defined by the United Nations in the 2030 Agenda [4].

Tackling food waste presents advantages for the climate, for food security and for agri-food system sustainability [3]. In order to meet current food industry sustainability and resource efficiency needs, valorizing food waste, by-products and surpluses has become both a challenge and a necessity. Often rich in nutrients and bioactive compounds, food waste and by-products hold immense potential to be transformed into valuable functional ingredients with applications in the food industry and other related areas [5–7]. This Special Issue aimed to explore processing technologies for obtaining functional ingredients from these underutilized resources, and to expand knowledge about the nutritional, technological and functional characteristics of the new ingredients and value-added products that can be obtained, exploring their wide-ranging applications across different sectors, from food and beverages to agriculture, packaging and beyond.

As a result, this Special Issue gathers eleven valuable scientific contributions in the form of original research works which could be categorized in two main groups: one dealing with processing and extraction techniques to obtain value-added products or ingredients, and another focused on new product development and other applications.

2. Article Overview

In this SI, various technologies are investigated to produce value-added products or ingredients from food waste and by-products. Maté et al. (Contribution 1) explore subcritical water extraction and hydrogen peroxide alkaline delignification to fractionate grape stalks, a lignocellulosic biomass from the winemaking process. Subcritical water extraction allowed for obtaining phenolic-rich extracts with more than 70% of the total phenolic content of the grape stalks, whereas a part of these compounds also remained in the solid fraction. After subcritical water extraction, the products displayed improved their polyphenolic contents and antioxidant activities compared to the original material. The

 https://doi.org/10.3390/foods14050847

solid residue fraction was subjected to a green bleaching process to purify cellulose. The first bleaching cycle led to a maximum reduction in the lignin content of the insoluble fractions of 75%, with a higher cellulose purity in the samples obtained at 170 °C and bleached with 4% alkaline hydrogen peroxide. The final lignin content in the bleached samples was still higher than 25%, thus further studies are proposed to optimize cellulose recovery.

In another paper, Bas-Bellver et al. (Contribution 2) propose a series of thermophysical and biological treatments to improve the antioxidant properties of broccoli stem products to eventually produce functional powdered ingredients. Ultrasounds, microwaves, autoclaving, pasteurization and lactic acid fermentation are applied as pretreatments to enhance the antioxidant properties of the ground broccoli residues, and the corresponding powders obtained after air- or freeze-drying. The pretreatments applied enhanced the antioxidant properties of broccoli wastes, particularly ultrasounds and pasteurization treatments, followed by fermentation and, finally, autoclaving and microwaving. Microscopic observations suggested that membrane permeabilization and cell breakage were related to this improvement. When applied to non-pretreated samples, drying improved the antioxidant properties of the flours; this effect was more significant in air-dried samples, since higher temperatures promote the formation of certain antioxidant compounds. However, combining the proposed pretreatments with dehydration was not successful in improving antioxidant properties, except for the ultrasonicated samples. It is concluded that pretreatments may be applied to enhance the antioxidant attributes of broccoli waste, but extending this to dried products was not demonstrated and requires further research.

Other than as a pre-treatment, ultrasounds are proposed to assist the extraction of different constituents from food wastes. On the one hand, Zhang et al. (Contribution 3) proposed ultrasounds for extracting proteins from *Dictyophora rubrovolvata* volva and modifying their techno-functional properties. A maximum 43% extraction rate could be achieved in the ultrasound-assisted process. Compared to conventional alkaline extraction, ultrasonication had an impact on proteins' morphology and molecular size; the secondary and tertiary structures of the extracted proteins were modified. Changes in the proteins' characteristics and inter-molecular forces led to significant improvements in techno-functional properties such as water and oil holding capacity, foaming capacity and stability, emulsion activity, and stability, thus leading to new applications for food product development. On the other hand, Leonarski et al. (Contribution 4) proposed an ultrasound-assisted extraction process to obtain an anthocyanin-rich extract from black rice bran. The obtained crude and resin-purified extracts were evaluated in terms of functional properties and biological activities by determining anthocyanin content, antioxidant activity, antidiabetic and antitumoral activities, cytotoxicity and oxidative stress. The process was successfully scaled up 20 times without negatively affecting anthocyanin recovery. The extracts exhibited antidiabetic and anticancer effects, showing promising potential. The authors suggest further research to verify whether the extracts can be proposed as alternative therapeutics to control hyperglycemia in diabetic patients or as a potential anticancer supplement.

Non-communicable diseases such as diabetes, cancer or obesity are closely related to human gut microbiota. Hence, improving gut microbiota diversity is critical for developing healthier diets. In this vein, Li et al. (Contribution 5) investigated the prebiotic activity of a novel polysaccharide extracted from Huangshui, a by-product of traditional Chinese Baijiu production. This polysaccharide was isolated by gradient ethanol precipitation and was found to mainly be composed of arabinose, xylose and glucose in a molar ratio which differed from those previously reported. The novel polysaccharide was used by gut microbiota during in vitro fecal fermentation and was found to regulate the composition and abundance of beneficial microorganisms such as Phascolarctobacterium, Parabacteroides, and Bacteroides. The prebiotic activity of the polysaccaride was also evidenced by the production of short chain fatty acids, including acetic, butyric, and propionic acid, during fermentation.

Postharvest losses also represent a waste of nutrients, and the resources used in their growth and harvesting. Addressing this, Hosseininejad et al. (Contribution 6) proposed the valorization of astringent 'Rojo brillante' persimmon through the development of energy bars formulated with dehydrated persimmons, walnuts, hazelnuts and chia seeds. The new formulated product showed higher levels of healthy fats, proteins, and fiber than other commercial energy bars. In vitro simulated digestion revealed a higher recovery index for soluble tannins than for carotenoids. The bars behaved satisfactorily during storage, since astringency was mitigated, and carotenoids and antioxidant activity remained stable. In this way, a discarded material was transformed into a sustainable nutritious option for the snack industry. Siczek et al. (Contribution 7) present another piece of research in which a by-product is used to formulate muesli bars. In this case, researchers aimed to evaluate the possibility of replacing walnuts with oil pomace in muesli bar recipes and assess whether the resulting product met quality standards. The pomace-enriched bars exhibited physicochemical properties comparable to those of control bars, with positive sensory evaluations. Walnut pomace is therefore presented as a sustainable ingredient, potentially expanding product diversity while reducing environmental impacts.

The beverage industry can also benefit from the valorization of food by-products. In this vein, sugarcane syrup, an industrial by-product, has potential for new product development in the fermented beverages sector. Lv et al. (Contribution 8) proposed the use of sugarcane syrup to partially replace barley in producing lager beer. Results evidenced that the fruit and hop aromas were more harmonious when 20% sugarcane juice was added to the basic wort. Under these conditions, the fermentation process was optimized. The beer brewed with sugarcane syrup met the basic specifications of beer, increasing the utilization of amino acids by the yeasts, and bringing new flavor compounds compared with the normal all-barley beer. It also showed greater resistance to aging. In regions where sugarcane is abundant, utilizing this versatile local raw material to replace barley in traditional brewing recipes offers the dual benefits of reduced production costs and more sustainable beer production.

Other than developing novel functional food products by the addition of upcycled ingredients, the value-added applications of food waste are vast. For instance, food by-products have potential as preservative agents due to their antioxidant and antimicrobial activities. In line with this, Martínez et al. (Contribution 9) presented a novel approach in which the preservative properties of melanin and melanin-free fractions of cephalopod ink were evaluated on canned golden seabream (*Sparus aurata*) fillets after 3 months of storage at room temperature. Samples treated with the ink at specific concentrations showed higher levels of free fatty acids and phospholipids compared to the control, although differences were not statistically significant. Other applications rely on improving food functionality by the indirect use of food wastes. For instance, Balanč et al. (Contribution 10) developed biodegradable nanofibers made from pumpkin leaf protein concentrate and gelatin through electrospinning. The nanofibers were designed as fast-dissolving systems for supplementing foods with nutraceuticals, specifically vitamin B12. The leaf protein concentrate not only facilitated a slower, sustained release of vitamin B12 over time, but also contributed to maintaining its thermal stability. A different approach is presented by Navajas-Porras et al. (Contribution 11), who investigated the use of chemically modified spent coffee grounds for the agronomic fortification of Dutch cucumbers. They focused on the effects on Zinc biofortification, antioxidant capacity and the production of short-chain fatty acids after in vitro digestion–fermentation. It is concluded that both the Zn content and chemical composition of cucumbers may vary significantly with growing conditions, thus determining their contribution to the dietary intake of nutrients and antioxidants.

3. Conclusions

In summary, this Special Issue "Functional Ingredients from Food Waste and By-Products: Processing Technologies, Functional Characteristics and Value-Added Applications" evidences the great interest in exploring new techniques for food waste and by-product valorization, as well as in developing appplications which contribute to produce healthier and more sustainable foods.

Author Contributions: Authors have contributed equally to this paper. All authors have read and agreed to the published version of the manuscript.

Acknowledgments: As the Guest Editors of the Special Issue "Functional Ingredients from Food Waste and By-Products: Processing Technologies, Functional Characteristics and Value-Added Applications", we would like to sincerely thank each and every author who contributed their work to this Special Issue and helped in its success. The authors would like also to thank the reviewers and Assistant Editors for their efforts.

Conflicts of Interest: The authors declare no conflicts of interest.

List of Contributors:

1. Maté, I.; Vargas, M.; Atarés, L.; Chiralt, A. Fractionation of Winemaking Grape Stalks by Subcritical Water Extraction to Obtain Added-Value Products. *Foods* **2024**, *13*, 3566.
2. Bas-Bellver, C.; Barrera, C.; Seguí, L. Impact of Thermophysical and Biological Pretreatments on Antioxidant Properties and Phenolic Profile of Broccoli Stem Products. *Foods* **2024**, *13*, 3585.
3. Zhang, Y.; Wei, S.; Xiong, Q.; Meng, L.; Li, Y.; Ge, Y.; Guo, M.; Luo, H.; Lin, D. Ultrasonic-Assisted Extraction of *Dictyophora rubrovolvata* Volva Proteins: Process Optimization, Structural Characterization, Intermolecular Forces, and Functional Properties. *Foods* **2024**, *13*, 1265.
4. Leonarski, E.; Kuasnei, M.; Santos, E.H.; Moraes, P.A.D.; Cesca, K.; Oliveira, D.d.; Zielinski, A.A.F. The Potential of Crude and Partially Purified Black Rice Bran Extracts Obtained by Ultrasound-Assisted Extraction: Anti-Glycemic, Cytotoxicity, Cytoprotective, and Antitumoral Effects. *Foods* **2024**, *13*, 597.
5. Li, M.; Su, J.; Wu, J.; Zhao, D.; Huang, M.; Lu, Y.; Zheng, J.; Li, H. The Prebiotic Activity of a Novel Polysaccharide Extracted from Huangshui by Fecal Fermentation In Vitro. *Foods* **2023**, *12*, 4406.
6. Hosseininejad, S.; Moraga, G.; Hernando, I. Valorizing Astringent 'Rojo Brillante' Persimmon Through the Development of Persimmon-Based Bars. *Foods* **2024**, *13*, 3748.
7. Siczek, P.; Libera, J. The Possibility of Using Oil Pomace as a Substitute for Walnuts (*Juglans regia* L.) in Muesli Bar Technology. *Foods* **2024**, *13*, 3807.
8. Lv, H.; Jia, Y.; Liu, C.; Xu, J.; Xie, C.; Li, K.; Huang, K.; Hang, F. A Preliminary Study on the Effect of Adding Sugarcane Syrup on the Flavor of Barley Lager Fermentation. *Foods* **2024**, *13*, 2339.
9. Martínez, B.; Trigo, M.; Rodríguez, A.; Aubourg, S.P. Influence of Cuttlefish-Ink Extract on Canned Golden Seabream (*Sparus aurata*) Quality. *Foods* **2024**, *13*, 1685.
10. Balanč, B.; Salević-Jelić, A.; Đorđević, V.; Bugarski, B.; Nedović, V.; Petrović, P.; Knežević-Jugović, Z. The Application of Protein Concentrate Obtained from Green Leaf Biomass in Structuring Nanofibers for Delivery of Vitamin B12. *Foods* **2024**, *13*, 1576.
11. Navajas-Porras, B.; Cervera-Mata, A.; Fernández-Arteaga, A.; Delgado-Osorio, A.; Navarro-Moreno, M.; Hinojosa-Nogueira, D.; Pastoriza, S.; Delgado, G.; Navarro-Alarcón, M.; Rufián-Henares, J.Á. Zn Biofortification of Dutch Cucumbers with Chemically Modified Spent Coffee Grounds: Zn Enrichment and Nutritional Implications. *Foods* **2024**, *13*, 1146.

References

1. Cherian, E.; Gurunathan, B. *Value Added Products from Food Waste*; Springer Nature: Cham, Switzerland, 2024. [CrossRef]
2. Sirohi, R.; Tarafdar, A.; Porto de Souza Vandenberghe, L.; Taherzadeh, M.J.; Pandey, A. *Sustainable Technologies for Food Waste Management*; CRC Press: Boca Raton, FL, USA, 2025. [CrossRef]
3. Katsarova, I. Waste Framework Directive: A More Sustainable Use of Natural Resources (Briefing). 2024. Available online: https://www.europarl.europa.eu/thinktank/en/document/EPRS_BRI(2023)757572 (accessed on 16 January 2025).
4. Hanson, C.; Flanagan, K.; Robertson, K.; Axmann, H.; Bos-Brouwers, H.; Broeze, J.; Kneller, C.; Maier, D.; Mcgee, C.; O'Connor, C.; et al. Reducing Food Loss and Waste: Ten Interventions to Scale Impact. World Resources Institute. 2019. Available online: https://www.wri.org/reducing-food-loss-and-waste-ten-interventions-scale-impact (accessed on 16 January 2025).
5. Capanoglu, E.; Nemli, E.; Tomas-Barberan, F. Novel approaches in the valorization of agricultural wastes and their applications. *J. Agric. Food Chem.* **2022**, *70*, 6787–6804. [CrossRef]
6. Radha; Prakash, S.; Kumari, N.; Sharma, N.; Puri, S.; Singh, J.; Mamta, T.; Pundir, A.; Kumar, M. Bioactives and Bioactivities from Food Byproducts. *Curr. Food Sci. Technol. Rep.* **2024**, *2*, 297–308. [CrossRef]
7. Fernandes, F.; Delerue-Matos, C.; Grosso, C. Unveiling the Potential of Agrifood By-products: A Comprehensive Review of Phytochemicals, Bioactivities and Industrial Applications. *Waste Biomass Valorization* **2024**. [CrossRef]

foods

MDPI

Article

The Potential of Crude and Partially Purified Black Rice Bran Extracts Obtained by Ultrasound-Assisted Extraction: Anti-Glycemic, Cytotoxicity, Cytoprotective, and Antitumoral Effects

Eduardo Leonarski [1], Mayara Kuasnei [1], Eloisa H. Santos [1], Paulo A. D. Moraes [2], Karina Cesca [1], Débora de Oliveira [1] and Acácio A. F. Zielinski [1,*]

[1] Department of Chemical Engineering and Food Engineering, Federal University of Santa Catarina (UFSC), Florianópolis 88010-970, SC, Brazil; eduardoleonarski@gmail.com (E.L.); mayara_kuasnei@hotmail.com (M.K.); ds.eloisa@gmail.com (E.H.S.); karinacesca@gmail.com (K.C.); debora.oliveira@ufsc.br (D.d.O.)

[2] Department of Chemistry, Federal University of Santa Catarina (UFSC), Florianópolis 88010-970, SC, Brazil; paulo.a.d.moraes@ufsc.br

* Correspondence: acacio.zielinski@ufsc.br; Tel.: +55-48-3721-2509

Abstract: Recovering anthocyanins from black rice bran is a way of valuing this byproduct, by obtaining an extract with biological potential. The objective of this study was to recover anthocyanins using ultrasound-assisted extraction. Some of the extract was partially purified, and both (crude and partially purified) extracts were evaluated for their anthocyanin content, antioxidant activity, antidiabetic and antitumoral activities, cytotoxicity, and oxidative stress. An increase in the laboratory scale was also achieved, making possible to increase the extraction volume up to 20 times without significantly changing the content of anthocyanins (1.85 mg C3G/g DW). It was found that the purified sample presented a 4.2 times higher value of total anthocyanins compared to the crude sample. The best IC_{50} values for the purified sample were verified by DPPH and ABTS (0.76 and 0.33 mg/mL). The best results for antidiabetic activity were obtained for the partially purified sample: 0.82 μM C3G for α-glucosidase and 12.5 μM C3G for α-amylase. The extracts demonstrated protection (~70%) when subjected to the oxidative stress of L929 cells. An antitumoral effect of 25–30% for both extracts was found in A459 cells. The crude and partially purified extracts of black rice have antidiabetic and anticancer effects and more studies are needed to explore their potential.

Keywords: *Oryza sativa* L.; purification; oxidative stress; antidiabetic potential

1. Introduction

Black rice processing produces about 10% rice bran, 14% broken rice, and 20% rice husk [1–3], which also has high functional value mainly due to its high anthocyanin content [4,5]. Black rice bran is a rich anthocyanin source, which has been reported to have antidiabetic effects by inhibiting the activities of α-amylase and α-glucosidase (reducing the risk of type 2 diabetes) [6,7]. Furthermore, it was also found that black rice bran extract has a cytoprotective effect on H_2O_2-induced oxidative stress in L292 cells, indicating that these extracts may have protective effects against oxidative reactions [5]. Therefore, the recovery of these extracts which are rich in anthocyanins may be interesting for evaluating their biological potential.

One way to recover these bioactive compounds is through extraction, using, preferably, eco-friendly methods and solvents [8]. The ultrasound method produces waves that achieve a greater penetration into the cellular material, requiring less time and less solvents, in addition to using less energy and allowing for the extraction of heat-labile compounds [9,10]. The expansion of the laboratory scale for UAE is important for the recovery of bioactive compounds, although few studies have been carried out with this objective [11]. This is because scaling up is not a trivial task that only increases the amount of biomass and solvent required; a more in-depth study is necessary, and, therefore, it is necessary to carry out tests from the laboratory scale (cells from 10 to 1000 mL) to the pilot scale (10 to 50 L), and up to industrial scale [9]. The extracts obtained during the process, in addition to being rich in anthocyanins, contain several other compounds, which are called crude extracts.

Crude anthocyanin extracts from black rice contain several other compounds (mainly sugars), which may affect their biological effects and possible applications. Therefore, the use of a macroporous resin (e.g., XAD-7HP, AB-8, HP-20, D-101, or X-5) is a good alternative to partially purify the anthocyanins from plant materials [12]. Through hydrophobic bonding and aromatic stacking, these synthetic resins adsorb to the phytochemicals from aqueous solutions and desorb in organic solvents such as methanol and ethanol [13]. The partial purification of phenolic extracts could improve their antioxidant activity and present an improvement in their anti-inflammatory effect, making them a more effective alternative for treating diseases [14].

Anthocyanin extraction from black rice bran by ultrasound and partial purification with Amberlite XAD-7HP resin could be an alternative method to obtain an anthocyanin-rich extract with high biological potential. Therefore, this study aims to recover anthocyanins from black rice bran using ultrasound-assisted extraction (UAE) and purify them using Amberlite XAD-7HP resin. Then, the crude and partially purified extracts' antioxidant activity, antidiabetic potential, cytotoxicity, and oxidative stress will be evaluated.

2. Materials and Methods

2.1. Sample Preparation

Black rice bran (BRB) was supplied by Ruzene (Pindamonhangaba, Brazil) and prepared according Leonarski et al. [5].

2.2. Ultrasound-Assisted Extraction (UAE)

At the lab scale, the extraction was carried out using an ultrasonic probe (Eco-Sonics, Ultronique Q3.0/37A, Indaiatuba, Brazil), a mass of 0.5 g of BRB, and a 15 mL volume of solvent. The following conditions parameters were used: a temperature of 50 °C, frequency power of 380 W, and a solvent ratio to 60% citric acid (0.1 mol/L) to 40% ethanol.

For the scale-up, with the same ultrasonic probe used at the lab scale, an experiment was carried out to increase the extraction volume from 15 to up to 300 mL (an increase of 20 times). For this, firstly, the real ultrasound power (P) was measured, considering that the real input power is converted into heat dissipated in the medium, which is determined by calorimetry, as in Equation (1), according to Carail et al. [15]:

$$P = m \cdot Cp \cdot (dT/dt) \quad (1)$$

where Cp is the heat capacity of the solvent at a constant pressure (J/g K), m is the mass of the solvent (g), and dT/dt is the temperature rise per second.

Using the power calculated in Equation (1), the consequent ultrasonic intensity (UI) was calculated for the ultrasonic probe, as shown in Equation (2).

$$UI = 4P/\pi D^2 \quad (2)$$

where UI is the ultrasonic intensity (W/cm^2), P is the ultrasound power (W) as calculated by Equation (1), and D is the internal diameter (cm) of the probe tip.

By calculating the real power (P) and the ultrasonic intensity (UI), it was possible to verify the required power increase when increasing the reactor size to carry out the scale-up. The instrumental parameters are presented in Table 1.

Table 1. Instrumental parameters for laboratory scale-up.

Frequency Power (W)	Real Frequency Power (W)		Ultrasound Intensity (W/cm²)	
	15 mL Reactor	300 mL Reactor	15 mL Reactor	300 mL Reactor
300	24.61	71.31	3.48	2.52
350	27.89	78.04	3.95	2.76
380	28.85	80.58	4.08	2.85
400	30.99	85.69	4.39	3.03
450	32.18	88.31	4.55	3.12

2.3. *Total Monomeric Anthocyanin (TMA)*

The total monomeric anthocyanin (TMA) analysis was performed according to the differential pH methodology [16], as described by Leonarski et al. [5]. The TMA concentrations were calculated using Equations (3) and (4):

$$A_. = (A_{520} - A_{700})_{pH\ 1.0} - (A_{520} - A_{700})_{pH\ 4.5} \quad (3)$$

$$TMA = (A_. \cdot MW \cdot DF \cdot 1000)/\varepsilon \quad (4)$$

where $A_.$ = absorbance, MW = molecular weight (cyanidin-3-glucoside: 449.2 g/mol), DF = dilution factor (15), and ε = the molar absorptivity of cyanidin (26,900).

2.3.1. Partial Purification of Anthocyanin-Rich Extracts

The steps for the extraction of anthocyanins, from black rice bran to freeze-drying, are shown in Figure 1. The anthocyanin-rich extracts obtained using the UAE technique were concentrated, eliminating all ethanol through vacuum rotary evaporation (Fisatom, model 801, São Paulo, SP, Brazil). Using 15 mL of ethyl acetate, 10 mL of concentrated extract was washed twice. A glass column (1.0 cm × 30 cm) was filled with 50 g of Amberlite XAD7HP resin (Sigma-Aldrich, Steinheim, Germany), and the extract was added. Elution was carried out with ultrapure water to remove sugars and aliphatic acids, using a peristaltic pump (Watson-Marlow 323 Series Drive, Falmouth, UK) with a flow rate of 6 mL/min, until reaching a final volume of 500 mL. With a hydroethanolic acid solution [containing 50% (*v*/*v*) ethanol and 1% of acetic acid or citric acid (*v*/*v*)], the purified extract was desorbed by elution. By rotary evaporation, the recovered solution was concentrated and lyophilized.

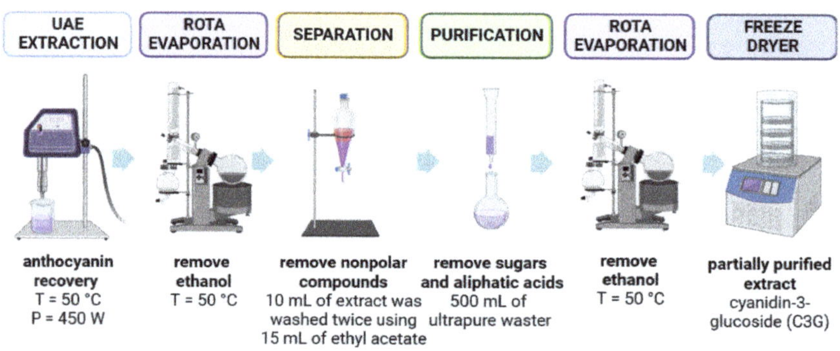

Figure 1. Extraction, partial purification, and freeze-drying steps to obtaining anthocyanins from BRB. Figure created with BioRender.com.

2.3.2. Determining the Individual Anthocyanins by HPLC-MS

The individual anthocyanins were determined by HPLC-MS and quantified by HPLC-PDA, as described by Leonarski et al. [5]. The lyophilized extracts were resuspended in formic acid 0.1% and filtered through a 0.22 μm filter. The samples (10 μL) were injected into a high-performance liquid chromatography system (model LCMS-2020, Shimadzu, Kyoto, Japan). The equipment consists of a photo-diode array detector (PDA), an LC-20AD binary pump, a SIL-20AC HT autosampler, a central controller, and a single-quadrupole MS detector (Shimadzu) with electrospray ionization (ESI). A Kromasil® C18 column (100 Å, 300 mm × 4.6 mm i.d.) was used. Cyanidin-3-glucoside (C3G) was quantified in mg/g.

2.4. Antioxidant Activity

The extracts' antioxidant potential was evaluated by DPPH and ABTS radical scavenging methods according to Brand-Williams et al. [17] and Re et al. [18], respectively, with adaptations to the microplate reader. The results were expressed as IC50 values (mg/mL).

2.5. Effects of Crude and Partially Purified Anthocyanin Extracts on α-Amylase and α-Glucosidase Inhibition

For the α-glucosidase assay, UAE extracts were diluted in phosphate buffer (0.1 mol/L, pH 6.9) and incubated at 25 °C for 10 min with 100 μL of α-glucosidase solution (0.5 U/mL). Then, 50 μL of pNPG (5 mmol/L in phosphate buffer) was added and the solution was incubated for 5 min at 25 °C. By adding 80 μL of sodium carbonate (0.2 mol/L), the reactions were stopped, and, finally, the absorbance was measured at 405 nm [19]. For the α-amylase assay, 40 μL of the extracts, diluted in buffer (0.1 M with 0.006 M NaCl, pH 6.9), was mixed with 150 μL of water, 400 μL of starch solution (0.5%), and 200 μL of enzyme solution (0.5 mg/mL), prepared in phosphate buffer (pH 6.8), and incubated at 25 °C for 15 min. The reaction was stopped using 400 μL of DNS reagent (3.5-dinitro salicyl) and the solution was maintained at 90 °C for 15 min, before, finally, the absorbances were read at 540 nm using a spectrophotometer [20]. The % inhibition of both assays was calculated according to Equation (5):

$$\%\text{inhibition} = [1 - (C - D)/(A - B)] \cdot 100 \quad (5)$$

where A = control (with enzyme and without sample), B = control blank (no enzyme and no sample), C = reaction (with enzyme and sample), D = reaction blank (without enzyme and with sample).

2.6. Cell Culture

2.6.1. Culture Conditioning

In Dulbecco's Modified Eagle Medium (DMEM), cell lines were cultured and supplemented with 10% fetal bovine serum (FBS) and 1% penicillin and streptomycin. The cells were maintained in a humidified environment at 37 °C with 5% CO_2. The cell culture medium was renewed every 48 h until the cells attained a confluence level of 75–90%. Subsequently, using Tryplex, the cells were dissociated, counted, and seeded.

2.6.2. Cytotoxicity Assay

Using a colorimetric method with MTS [3-(4,5-dimethylthiazol-2-yl)-5-(3-carboxymethoxyphenyl)-2-(4-sulfophenyl)-2H-tetrazolium colorimetric assay], the cytotoxicity of the extracts was evaluated by measuring the metabolic activity of a normal fibroblast cell (L929) [21]. After cell adhesion (1×10^4 cells/well) for 24 h in 96-well plates, extracts were added (0.5, 1, 2.5, 5, 7.5, 10, and 20 μM) and incubated for 24 h and 48 h. By spectrophotometry (490 nm, Molecular Devices, CA, USA), optical density was evaluated, and cell viability was determined with respect to the control (100%) [22].

2.6.3. Hydrogen Peroxide-Induced Oxidative Stress in L929 Fibroblast Cells

At a density of 1×10^4 cells/well, L929 fibroblast cells were seeded with DMEM medium, supplemented with 10% FBS, and incubated with 5% CO_2 at 37 °C overnight. Then, 0.5, 1, 2.5, 5, 7.5, 10, and 20 μM of extracts rich in anthocyanins were added to these wells concomitantly with 1.0 mM H_2O_2 to treat exposed cells, and cells incubated for 24 h [23]. Cell viability was evaluated using an MTT assay as described in Leonarski et al. [5].

2.6.4. Antitumoral Activity

Using an MTS colorimetric assay, cell viability was measured. In 96-well plates, mouse glioma (GL261) and lung adenocarcinoma (A549) cells were seeded at a density of 1×10^4 cells/well and placed in a CO_2 incubator for 24 h at 37 °C for attachment. After 24 h, the cells were treated with 2.5 μM of extracts and doxorubicin (5 μM) as a positive control. Then, cells were incubated at 37 °C in CO_2 for 24 h and 48 h. Then, the medium was removed, and the cells were washed twice with PBS. In total, 100 μL of medium and 20 μL of MTS reagent were added and incubated for 2 h at 37 °C with 5% CO_2. The cells' optical density (OD) was measured on plates at 490 nm [24].

2.7. Statistical Analysis

The dataset was evaluated by a one-way analysis of variance (ANOVA). Using *t*-test or Tukey test, significant differences were determined at a probability level of less than 5% ($p < 0.05$). All statistical procedures were performed using Statistica v. 13.5 software (TIBCO Software Inc., Palo Alto, CA, USA). The results were presented as the mean ± standard deviation.

3. Results and Discussion

3.1. Laboratory Scale-up

The use of new technologies for food processing, such as UAE, aims to save energy, reduce time, and improve product quality and shelf life [15]. In studies by Carail et al. [15] and Gille et al. [25], the authors verified the use of UAE for carotene extraction, using Equations (1) and (2) to calculate the ultrasonic intensity (UI) to verify its effect on possible carotene degradation. In our study, these equations were used with the main objective of scaling up the process.

For scale-up (using the 300 mL reactor), the use of 450 W achieves a maximum ultrasound intensity (UI) equal that of 300 W for the 15 mL reactor (Table 1). Therefore, for a 15 mL vessel, the extraction of anthocyanins from black rice bran at 50 °C, using a 60:40 ratio of 0.1 mol/L citric acid–ethanol and a power of 300 W, obtained 1.97 mg C3G/g DW of anthocyanins (anthocyanin recovery (AR) = 67.7%). For the laboratory scale-up, the same conditions of time, temperature, solvent, and a power of 450 W were used for volumes of 150 mL (10× increase) and 300 mL (20× increase), obtaining 1.94 mg C3G/g DW (AR = 66.7%) and 1.82 mg C3G/g DW (AR = 62.7%) of anthocyanins, respectively.

In a study by Belwal et al. [11], in which anthocyanins were extracted from *Pyrus communis* 'Starkrimson' fruit peel, the authors scaled up from 30 to 150 mL (5× increase), 450 mL (15× increase), and 3000 mL (100× increase). The authors obtained comparisons with the 30 mL reactor, where were around 92% for the 150 mL reactor, 88% for the 450 mL reactor, and 83% when increasing the reactor to 3 L. In our study, compared with the extraction in the 15 mL reactor, there was no significant difference for extractions in a reactor that had a 150- or 300-mL volume, which recovered around 98 and 92.5% of anthocyanins, respectively.

To increase the scale, whether to a pilot or industrial scale, we needed to use equipment that has a greater variation in power. In the study by Chen et al. [26], the authors carried out an increase in scale for anthocyanins extracted from purple corn bran, from 50 mL to 42 L, using ultrasound, obtaining a small decrease of 5.46% in the anthocyanins extracted compared to the laboratory scale. According to the authors, the results obtained suggest that scaling up to an industrial scale of ultrasound extraction is easily achieved.

3.2. Partial Purification of Anthocyanins Extracted from Black Rice Bran

Amberlite XAD-7HP stands out in terms of separating anthocyanins of all the resins available for partial purification. In studies by Das et al. [13], Heinonen et al. [27], and Chen et al. [12], of the several resins used for purifying anthocyanins from different sources (purple rice bran, purple-fleshed potato, and mulberry, respectively), the one that stood out was Amberlite XAD-7HP, which presented better adsorption and desorption characteristics compared to other resins. The efficient elution of the solute adsorbed to the resin is important to guarantee the use of the resin over multiple cycles [28].

For the partial purification of the extract, two organic acids (acetic and citric acid) were used, in a 1% volume, to desorb the anthocyanin. According to Chang et al. [29], citric acid solution, as an eluent, is preferred to hydrochloric acid, as it avoids traces of this acid in the product after purification. Among the acids used as an eluent, acetic acid showed better results, recovering about 57.2% of the total anthocyanins after purification. In contrast, with citric acid that recovery is 36.8% of the total anthocyanins. Adding acid during anthocyanin elution is necessary for these compounds to maintain a stable form (flavylium cations) and prevent degradation [30]. In their flavylium cation form, anthocyanins have a lower affinity for the resin; therefore, the addition of acetic acid possibly led to a greater formation of flavylium cations, which resulted in a higher purification yield.

In a study by Contreras et al. [31], a raw extract obtained from native black beans after purification with Amberlite XAD-7HP, using acidified ethanol/water 70/30 v/v (0.3% formic acid) as eluent, reached a 53.8% recovery of anthocyanins, a result close to that reported in our study using acetic acid (1%). In the study by Das et al. [13], the authors reported an efficiency in the recovery of anthocyanins extracted from purple rice bran of 41.5%, using 95% ethanol for adsorption; these values are closer to those reported when using citric acid.

In a study by Zhao et al. [32], the authors observed that most anthocyanins from black peanut skin were eluted from the column early due to their high polarity and low adsorption capacity to Amberlite XAD-7HP resin. The use of acidified ethanol (40–75%, v/v) to recover anthocyanins in the purification of Amberlite XAD-7HP can be considered an efficient strategy [12,27,32].

3.3. Comparison of Crude and Partially Purified Anthocyanin Extracts

The crude anthocyanin extract achieved 1.06 mg/g of TMA, while the partially purified extract increased the anthocyanin content by 4.2-fold (4.44 mg/g) (Table 2). The same behavior was observed for cyanidin-3-glucoside, for which the crude extract presented 1.03 mg/g (about 98% of the total anthocyanins) and purified extract presented 3.97 mg/g (corresponding to 89.5% of total anthocyanins).

Table 2. Analysis of the anthocyanin-rich extract of black rice bran before and after purification.

Analysis	Extracts	
	Crude	Partially Purified
TMA (mg/g)	1.06 ± 0.04^{b}	4.44 ± 0.60^{a}
C3G (mg/g)	1.03 ± 0.01^{b}	3.97 ± 0.02^{a}
DPPH IC_{50} (mg/mL)	1.57 ± 0.03^{a}	0.67 ± 0.01^{b}
ABTS IC_{50} (mg/mL)	1.26 ± 0.01^{a}	0.33 ± 0.01^{b}

Results show the mean ± standard deviation. TMA: total monomeric anthocyanins; C3G: cyanidin-3-glucoside. Different letters indicate significant differences by *t*-Test ($p < 0.05$).

In the study by Das et al. [13], XAD resin was used to purify purple rice bran extracts, verifying a 5.5-fold increase for the purified extract, while, for C3G, the increase was 6.6-fold. In a study by Jeyaraj et al. [33], in which the Amberlite resins XAD-16 and C18-OPN were used to purify an extract of *Clitoria ternatea* flower, the concentration of anthocyanins was found to be approximately 4.5–4.7 times higher; values close to those reported in our study.

It was also observed that the purification process improved in vitro antioxidant activity (DPPH and ABTS). It can be seen in Table 2 that the IC_{50} for the crude extract was 1.57 and 1.26 mg/mL for the DPPH and ABTS radicals, respectively. These values decreased to 0.67 and 0.33 mg/mL for the DPPH and ABTS radicals in the partially purified extract, respectively. This increase in antioxidant activity is associated with the C3G concentration discussed previously.

In a study by Chumchoochart and Sutthanut [34], the authors reported an IC_{50} value for DPPH equal to 1.69 mg/mL, and in a study by Singha et al. [35], the authors reported values between 0.10 and 1.25 mg/mL; both corroborate those found in our study. Sansenya and Nanok [36] reported purified fractions of black rice IC_{50} values between 0.35 and 0.51 mg/mL for the ABTS radical, a value close to that of the purified sample presented in our study.

3.4. *Effects of Crude and Partially Purified Extracts on In Vitro α-Amylase and α-Glucosidase Inhibition*

Figure 2 shows the black rice bran extracts' in vitro enzymatic inhibition values for α-glucosidase and α-amylase. Partial purification improved significantly ($p < 0.05$) the inhibitory effect of the extract. For the inhibition of α-glucosidase, an IC_{50} value of 3.23 μM (1.45 mg/mL) was obtained for the crude extract and 0.82 μM (0.37 mg/mL) for the partially purified extract, improving its activity by approximately 4-fold after purification. Shimoda et al. [37] reported an IC_{50} value for α-glucosidase of 0.41 mg/mL for a purple rice extract, while Choi et al. [38] presented, for C3G (the main compound from black rice extract), an IC_{50} of 13.7 μM for α-glucosidase.

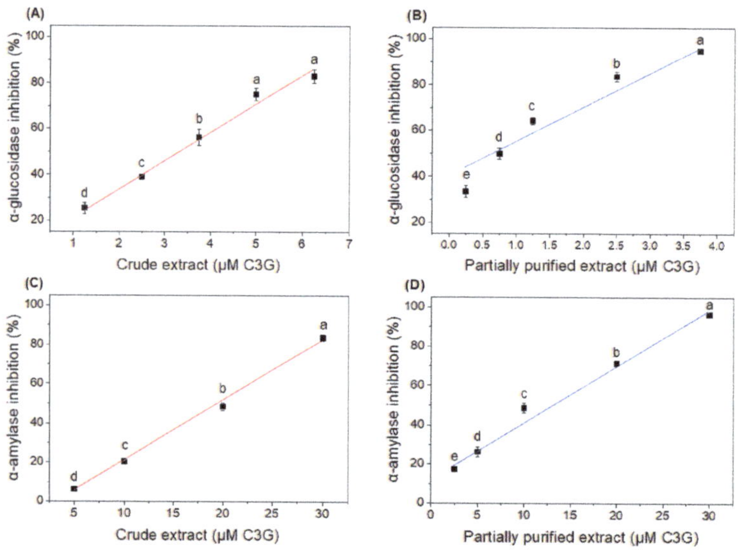

Figure 2. Inhibitory effects of (**A**) crude and (**B**) partially purified anthocyanin-rich extracts on α-glucosidase, and of (**C**) crude and (**D**) purified anthocyanin-rich extracts on α-amylase. Different letters indicate significant differences by Tukey test ($p < 0.05$).

For α-amylase inhibition (Figure 2C,D), the IC_{50} was found to be 19.31 μM (8.67 mg/mL) for the crude extract, while for the partially purified extract it was 12.5 μM (5.60 mg/mL), meaning they differed significantly from each other ($p < 0.05$). Values close to the IC_{50} for α-amylase inhibition were reported by Aalim et al. [39] for black rice grains (8.36 mg/mL).

According to Choi et al. [38], the intestinal enzymes α-glucosidase and pancreatic α-amylase are responsible for the hydrolysis of various carbohydrates (starch, glycogen, sucrose, etc.), and the inhibition of these enzymes leads to a delay in the rise of blood glucose levels. The degradation of this starch would lead to high postprandial hyperglycemia [40]. This delay in hydrolysis through enzyme inhibition is a therapeutic approach to controlling postprandial hyperglycemia in pre-diabetes, diabetes, and obesity [41,42].

The inhibition of α-amylase or α-glucosidase enzymes occurs because anthocyanins enter their active site and reduce their catalytic action through hydrogen bonding [43]. In a study by Sui et al. [7], it was found that C3G made seven hydrogen bonds with porcine pancreatic α-amylase and had the highest inhibition activity among the four anthocyanins studied (cyanidin-3-glucoside, cyanidin-3,5-glucoside, cyanidin-3-rutinoside, and peonidin-3-glucoside).

Drugs used to treat diabetes can have undesirable effects, such as causing hypoglycemia in higher doses, liver problems, lactic acidosis, and diarrhea [42]. Therefore, many studies have suggested the use of natural compounds that can replace conventional medicines and consequently lead to a reduction in side effects [38,42].

3.5. Cell Culture

3.5.1. Cytotoxicity

A cytotoxicity assay was performed using L929 cells; the results are presented in Figure 3A. For crude extracts, only 20 μM presented a cell viability lower than 70%, while, for the partially purified extracts, volumes higher than 2.5 μM presented a cell viability lower than 70%. According to ISO 10993-5:2009 [21], an L929 cell viability higher than 70% does not show cytotoxicity. The semi-purified extract may have concentrations of other compounds that were not removed in previous steps, which could have affected its cytotoxicity, such as the acetic acid used to elute the extract, or even traces of other solvents. However, further studies must be carried out to identify these compounds.

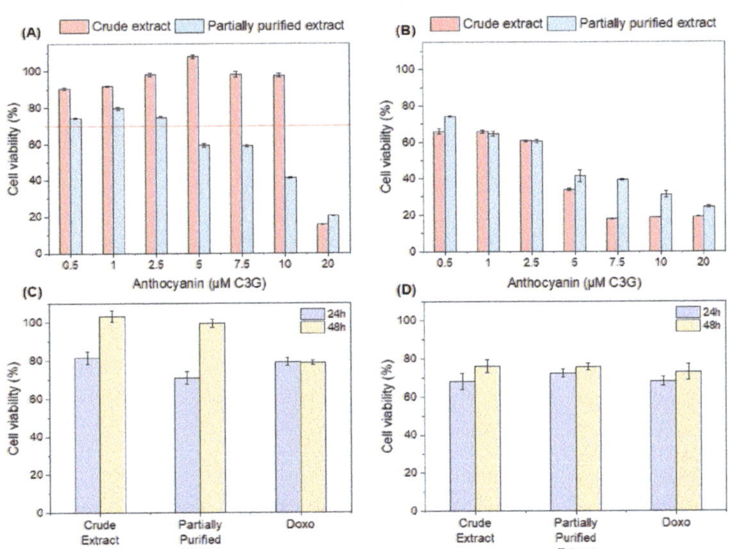

Figure 3. (**A**) Cytotoxicity of L929 cells treated with the crude and partially purified extracts. The dotted line corresponds to 70% cell viability (below 70%, the extract is considered cytotoxic); (**B**) viability of L929 cells treated with the crude and partially purified extracts with concomitant H_2O_2 exposure; (**C**) antitumoral activity (%) of crude and partially purified extracts in mouse glioma (GL261) cells, and (**D**) in lung adenocarcinoma (A549) cells.

Sangkitikomol et al. [44] verified the cytotoxicity of a black rice extract in HepG2 cells at concentrations above 800 μg/mL. In Aprodu et al.'s [45] study, at concentrations greater than 800 μg/mL of microencapsulated black rice extracts, a significant decrease was observed but without cytotoxicity (a cell viability of L929 cells of around 80%). In a study by Wang et al. [46], it was verified using L929 cells that the anthocyanins from their blueberry extract showed cytotoxicity (viability less than 70%) at a concentration of 800 μg/mL.

Toxicity analysis is important to verify the concentration at which extracts may be harmful in their application. Alongside this, oxidative stress analysis and antitumor activity were carried out to verify the biological potential of these extracts.

3.5.2. Effects of Crude and Partially Purified Extracts on Cell Viability in H_2O_2-Induced L929 Cells

Oxidative stress was achieved in L929 cells using hydrogen peroxide (H_2O_2). At the concentration of 1 mM H_2O_2, cell viability was approximately 11.2%. By adding the crude and partially purified extracts, the cytoprotection of the L929 cells was verified (Figure 3B). At low concentrations (0.5, 1, and 2.5 μM), around a 60–74% cytoprotection of the L929 cells was observed, and this was higher for the partially purified sample at a concentration of 0.5 μM, significantly differentiating it from the crude extract ($p < 0.05$). With increasing concentrations, there was a decrease in the cytoprotection of both samples, which was more pronounced for the crude extract. In this case, it was shown that small concentrations of the extracts are efficient at carrying out cell cytoprotection.

Ereminas et al. [47] reported a cytoprotective effect of C3G on rat C6 glial cells at concentrations of 5, 10, and 20 μM, ranging from 58 to 65% (H_2O_2 100 μM) and not differing significantly from each other. In a study by Zhang et al. [48], C3G from Chinese Bayberry showed a cytoprotection of around 70% in β cells (INS-1) (with 1 mM H_2O_2) at concentrations of 0.5 and 1 μM, which decreased significantly ($p < 0.05$) to around 60% when the concentration was increased to 5 μM. According to Tan et al. [49], the cytoprotective effect of C3G is death receptor-dependent, as is obtained by regulating two apoptotic pathways: the mitochondrial pathways and the external pathway.

For anthocyanin-rich extracts from black rice bran obtained by pressurized liquid extraction (PLE) and heat-stirring extraction (HSE), a cytoprotection of around 80% was also evidenced in L929 cells when using 250 μg/mL [5]. Palungwachira et al. [50] showed an improvement (around 25% for 10 μg/mL of extract and 10% for 25 μg/mL) in the cytoprotection of rat dermal fibroblast (RDF) cells (with 0.6 mM H_2O_2) by black rice extracts purified with a C18 Sep-Pak cartridge compared to the crude extract.

Due to their antioxidant potential, their cytoprotective effect comes from the anthocyanins present in the extracts [5]. This activity is related to the capture of the free radicals (H_2O_2) responsible for tissue damage; therefore, the extract has the potential for application in wound healing [23,51]. However, more studies are needed to explore this potential in full.

3.5.3. Antitumoral Activity

Antitumoral activity (or antiproliferative activity) assays were performed using mouse glioma (GL261) cells and lung adenocarcinoma (A549) cells (Figure 3C,D). A concentration of 2.5 μM was used for both extracts (a concentration that did not show cytotoxicity according to Section 3.5.1), and 5 μM doxorubicin (Doxo) was used as the standard antibiotic.

In terms of antitumoral activity in GL261 cells, within 24 h the samples showed a decrease in metabolic activity of around 20–30%, a value similar to the antibiotic Doxo, and did not differ from each other. However, within 48 h, the cells began to increase again, recovering to their initial state (close to 100%) for both extracts, while, for Doxo, the activity remained at around 80%. Therefore, although both extracts showed initial promise, over time the concentration was not sufficient to cause a significant effect on the metabolic activity of GL261 cells.

For A549 cells, it was found that both extracts presented effects between approximately 68 and 76%, while the antibiotic Doxo presented similar values, between 68% and 73%, with no sample differing from the others. Although the result was not as expressive, similar results were obtained for the antibiotic. In a study by Xue et al. [52], when evaluating the antitumor activity in A459 cells of raspberry wine residue extract at a concentration of 2.5 μM, the authors showed a decrease in cell viability of between approximately 15% and 28% within 48 h, results close to those reported in our study. Therefore, our extracts have the potential for application against A549 cells.

Further studies have shown that C3G (the main compound of black rice bran extracts) inhibited the proliferation, migration, and invasion and promoted the apoptosis of A549 cells, processes that involve cancer metastasis [53,54]. The negative regulation of TP53I3 and the inhibition of the PI3K/AKT/mTOR pathway carried out by the anthocyanin C3G are responsible for inhibiting the proliferation, migration, and invasion and also facilitating the apoptosis of A459 cells [53]. Although the extracts have potential in treating lung adenocarcinoma, more studies are needed to highlight their possible mechanisms and effectiveness.

4. Conclusions

It was possible to increase the laboratory scale for ultrasound-assisted extraction (UAE) by 20 times without significantly affecting the recovery of anthocyanins from black rice bran (around 63% were recovered). Furthermore, their partial purification using a microporous resin (Amberlite XAD-7HP) resulted in an extract with a higher total level of anthocyanins (by 4.2-fold), and greater antioxidant power (DPPH and ABTS). In addition, the partially purified anthocyanin extracts presented an antidiabetic effect greater than that of the crude extract (being approximately 4 times higher for α-glucosidase and 0.65 for α-amylase). No cytotoxicity was found towards L929 cells for the crude (concentration $\leq$ 10 μM) or partially purified extracts (concentration $\leq$ 2.5 μM). For oxidative stress, low concentrations of both extracts (0.5, 1, and 2.5 μM) were sufficient to protect about 70% of L929 cells. Around a 25–30% antitumor activity was found in lung adenocarcinoma (A549) cells for both extracts. Therefore, further studies on the extract should be conducted to verify whether this compound can be used as an alternative therapeutic approach to control hyperglycemia in diabetic patients or as a potential anti-cancer supplement.

Author Contributions: Conceptualization, E.L., M.K., E.H.S. and K.C.; methodology, E.L., M.K., E.H.S. and K.C.; software, E.L. and A.A.F.Z.; validation, E.L. and A.A.F.Z.; formal analysis, E.L., M.K., E.H.S., P.A.D.M. and K.C.; investigation, E.L., M.K., E.H.S. and K.C.; resources, D.d.O. and A.A.F.Z.; data curation, E.L.; writing—original draft preparation, E.L., M.K., E.H.S. and K.C.; writing—review and editing, D.d.O. and A.A.F.Z.; visualization, E.L., D.d.O. and A.A.F.Z.; supervision, D.d.O. and A.A.F.Z.; project administration, D.d.O. and A.A.F.Z. All authors have read and agreed to the published version of the manuscript.

Funding: This research was funded by Brazilian agencies, CNPq (Conselho Nacional de Desenvolvimento Científico e Tecnológico) and CAPES-PRINT (Project numbers 88887.310560/2018-00 and 88887.310727/2018-00).

Institutional Review Board Statement: Not applicable.

Informed Consent Statement: Not applicable.

Data Availability Statement: The data are contained within this article.

Conflicts of Interest: The authors declare no conflicts of interest.

References

1. Moraes, C.A.M.; Fernandes, I.J.; Calheiro, D.; Kieling, A.G.; Brehm, F.A.; Rigon, M.R.; Berwanger Filho, J.A.; Schneider, I.A.H.; Osorio, E. Review of the Rice Production Cycle: By-Products and the Main Applications Focusing on Rice Husk Combustion and Ash Recycling. *Waste Manag. Res.* **2014**, *32*, 1034–1048. [CrossRef]
2. Nagrale, S.D.; Hajare, H.; Modak, P.R. Utilization of Rice Husk Ash. *Int. J. Eng. Res. Appl.* **2012**, *2*, 1–5.
3. Van Hoed, V.; Depaemelaere, G.; Ayala, J.V.; Santiwattana, P.; Verhé, R.; De Greyt, W. Influence of Chemical Refining on the Major and Minor Components of Rice Bran Oil. *JAOCS J. Am. Oil Chem. Soc.* **2006**, *83*, 315–321. [CrossRef]

4. Das, A.B.; Goud, V.V.; Das, C. Extraction of Phenolic Compounds and Anthocyanin from Black and Purple Rice Bran (*Oryza sativa* L.) Using Ultrasound: A Comparative Analysis and Phytochemical Profiling. *Ind. Crops Prod.* **2017**, *95*, 332–341. [CrossRef]
5. Leonarski, E.; Kuasnei, M.; Moraes, P.A.D.; Cesca, K.; de Oliveira, D.; Zielinski, A.A.F. Pressurized Liquid Extraction as an Eco-Friendly Approach to Recover Anthocyanin from Black Rice Bran. *Innov. Food Sci. Emerg. Technol.* **2023**, *86*, 103372. [CrossRef]
6. Ji, Y.; Liu, D.; Jin, Y.; Zhao, J.; Zhao, J.; Li, H.; Li, L.; Zhang, H.; Wang, H. In Vitro and in Vivo Inhibitory Effect of Anthocyanin-Rich Bilberry Extract on α-Glucosidase and α-Amylase. *LWT* **2021**, *145*, 111484. [CrossRef]
7. Sui, X.; Zhang, Y.; Zhou, W. Bread Fortified with Anthocyanin-Rich Extract from Black Rice as Nutraceutical Sources: Its Quality Attributes and in Vitro Digestibility. *Food Chem.* **2016**, *196*, 910–916. [CrossRef]
8. Leonarski, E.; Kuasnei, M.; Cesca, K.; de Oliveira, D.; Zielinski, A.A.F. Black Rice and Its By-Products: Anthocyanin-Rich Extracts and Their Biological Potential. *Crit. Rev. Food Sci. Nutr.* **2023**, 1–19. [CrossRef]
9. Chemat, F.; Rombaut, N.; Sicaire, A.G.; Meullemiestre, A.; Fabiano-Tixier, A.S.; Abert-Vian, M. Ultrasound Assisted Extraction of Food and Natural Products. Mechanisms, Techniques, Combinations, Protocols and Applications. A Review. *Ultrason. Sonochem.* **2017**, *34*, 540–560. [CrossRef]
10. Moreira, S.A.; Alexandre, E.M.C.; Pintado, M.; Saraiva, J.A. Effect of Emergent Non-Thermal Extraction Technologies on Bioactive Individual Compounds Profile from Different Plant Materials. *Food Res. Int.* **2019**, *115*, 177–190. [CrossRef]
11. Belwal, T.; Devkota, H.P.; Hassan, H.A.; Ahluwalia, S.; Ramadan, M.F.; Mocan, A.; Atanasov, A.G. Phytopharmacology of Acerola (*Malpighia* spp.) and Its Potential as Functional Food. *Trends Food Sci. Technol.* **2018**, *74*, 99–106. [CrossRef]
12. Chen, Y.; Zhang, W.; Zhao, T.; Li, F.; Zhang, M.; Li, J.; Zou, Y.; Wang, W.; Cobbina, S.J.; Wu, X.; et al. Adsorption Properties of Macroporous Adsorbent Resins for Separation of Anthocyanins from Mulberry. *Food Chem.* **2016**, *194*, 712–722. [CrossRef]
13. Das, A.B.; Goud, V.V.; Das, C. Adsorption/Desorption, Diffusion, and Thermodynamic Properties of Anthocyanin from Purple Rice Bran Extract on Various Adsorbents. *J. Food Process Eng.* **2018**, *41*, e12834. [CrossRef]
14. Hernández, D.F.; Mojica, L.; Berhow, M.A.; Brownstein, K.; Lugo Cervantes, E.; Gonzalez de Mejia, E. Black and Pinto Beans (*Phaseolus vulgaris* L.) Unique Mexican Varieties Exhibit Antioxidant and Anti-Inflammatory Potential. *Food Res. Int.* **2023**, *169*, 112816. [CrossRef]
15. Carail, M.; Fabiano-Tixier, A.-S.; Meullemiestre, A.; Chemat, F.; Caris-Veyrat, C. Effects of High Power Ultrasound on All-E-β-Carotene, Newly Formed Compounds Analysis by Ultra-High-Performance Liquid Chromatography–Tandem Mass Spectrometry. *Ultrason. Sonochem.* **2015**, *26*, 200–209. [CrossRef]
16. Giusti, M.M.; Wrolstad, R.E. Characterization and Measurement of Anthocyanins by UV-Visible Spectroscopy. *Curr. Protoc. Food Anal. Chem.* **2001**, 1–13. [CrossRef]
17. Brand-Williams, W.; Cuvelier, M.E.; Berset, C. Use of a Free Radical Method to Evaluate Antioxidant Activity. *LWT Food Sci. Technol.* **1995**, *28*, 25–30. [CrossRef]
18. Re, R.; Pellegrini, N.; Proteggente, A.; Pannala, A.; Yang, M.; Rice-Evans, C. Antioxidant Activity Applying an Improved ABTS Radical Cation Decolorization Assay. *Free Radic. Biol. Med.* **1999**, *26*, 1231–1237. [CrossRef]
19. Barik, S.K.; Russell, W.R.; Moar, K.M.; Cruickshank, M.; Scobbie, L.; Duncan, G.; Hoggard, N. The Anthocyanins in Black Currants Regulate Postprandial Hyperglycaemia Primarily by Inhibiting α-Glucosidase while Other Phenolics Modulate Salivary α-Amylase, Glucose Uptake and Sugar Transporters. *J. Nutr. Biochem.* **2020**, *78*, 108325. [CrossRef]
20. Ali, H.; Houghton, P.J.; Soumyanath, A. α-Amylase Inhibitory Activity of Some Malaysian Plants Used to Treat Diabetes; with Particular Reference to *Phyllanthus amarus*. *J. Ethnopharmacol.* **2006**, *107*, 449–455. [CrossRef]
21. *ISO 10993-5:2009*; Biological Evaluation of Medical Devices—Part 5: Tests for in Vitro Cytotoxicity. International Organization for Standardization: Geneva, Switzerland, 2009.
22. Aguilar, A.E.M.; Fagundes, A.P.; Macuvele, D.L.P.; Cesca, K.; Porto, L.; Padoin, N.; Soares, C.; Gracher Riella, H. Green Synthesis of Nano Hydroxyapatite: Morphology Variation and Its Effect on Cytotoxicity against Fibroblast. *Mater. Lett.* **2021**, *284*, 129013. [CrossRef]
23. Pitz, H.D.S.; Pereira, A.; Blasius, M.B.; Voytena, A.P.L.; Affonso, R.C.L.; Fanan, S.; Trevisan, A.C.D.; Ribeiro-Do-Valle, R.M.; Maraschin, M. In Vitro Evaluation of the Antioxidant Activity and Wound Healing Properties of Jaboticaba (*Plinia peruviana*) Fruit Peel Hydroalcoholic Extract. *Oxid. Med. Cell. Longev.* **2016**, *2016*, 3403586. [CrossRef]
24. Sousa, M.H.O.; Morgan, J.M.S.; Cesca, K.; Flach, A.; de Moura, N.F. Cytotoxic Activity of *Cunila angustifolia* Essential Oil. *Chem. Biodivers.* **2020**, *17*, 3–7. [CrossRef]
25. Gille, A.; Trautmann, A.; Posten, C.; Briviba, K. Bioaccessibility of Carotenoids from *Chlorella Vulgaris* and *Chlamydomonas Reinhardtii*. *Int. J. Food Sci. Nutr.* **2016**, *67*, 507–513. [CrossRef]
26. Chen, L.; Yang, M.; Mou, H.; Kong, Q. Ultrasound-Assisted Extraction and Characterization of Anthocyanins from Purple Corn Bran. *J. Food Process. Preserv.* **2018**, *42*, e13377. [CrossRef]
27. Heinonen, J.; Farahmandazad, H.; Vuorinen, A.; Kallio, H.; Yang, B.; Sainio, T. Extraction and Purification of Anthocyanins from Purple-Fleshed Potato. *Food Bioprod. Process.* **2016**, *99*, 136–146. [CrossRef]
28. Kang, Y.J.; Jung, S.W.; Lee, S.J. An Optimal Extraction Solvent and Purification Adsorbent to Produce Anthocyanins from Black Rice (*Oryza sativa* Cv. Heugjinjubyeo). *Food Sci. Biotechnol.* **2014**, *23*, 97–106. [CrossRef]
29. Chang, X.L.; Wang, D.; Chen, B.Y.; Feng, Y.M.; Wen, S.H.; Zhan, P.Y. Adsorption and Desorption Properties of Macroporous Resins for Anthocyanins from the Calyx Extract of Roselle (*Hibiscus sabdariffa* L.). *J. Agric. Food Chem.* **2012**, *60*, 2368–2376. [CrossRef]

30. Wathon, M.H.; Beaumont, N.; Benohoud, M.; Blackburn, R.S.; Rayner, C.M. Extraction of Anthocyanins from *Aronia Melanocarpa* Skin Waste as a Sustainable Source of Natural Colorants. *Color Technol.* **2019**, *135*, 5–16. [CrossRef]
31. Contreras, J.; Alcázar-Valle, M.; Lugo-Cervantes, E.; Luna-Vital, D.A.; Mojica, L. Mexican Native Black Bean Anthocyanin-Rich Extracts Modulate Biological Markers Associated with Inflammation. *Pharmaceuticals* **2023**, *16*, 874. [CrossRef]
32. Zhao, Z.; Wu, M.; Zhan, Y.; Zhan, K.; Chang, X.; Yang, H.; Li, Z. Characterization and Purification of Anthocyanins from Black Peanut (*Arachis hypogaea* L.) Skin by Combined Column Chromatography. *J. Chromatogr. A* **2017**, *1519*, 74–82. [CrossRef] [PubMed]
33. Jeyaraj, E.J.; Lim, Y.Y.; Choo, W.S. Antioxidant, Cytotoxic, and Antibacterial Activities of *Clitoria ternatea* Flower Extracts and Anthocyanin-Rich Fraction. *Sci. Rep.* **2022**, *12*, 14890. [CrossRef] [PubMed]
34. Chumchoochart, W.; Sutthanut, K. Anti-Obesity Potential of Glutinous Black Rice Bran Extract: Anti-Adipogenesis and Lipolysis Induction in 3T3-L1 Adipocyte Model. *Songklanakarin J. Sci. Technol.* **2020**, *42*, 284–291. [CrossRef]
35. Marathe, S.J.; Shah, N.N.; Bajaj, S.R.; Singhal, R.S. Esterification of Anthocyanins Isolated from Floral Waste: Characterization of the Esters and Their Application in Various Food Systems. *Food Biosci.* **2021**, *40*, 100852. [CrossRef]
36. Sansenya, S.; Nanok, K. A-glucosidase, A-amylase Inhibitory Potential and Antioxidant Activity of Fragrant Black Rice (Thai Coloured Rice). *Flavour Fragr. J.* **2020**, *35*, 376–386. [CrossRef]
37. Shimoda, H.; Aitani, M.; Tanaka, J.; Hitoe, S. Purple Rice Extract Exhibits Preventive Activities on Experimental Diabetes Models and Human Subjects. *Rice Res. Open Access* **2015**, *3*, 137. [CrossRef]
38. Choi, K.; Choi, S.-I.; Park, M.H.; Han, J.-S. Cyanidin-3-O-Glucoside Ameliorates Postprandial Hyperglycemia in Diabetic Mice. *J. Life Sci.* **2017**, *27*, 32–37. [CrossRef]
39. Aalim, H.; Wang, D.; Luo, Z. Black Rice (*Oryza sativa* L.) Processing: Evaluation of Physicochemical Properties, in Vitro Starch Digestibility, and Phenolic Functions Linked to Type 2 Diabetes. *Food Res. Int.* **2021**, *141*, 109898. [CrossRef]
40. Akkarachiyasit, S.; Yibchok-Anun, S.; Wacharasindhu, S.; Adisakwattana, S. In Vitro Inhibitory Effects of Cyandin-3-Rutinoside on Pancreatic α-Amylase and Its Combined Effect with Acarbose. *Molecules* **2011**, *16*, 2075–2083. [CrossRef] [PubMed]
41. Alam, F.; Shafique, Z.; Amjad, S.T.; Bin Asad, M.H.H. Enzymes Inhibitors from Natural Sources with Antidiabetic Activity: A Review. *Phyther. Res.* **2019**, *33*, 41–54. [CrossRef]
42. Tundis, R.; Loizzo, M.R.; Menichini, F. *Natural Products As-Amylase and-Glucosidase Inhibitors and Their Hypoglycaemic Potential in the Treatment of Diabetes: An Update*; Bentham Science Publishers: Sharjah, United Arab Emirates, 2010; Volume 10.
43. Hui, X.; Wu, G.; Han, D.; Stipkovits, L.; Wu, X.; Tang, S.; Brennan, M.A.; Brennan, C.S. The Effects of Bioactive Compounds from Blueberry and Blackcurrant Powders on the Inhibitory Activities of Oat Bran Pastes against α-Amylase and α-Glucosidase Linked to Type 2 Diabetes. *Food Res. Int.* **2020**, *138*, 109756. [CrossRef]
44. Sangkitikomol, W.; Tencomnao, T.; Rocejanasaroj, A. Effects of Thai Black Sticky Rice Extract on Oxidative Stress and Lipid Metabolism Gene Expression in Hepg2 Cells. *Genet. Mol. Res.* **2010**, *9*, 2086–2095. [CrossRef]
45. Aprodu, I.; Milea, S.A.; Anghel, R.M.; Enachi, E.; Barbu, V.; Crăciunescu, O.; Râpeanu, G.; Bahrim, G.E.; Oancea, A.; Stănciuc, N. New Functional Ingredients Based on Microencapsulation of Aqueous Anthocyanin-Rich Extracts Derived from Black Rice (*Oryza sativa* L.). *Molecules* **2019**, *24*, 3389. [CrossRef] [PubMed]
46. Wang, E.; Liu, Y.; Xu, C.; Liu, J. Antiproliferative and Proapoptotic Activities of Anthocyanin and Anthocyanidin Extracts from Blueberry Fruits on B16-F10 Melanoma Cells. *Food Nutr. Res.* **2017**, *61*, 1325308. [CrossRef] [PubMed]
47. Ereminas, G.; Majiene, D.; Sidlauskas, K.; Jakstas, V.; Ivanauskas, L.; Vaitiekaitis, G.; Liobikas, J. Neuroprotective Properties of Anthocyanidin Glycosides against H_2O_2-Induced Glial Cell Death Are Modulated by Their Different Stability and Antioxidant Activity in Vitro. *Biomed. Pharmacother.* **2017**, *94*, 188–196. [CrossRef] [PubMed]
48. Zhang, B.; Kang, M.; Xie, Q.; Xu, B.; Sun, C.; Chen, K.; Wu, Y. Anthocyanins from Chinese Bayberry Extract Protect β Cells from Oxidative Stress-Mediated Injury via HO-1 Upregulation. *J. Agric. Food Chem.* **2011**, *59*, 537–545. [CrossRef] [PubMed]
49. Tan, J.; Li, P.; Xue, H.; Li, Q. Cyanidin-3-Glucoside Prevents Hydrogen Peroxide (H_2O_2)-Induced Oxidative Damage in HepG2 Cells. *Biotechnol. Lett.* **2020**, *42*, 2453–2466. [CrossRef] [PubMed]
50. Palungwachira, P.; Tancharoen, S.; Dararat, P.; Nararatwanchai, T. Anthocyanins Isolated from *Oryza sativa* L. Protect Dermal Fibroblasts from Hydrogen Peroxide-Induced Cell Death. *J. Nat. Sci. Biol. Med.* **2020**, *11*, 45–54. [CrossRef]
51. Xu, L.; Choi, T.H.; Kim, S.; Kim, S.H.; Chang, H.W.; Choe, M.; Kwon, S.Y.; Hur, J.A.; Shin, S.C.; Chung, J., II; et al. Anthocyanins from Black Soybean Seed Coat Enhance Wound Healing. *Ann. Plast. Surg.* **2013**, *71*, 415–420. [CrossRef]
52. Xue, H.; Tan, J.; Li, Q.; Tang, J.; Cai, X. Ultrasound-Assisted Enzymatic Extraction of Anthocyanins from Raspberry Wine Residues: Process Optimization, Isolation, Purification, and Bioactivity Determination. *Food Anal. Methods* **2021**, *14*, 1369–1386. [CrossRef]
53. Chen, X.; Zhang, W.; Xu, X. Cyanidin-3-Glucoside Suppresses the Progression of Lung Adenocarcinoma by Downregulating TP53I3 and Inhibiting PI3K/AKT/MTOR Pathway. *World J. Surg. Oncol.* **2021**, *19*, 232. [CrossRef] [PubMed]
54. Lu, J.N.; Panchanathan, R.; Lee, W.S.; Kim, H.J.; Kim, D.H.; Choi, Y.H.; Kim, G.S.; Shin, S.C.; Hong, S.C. Anthocyanins from the Fruit of *Vitis coignetiae Pulliat* Inhibit TNF-Augmented Cancer Proliferation, Migration, and Invasion in A549 Cells. *Asian Pacific J. Cancer Prev.* **2017**, *18*, 2919–2923. [CrossRef]

foods

MDPI

Article

Zn Biofortification of Dutch Cucumbers with Chemically Modified Spent Coffee Grounds: Zn Enrichment and Nutritional Implications

Beatriz Navajas-Porras [1], Ana Cervera-Mata [2], Alejandro Fernández-Arteaga [3], Adriana Delgado-Osorio [1], Miguel Navarro-Moreno [1], Daniel Hinojosa-Nogueira [1], Silvia Pastoriza [1], Gabriel Delgado [2], Miguel Navarro-Alarcón [1] and José Ángel Rufián-Henares [1,4,*]

[1] Departamento de Nutrición y Bromatología, Instituto de Nutrición y Tecnología de Alimentos, Centro de Investigación Biomédica, Universidad de Granada, 18011 Granada, Spain; beatriznavajas@ugr.es (B.N.-P.); adrianadelgado@ugr.es (A.D.-O.); miguelnav@correo.ugr.es (M.N.-M.); dhinojosa@ugr.es (D.H.-N.); spdelacueva@ugr.es (S.P.); nalarcon@ugr.es (M.N.-A.)

[2] Department of Soil Science and Agricultural Chemistry, Faculty of Pharmacy, University of Granada, 18011 Granada, Spain; anacervera@ugr.es (A.C.-M.); gdelgado@ugr.es (G.D.)

[3] Department of Chemical Engineering, University of Granada, 18071 Granada, Spain; jandro@ugr.es

[4] Instituto de Investigación Biosanitaria Ibs.Granada, Universidad de Granada, 18014 Granada, Spain

* Correspondence: jarufian@ugr.es; Tel.: +34-958-24-28-41; Fax: +34-958-24-95-77

Abstract: Spent coffee grounds (SCGs) are a food waste with a large generation around the world. However, their utilization as a soil organic amendment is difficult due to their phytotoxic effect. In the present work, the impact of agronomic biofortification on Dutch cucumbers was studied by using different chemically modified SCGs, analyzing their effects on Zn content, the release of antioxidant capacity and the production of short-chain fatty acids after in vitro digestion–fermentation. The results indicated variations in the Zn content and chemical composition of cucumbers according to the treatment groups. The functionalized with Zn and activated SCGs were able to increase Zn levels in cucumbers. Meanwhile, the activated hydrochar obtained at 160 °C and the activated and functionalized with Zn SCGs showed the highest Zn supply per serving. Differences in the antioxidant capacity and short-chain fatty acid production were observed between the groups. It is concluded that the growing conditions and the presence of Zn may significantly influence the contribution of these cucumbers to the dietary intake of nutrients and antioxidants, which could have important implications for human health and nutrition.

Keywords: cucumber; Zn; spent coffee grounds (SCGs); hydrochar; in vitro digestion; in vitro fermentation; antioxidant capacity; short-chain fatty acids

1. Introduction

Spent coffee grounds (SCGs) are the main by-product obtained during coffee brewing, mainly in coffee shops, with a worldwide production of around 15 million tons/year [1]. SCGs have a positive impact on soil quality, so that they can improve crop growth in the short term by increasing organic matter levels, soil fertility and carbon stock [2]. However, some compounds present in the dregs can be toxic and can reduce microbial diversity when they accumulate, so their application should be moderate. As an alternative, biochar is being used, which comprises stable carbonaceous materials that can remain in the soil for hundreds of years, improving crop fertility and plant growth [3]. It also has environmental benefits, as it can reduce greenhouse gas emissions and promotes the sustainable management of organic waste. Hydrochar, on the other hand, is similar to biochar, but is produced from hydrolysis instead of pyrolysis [4].

To improve the physical, chemical and biological properties of crops and soil quality, organic amendment (organic matter added to croplands) is carried out. This practice is commonly used for its positive influence on soil's biodiversity and biochemical parameters and provision of essential nutrients for crop growth, among others [5]. Despite this, organic amendment also has some disadvantages, as the quality and composition of amendments may vary, being limited in certain regions, or supplying nutrients more slowly than other fertilizers [6]. Due to this, alternatives to organic matter have been proposed for use in crop improvement.

In the context of organic amendment, agronomic biofortification is a procedure that aims to increase the content of essential nutrients in crops and combat malnutrition in populations that depend on crops as their main food source [7]. Biofortification can increase the mineral content in the edible portion of the crop, improving the nutritional value of food. In addition, micronutrient fertilization can have a positive impact on other nutritional parameters of crops, such as the protein content, amino acids, phenolic compounds, chlorophyll, carotenoids and essential oils [8]. One of the most used elements for agronomic biofortification is Zn, since it is an essential mineral for humans as well as plants [7,9].

Cucumber (*Cucumis sativus* L.) belongs to the Cucurbitaceae family and is one of the most cultivated vegetables in the world. Despite its high water content, it contains bioactive compounds such as tannins, terpenoids, saponins, cardiac glycosides and dietary fiber that provide remarkable antioxidant capacity [10]. In this sense, dietary fiber (and attached phenolic compounds) is poorly digested by human beings, being transformed by the gut microbiota in the colon into simpler metabolites with enhanced absorption and bioactivity [11]. The main metabolites of dietary fiber are short-chain fatty acids (SCFAs) such as acetic, propionic and butyric acids, with demonstrated health properties [12]. In addition, the gut microbiota also releases antioxidant compounds from the food matrix [13].

Considering all the above stated information, the aim of this study was to evaluate the effect of agronomic biofortification of Dutch cucumbers by using different soil treatments based on SCGs: activated spent coffee grounds (ASCGs), Zn-functionalized and activated coffee grounds (ASCG-Zn), activated hydrochar of SCGs at 160 °C (AH160) and Zn-functionalized and activated SCG hydrochar (AH160-Zn). Modifications in the Zn content, as well as the release of antioxidant capacity and production of SCFAs after human digestion–fermentation, were assessed. Finally, the contribution of the consumption of cucumbers to the daily intake of Zn and polyphenols in the Spanish diet was also calculated. It is important to highlight that mineral biofortification is an innovative and relevant strategy due to the importance of minerals in health, the prevalence of mineral deficiencies in the diet and its potential as a sustainable agronomic solution [14]. Thus, the novelty of this paper compared to existing ones relies on three aspects: (i) the use of a food waste (spent coffee grounds) as a source of bio-chelates for mineral biofortification; (ii) the study of the effect of Zn biofortification on other nutritional parameters (such as total phenolic content and release of antioxidant capacity and short-chain fatty acids after in vitro digestion–fermentation); and (iii) the evaluation of the contribution to daily intake of the biofortified cucumbers on Zn and polyphenols intake.

2. Materials and Methods

2.1. SCGs Bio-Products

SCGs were acquired from the cafeteria of the Faculty of Pharmacy (University of Granada). They were spread into a thin layer and dried at room temperature (18–21 °C) to remove residual moisture for 7 days. Four soil treatments (based on SCGs) were obtained applying the same methodology described by Lara-Ramos et al. [3]: SCGs hydrochar at 160 °C (H160), activated SCGs (ASCGs) and hydrochar (AH160), and their corresponding Zn bio-chelates activated and functionalized (ASCG-Zn, AH160-Zn). The Zn content of the bio-chelates was determined using mineralization with HNO_3 (65%) and H_2O_2 at 185 °C for 20 min in a microwave digestor (Multiwave 5000 with Rotor 24HVT50, Anton Paar GmbH, Graz, Austria). Mineral ions in the previous extracts were measured using inductively coupled plasma mass spectrometry (ICP-MS/MS; Agilent 8900 Triple Quadrupole ICP-

MS/MS, Agilent Technologies Inc., Santa Clara, CA, USA). The final Zn content of the bio-products used in this study, expressed in mg/kg, was as follows: ASCGs: 6.5 ± 0.5; AH160: 15.5 ± 0.1; ASCG-Zn: 11,123 ± 242; AH160-Zn: 11,799 ± 241.

2.2. Greenhouse Experiment

The experiment was conducted in a greenhouse of the "Fundación Cajamar" Experimental Station located in Almería, Spain (36°47′23″ N, 2°43′13″ W; 155 m a.s.l.) from October 2022 to January 2023. The site is located in a semi-arid subtropical Mediterranean climate with an average minimum temperature of 12.2 °C and an average maximum temperature of 16.1 °C corresponding to the aforementioned period; with these oscillations in external temperature the mean temperature in the greenhouse was 23.0 °C (ranging from 19.2 to 25.6 °C). The average annual precipitation is 220.5 mm while the mean solar radiation is 4576 Wh/m^2 day. The plot used had dimensions of 30 m long and 8.5 m wide, with a total useful area of 255 m^2.

The soil of the experiment was previously characterized by the staff of the experimental station. According to the internal report the soil has the following properties: sand 69.79%, silt 17.57%, clay 12.64%, pH 7.4, electrical conductivity at 25 °C (EC25) 2.05 dS/m, organic carbon (OC) 1.35%, carbonates as $CaCO_3$ 29.2%, C/N 8.06, total N 0.17%, available P 318.1 mg/kg.

The assayed bio-products at doses of 0.2% were as follows: ASCGs, AH160, ASCG-Zn, and AH160-Zn. These bio-products were added at 0.2%, since this is the proper dosage to avoid phytotoxicity for plants [3]. The 0.2% dosage corresponds to different amounts of Zn added (mg/kg) with the bio-chelates: ASCG: 0.011; AH160: 0.026; ASCG-Zn: 19; AH160-Zn: 20. Two controls were established: soil without any bio-product (Control) and a commercial Zn chelate at a concentration of 10 mg/kg soil (Control-Zn). The commercial chelate Zn ethylenediaminetetraacetate (EDTA-Zn, 14% *w:w*) was supplied by Trade Corporation International, S.A.U. As a summary, the study involved a total of six treatments (Table 1). The experiment was set up with four replicates in split-plots design. Two plots were used, each plot containing one replicate per treatment randomly distributed, composed of 4 plants. Overall, each treatment consisted of 8 plants (n = 8) per plot.

Table 1. Treatment details of the greenhouse experiment.

N°	Treatment	Description
1	Control	No bio-product
2	Control-Zn	Commercial chelate (EDTA-Zn, 14%)
3	ASCGs	Activated spent coffee grounds (SCGs)
4	AH160	Activated hydrochar obtained from SCGs at 160 °C
5	ASCG-Zn	Activated SCGs functionalized with Zn
6	AH160-Zn	Activated hydrochar obtained from SCGs at 160 °C functionalized with Zn

The planting points were distributed as follows: 1.5 m between crop lines and 0.5 m between plants within each crop line, meaning a planting density of 1.33 plants/m^2. For the placement of the bio-products, a 20 cm deep × 15 cm diameter hole, equivalent to 4 kg of soil, was made in each planting point. The soil of all holes corresponding to the same treatment was extracted and homogenized for 5 min with the corresponding bio-product in a concrete mixer. The holes were then individually re-filled with the soil–bio-product mixture. The holes of the controls had the same soil treatment. The commercial chelate was dissolved in distilled water and poured at the surface.

Cucumber seedlings (*Cucumis sativus* L. var. 'Almería' cv. 'Huracán') that were 21 days old were acquired from the commercial greenhouse Acrena S.A.T 251 (Almería, Spain). One seedling per planting point was planted in 8 cm soil depth after 6 days of the soil preparation.

2.3. Crop Maintenance

The experiment was managed organically, using fertilizers authorized for organic cultivation by applying the drip irrigation method. Each row of cucumber was irrigated using a polyethylene pipe. The emitters were spaced at a 50 cm interval with a daily irrigation rate of 3 L/(h.m^2) from 8:00 till 11:00. For the first 15 days after transplant, only water was used for irrigation; from day 16 till the end of the experiment the cultivation was irrigated with a nutrient solution containing 10 mM/L of N, 1.41 mM/L of P and 5.56 mM/L of K based on an organic fertilizer with a richness of 3.5% N, 2.5% P_2O_5 and 6.5% K_2O diluted to a concentration of 3.2‰. The concentration of nitrate, potassium, calcium and sodium in the extracted soil solution was monitored biweekly using suction probes to maintain a nitrate concentration in a range of 3 to 10 mM/L and a K/Ca ratio between 0.5 and 1. These concentrations were measured using rapid analysis ionometers (LAQUAtwin, Horiba, Japan).

2.4. Plant Sampling and Processing

Cucumbers were harvested after 105 days of cultivation. Then, the fruits were cleaned, chopped and divided for antioxidant capacity of Zn analysis. Those samples for antioxidant capacity and total phenolic content were stored at −80 °C until in vitro digestion–fermentation. The samples for Zn analysis were dried in aluminum trays at room temperature for 48 h for water loss, then placed in an oven at 50 °C for 72 h and weighed again (dry weight). The dried material was milled and stored at room temperature until analysis.

2.5. Zn Determination

Cucumbers were harvested following a 105-day cultivation period. Subsequently, the fruits underwent cleaning, chopping and division for antioxidant capacity and Zn analysis. Samples designated for antioxidant capacity assessment were preserved at −80 °C until subjected to in vitro digestion–fermentation. Meanwhile, samples intended for Zn analysis underwent a drying process in aluminum trays at room temperature for 48 h to eliminate water content. They were then placed in an oven at 50 °C for 72 h to achieve a constant dry weight. The dried material was subsequently ground into powder and stored at room temperature till analysis.

For Zn determination, 0.200 g of the homogenized and dried sample was weighed using a precision balance (Ohaus, model PA224C, Europe GmbH, Greifensee, Switzerland) in borosilicate tubes. To these tubes, 3 mL of 69% HNO_3 (TraceSELECT, Honeywell, Fluka, France), 0.5 mL of 30% H_2O_2 (Merck Suprapur, Darmstadt, Germany) and 0.5 mL of Milli-Q reagent-grade water were added. Prior to mineralization, 3 mL of an 11.5% HNO_3 solution was added to the Teflon cups of the microwave digester, where the borosilicate tubes containing the samples were placed. After optimization of a suitable time-temperature program (20 min at a temperature range of 150 to 185 °C), mineralization of the samples was carried out.

Post-mineralization, the samples were diluted to 50 mL with Milli-Q reagent-grade water to prepare the analytical solution. Zn content in the final analytical dissolution was measured using ICP-MS/MS. Calibration curves were prepared using serial dilutions from a standard solution of 1000 mg/L of zinc in HNO_3 at 1% (ppm; Merck; Darmstadt, Germany). The "Internal Standard Kit (Ge, Ir, Rh Sc; ISC Science, batch 20210712)" was utilized for correcting counts per second (CPS) for the analyzed atomic mass of Zn (^{66}Zn) to that of ^{72}Ge. Measurements were conducted using the linear calibration method and performed in triplicate for each sample.

Prior to Zn concentration determination using the ICP/MS technique, the analytical parameters of the procedure were validated. The limit of detection (LOD) for Zn was determined to be 0.42 µg/L. Accuracy and precision of the method were evaluated using reference standards certified in Zn, with recovery percentages ranging from 98.3 to 102.1% and coefficients of variation averaging 3.56%. These results demonstrate the suitability of the mineralization and analysis technique for measuring Zn in cucumber samples. The results are expressed as mg Zn/100 g of fresh weight.

To evaluate the efficiency of the different bio-products used, the Zn utilization efficiency (UE) was calculated according to Zhao et al. [15]:

$$UE\ (\%) = (\text{Uptake in treatment-uptake in control})/\text{Micronutrient added} \times 100$$

2.6. In Vitro Digestion and Fermentation

Samples were subjected to in vitro gastrointestinal fermentation and in vitro fermentation in triplicate, following previously described protocols [16]. First, an oral phase was carried out by adding cucumbers to falcon tubes along with simulated salivary fluid (1:1, *w*/*v*) consisting of salts (KCl, KH_2PO_4, $NaHCO_3$, NaCl, $MgCl_2(H_2O)_6$, $NH_4(CO_3)_2$, $CaCl_2(H_2O)_2$ and HCl) and α-amylase (75 U/mL). The falcon tubes were kept for 2 min at 37 °C under oscillation.

Subsequently, the gastric phase was performed. For this, 5 mL of simulated gastric fluid was added, simulating the content of gastric juices in salts (same salts as in the oral phase) and pepsin (2000 U/mL). On this occasion, the mixture was kept at 37 °C for 120 min, at pH 3 and in oscillation. Finally, the intestinal phase was carried out, in which 10 mL of simulated intestinal fluid was added, simulating the content of intestinal juices in salts (same salts as in the oral and gastric phase), bile salts and enzymes (specifically 67.2 mg/mL pancreatin was used). The mixture was kept at 37 °C for 120 min in oscillation as in the gastric phase, but this time at pH 7. Once the intestinal phase was finished, the tubes were placed on ice and the enzymatic reactions were stopped. The tubes were then centrifuged for 10 min at 3500 rpm. The resulting solid pellet served as an in vitro fermentation substrate, and represents the undigested portion entering the large intestine. The supernatant, which represents the fraction available for absorption in the small intestine, was stored in 1 mL tubes at −80 °C until analysis.

In vitro fermentation was carried out using faecal samples from five healthy donors with an average Body Mass Index of 21.3 who had abstained from antibiotics during the three months prior to the trial. The eligibility criteria of the European project Stance4Health were used for healthy donors [17]. The informed consent document was signed by faecal donors. That form included all of the information of the study as well as the exclusion and inclusion criteria. The study was conducted according to the guidelines of the Declaration of Helsinki. It was approved by the Ethics Committee of the University of Granada (protocol code 1080/CEIH/2020). The faeces were pulled to reduce inter-individual variability. The fermentation was carried out for 20 h at 37 °C under agitation. After completion of the in vitro fermentation, the samples were placed on ice, as after digestion, to stop microbial reactions, and then centrifuged at 3500 rpm for 10 min. The supernatant, representing the fraction available for absorption in the large intestine, was stored at −80 °C until analysis. The solid pellet, representing the unfermented portion, was properly disposed of.

The result of the in vitro gastrointestinal digestion and fermentation was two fractions: the digestion supernatant, which corresponds to absorption in the small intestine, and the fermentation supernatant, which corresponds to absorption in the large intestine.

2.7. Antioxidant Tests

The antioxidant capacity of the supernatants obtained after digestion and fermentation were studied. The sum of the antioxidant capacity of both was considered the total antioxidant capacity [18].

2.7.1. Trolox Equivalent Antioxidant Capacity Referred to Reducing Capacity ($TEAC_{FRAP}$ Assay)

The procedure used to determine the ability of the samples to reduce ferric iron was based on that described by Benzie and Strain [19] and modified for use with a microplate reader (Cytation 5, Agilent Technologies Inc., Santa Clara, CA, USA). Briefly, 280 μL of freshly prepared FRAP reagent was combined with 20 μL of digestion or fermentation supernatant in a 96-well plate. The antioxidant response was observed for half an hour. Trolox was used to create a calibration curve at a concentration of 0.01 to 4 mg/mL. The results are expressed using mmol Trolox equivalent/kg fresh weight.

2.7.2. Trolox Equivalent Antioxidant Capacity against DPPH Radicals ($TEAC_{DPPH}$ Assay)

The technique was modified to work with a microplate reader (Cytation 5, Agilent Technologies Inc., Santa Clara, CA, USA) and followed the protocol of Yen and Chen [20]. Briefly, 280 µL of DPPH reagent was combined with 20 µL of digestion supernatant or fermentation supernatant in duplicate. To make this reagent, 7.4 mg DPPH per 100 mL methanol was used. The antioxidant response was observed for a full hour. Trolox was used to create a calibration curve, with concentrations ranging from 0.01 to 4 mg/mL. Results are given in mmol Trolox equivalent/kg fresh weight.

2.7.3. Trolox Equivalent Antioxidant Capacity against ABTS+ Radicals ($TEAC_{ABTS}$ Assay)

The ABTS assay was conducted following the method described by Ozgen et al. [21] with minor adjustments. A stock solution of 7 mM ABTS was mixed with 2.45 mM potassium persulfate to generate ABTS+, which was then left to incubate at room temperature for 12–16 h in the absence of light before utilization. The ABTS working solution, which remained stable for two days, was diluted with a mixture of ethanol and water (50:50) to achieve an absorbance of 0.70 ± 0.02 at 730 nm. After a 20 min incubation period, absorbance measurements were taken using a Cytation 5 microplate reader (Agilent Technologies Inc., Santa Clara, CA, USA). Trolox dilutions ranging from 0.15 to 1.15 mM were employed for calibration purposes. Results are reported in mmol Trolox equivalent/kg fresh weight.

2.7.4. Total Phenolic Content: Folin–Ciocalteu (FC) Assay

A microplate reader (Cytation 5, Agilent Technologies Inc., Santa Clara, CA, USA) was used to estimate the total phenolic content. The procedure of Singleton and Rossi [22] was followed, with some modifications. The FC assay was carried out. The results are expressed as gallic acid equivalents (GAE) per kg of fresh weight.

2.8. Short-Chain Fatty Acids Analysis

Following a previous study [23], the analysis of SCFAs was performed using ultra-high performance liquid chromatography (UHPLC). There was no need for sample pre-treatment prior to injection. SCFAs standards, prepared in the mobile phase at concentrations ranging from 5 to 10,000 ppm, were quickly created. The mobile phase was supplied at a flow rate of 0.250 mL/min and consisted of a mixture of two solutions. The mobile phase consisted of aqueous acetonitrile (1%) and ultrapure water (99%), both acidified with 1% formic acid. After fermentation, 1 mL of the supernatant was centrifuged, filtered through a 0.22 µm nylon filter and then transferred to a vial for UHPLC analysis. The UV-Vis photodiode array detector (PDA) was set at 210 nm and the column was a reversed-phase Accucore™ C18 (ThermoFisher Scientific, Waltham, MA, USA) with a particle size of 2.6 µm and 150 mm in length. The analysis was performed twice and the information shown represents the mean values of the millimolar (mM) concentration of each SCFA.

2.9. Calculations of Daily Antioxidant and Zn Intakes

The contribution of each cucumber group to the overall dietary intake of polyphenols and Zn was assessed individually according to treatment variations and plot distinctions. This assessment took into account several factors: (i) the average daily cucumber consumption in Spain and portion size [24]; (ii) the total phenolic content of the samples previously determined after the FC test; (iii) the Zn content within each cucumber group. In addition, the total phenolic and Zn content of each food was standardized to the typical serving size in Spain [25] and resembled previously documented findings on polyphenol [26] and Zn [27] intake.

2.10. Statistical Analysis

First, the normality of the samples was studied using the Shapiro–Wilk test. Then, a one-way analysis of variance (ANOVA) was performed for statistical differences between treatments which were assessed using Tukey's test. Student's *t*-test was also performed for graphical plots. The significance level was set at 95% ($p < 0.05$). Statistical analysis was carried out in triplicate and performed using control (non-bio-product) cucumbers as the reference group, as well as comparisons between all cucumber groups with different treatments. Statgraphics Plus, version 5.1, as well as SPSS 26.0 for Windows (IBM SPSS Inc., New York, NY, USA), was used.

3. Results

3.1. Zn Content

The results of Zn content for each group of cucumbers showed statistically significant differences ($p < 0.05$) in both plots (Figure 1a,b). First, comparisons were made with the control group (Figure 1a). In cucumbers from plot 1, the AH160, AH160-Zn and ASCG-Zn groups showed higher Zn levels ($p < 0.05$) than those of the control group. The treatment that increased to a greater extent the Zn content in cucumbers was ASCG-Zn, reaching levels of 0.13 mg/100 g fresh weight. The opposite was found for cucumbers of the ASCGs and Control-Zn groups, which reported lower Zn content ($p < 0.05$) than the control group. When comparisons were made between the different groups (Table 2), in addition to those already mentioned with the control group, statistically significant differences ($p < 0.05$) were found between all groups.

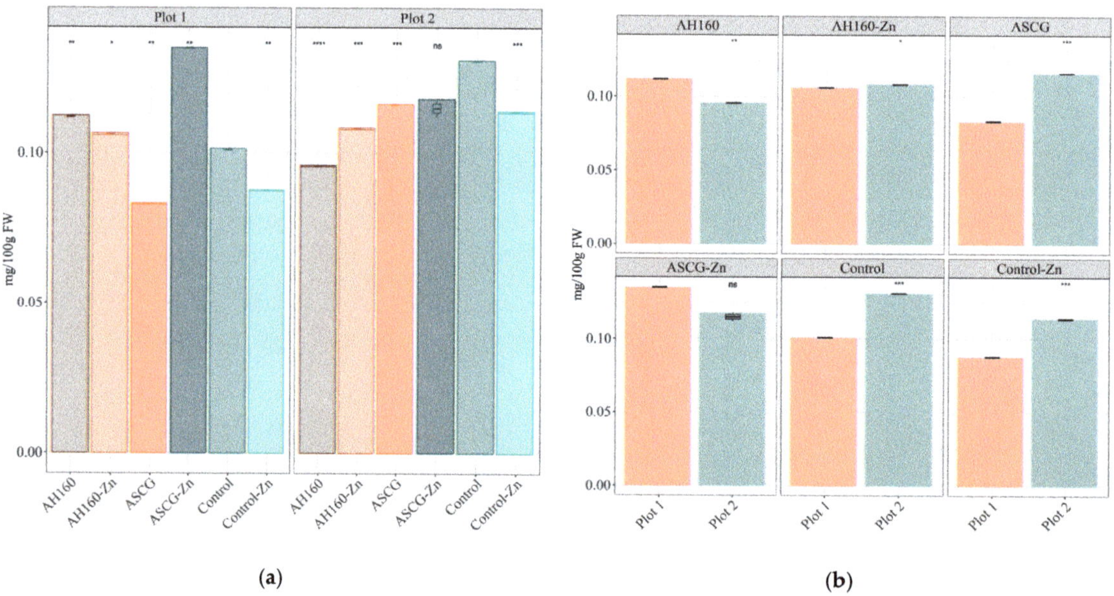

Figure 1. Differences in the Zn content of cucumbers. (**a**) Differences among treatment groups. Statistical analysis was performed using Student's *t* test using the control group as the reference group. Statistical labels: *: $p < 0.05$, **: $p < 0.01$, ***: $p < 0.001$, ****: $p < 0.0001$, *ns*: not significant. (**b**) Differences between plots for each group. Statistical analysis was performed using Student's *t*-test using Plot 1 as the reference group. Statistical labels: *: $p < 0.05$, **: $p < 0.01$, ***: $p < 0.001$, ns: not significant.

Table 2. Differences in Zn content and antioxidant capacity (FRAP, DPPH, ABTS and FC assays) between cucumber groups. Differences in the two plots were studied separately. Statistical analysis was performed using ANOVA and Tukey's test. Statistical labels: *: $p < 0.05$, ns: not significant.

	Zn Content	FRAP Assay	DPPH Assay	ABTS Assay	FC Assay
		Plot 1			
AH160-Zn/AH160	*	ns	ns	ns	ns
ASCGs/AH160	*	ns	ns	ns	ns
ASCG-Zn/AH160	*	ns	ns	*	ns
Control-Zn/AH160	*	ns	ns	*	ns
ASCGs/AH160-Zn	*	ns	ns	ns	ns
ASCG-Zn/AH160-Zn	*	ns	ns	*	ns
Control-Zn/AH160-Zn	*	ns	ns	*	ns
ASCG-Zn/ASCGs	*	ns	ns	*	ns
Control-Zn/ASCGs	*	ns	ns	*	ns
Control-Zn/ASCG-Zn	*	ns	ns	*	ns
	Zn Content	FRAP assay	DPPH Assay	ABTS Assay	FC Assay
		Plot 2			
AH160-Zn/AH160	*	ns	ns	ns	*
ASCGs/AH160	*	ns	ns	ns	ns
ASCG-Zn/AH160	*	ns	ns	ns	*
Control-Zn/AH160	*	ns	ns	ns	*
ASCGs/AH160-Zn	*	ns	ns	ns	*
ASCG-Zn/AH160-Zn	*	ns	ns	*	*
Control-Zn/AH160-Zn	ns	ns	ns	ns	ns
ASCG-Zn/ASCGs	ns	ns	ns	ns	ns
Control-Zn/ASCGs	ns	ns	ns	ns	*
Control-Zn/ASCG-Zn	ns	ns	ns	*	*

In the case of plot 2, all cucumbers from the different groups (except the ASCG-Zn group) had lower Zn content ($p < 0.05$) than those from the control group, which reported higher Zn levels (reaching concentrations of 0.13 mg/100 g fresh weight). When comparisons were made between the different groups (Table 2), statistically significant differences ($p < 0.05$) were found between the AH160 group and all other groups. Almost the same was found for the AH160-Zn group: statistically significant differences between this group and all the other groups were found, except with control-Zn group.

Finally, differences between the two plots were studied for each cucumber group (Figure 1b). Statistically significant differences were found for all groups except for the ASCG-Zn group. Cucumbers from plot 2 had more Zn ($p < 0.05$) than those from plot 1 for the AH160-Zn, ASCGs, control and control-Zn groups, while the opposite was found for the AH160 group.

3.2. *Antioxidant Capacity*

Regarding the antioxidant capacity released after in vitro digestion and fermentation, comparisons were made with those cucumbers belonging to the control group. For the FRAP method, no statistically significant differences were found (Figure 2a). In addition, comparisons were made between all cucumber groups taking into account the total antioxidant capacity. For neither of the two plots were statistically significant differences found between both groups (Table 2).

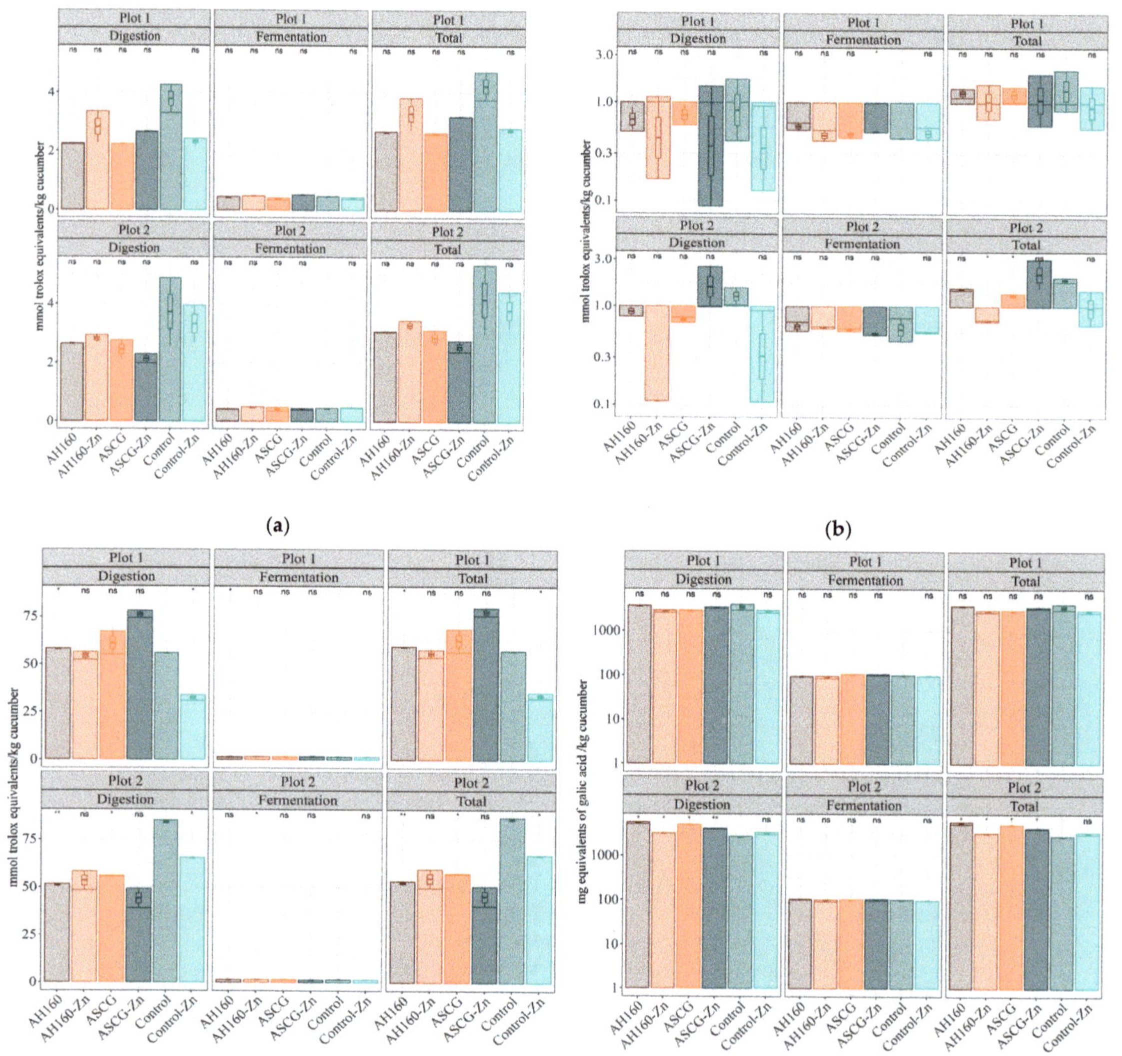

Figure 2. Antioxidant capacity of the different cucumber groups (obtained after in vitro digestion, in vitro fermentation and total antioxidant capacity). (**a**) The Trolox equivalent antioxidant capacity referred to reducing capacity ($TEAC_{FRAP}$). (**b**) The Trolox equivalent antioxidant capacity against DPPH radicals ($TEAC_{DPPH}$). (**c**) The Trolox equivalent antioxidant capacity against ABTS radicals ($TEAC_{ABTS}$). (**d**) The total phenolic content (Folic–Ciocalteu). Results were log10 transformed to improve visualization. Statistical analysis was performed using Student's *t* test using the control group as the reference group. Statistical labels: *: $p < 0.05$, **: $p < 0.01$, ns: not significant.

In the case of the DPPH method (Figure 2b), comparing all groups with respect to the control, statistically significant differences ($p < 0.05$) were found in the in vitro fermentation fraction for samples from plot 1, where ASCG-Zn cucumbers showed higher antioxidant capacity than those of the control group. In plot 2, statistically significant differences were found for total antioxidant capacity, where ASCGs and AH160-Zn cucumbers showed significantly ($p < 0.05$) lower antioxidant capacity than the control group. When studying the comparisons between the different groups (Table 2) there were no significant differences in their antioxidant capacity.

For the ABTS method, when comparisons were made with the control group (Figure 2c), significant differences were found for both factions (in vitro digestion, in vitro fermentation) and total antioxidant capacity. In plot 1, statistically significant differences ($p < 0.05$) were obtained between the AH160 cucumbers and the control group, as well as between the control-Zn and the control group (in the digestion fraction and total antioxidant capacity); the antioxidant capacity of AH160 was higher and that of the control-Zn group. In the in vitro fermentation fraction, the antioxidant capacity of the AH160 group was also significantly ($p < 0.05$) higher than that of the control group. On the other hand, for the samples from plot 2, the antioxidant capacity of the AH160, ASCGs and control-Zn groups were lower ($p < 0.05$) than that of the control group (for the in vitro digestion fraction and total antioxidant capacity). In the in vitro fermentation fraction, only the antioxidant capacity of the AH160-Zn group was significantly lower ($p < 0.05$) than that of the control group. When multiple comparisons (ANOVA) were performed among all cucumber groups (Table 2), statistically significant differences were also found. In addition to those already stated for plot 1 with respect to the control group, it was found that the antioxidant capacity of ASCG-Zn cucumbers was significantly higher than that of the AH160, AH160-Zn, ASCG and control-Zn groups ($p < 0.05$). On the other hand, the control-Zn group showed lower antioxidant capacity than AH160 and ASCG groups ($p < 0.05$). For plot 2, again the antioxidant capacity of the ASCG-Fe group was higher than that of the AH160 and control-Zn groups ($p < 0.05$).

Finally, for the FC method, no significant differences were found in plot 1 (Figure 2d) when comparisons were made with the control group. In plot 2, the antioxidant capacity of the AH160, AH160-Zn, ASCGs and ASCG-Zn groups was higher ($p < 0.05$) than that of the control group (in the in vitro fraction and total antioxidant capacity). In the in vitro fermentation fraction, no significant differences were found. In the ANOVA analysis (Table 2) no significant differences were found in plot 1. For plot 2, in addition to the aforementioned differences with respect to the control group, it was found a higher antioxidant capacity in AH160 cucumbers compared to the other groups ($p < 0.05$), except for the ASCG group (no differences were found). In addition, the antioxidant capacity of the AH160-Zn group was lower ($p < 0.05$) than that of the ASCG and ASCG-Zn groups. The control-Zn group had a lower antioxidant capacity than the ASCGs and ASCG-Zn groups.

Differences in Antioxidant Capacity in the In Vitro Digestion–Fermentation Fractions

Figure 3 depicts the antioxidant capacity of both fractions obtained after in vitro digestion and fermentation. In general, the antioxidant capacity after in vitro fermentation decreased considerably compared to that obtained after in vitro digestion. This occurred for both ABTS and FC methods and in both plots. For the FRAP method, the fraction belonging to in vitro fermentation did not reach 25% for any of the treated groups of cucumbers, nor for any of the plots. In the case of the DPPH method, it was observed that the contribution to the total antioxidant capacity of the in vitro fermentation fraction was higher than for the other methods, although none of the treated cucumber groups reached 50% for plot 1; in addition, the AH160-Fe group exceeded 75% in plot 2. Therefore, in vitro digestion contributes more to the release of total antioxidant capacity.

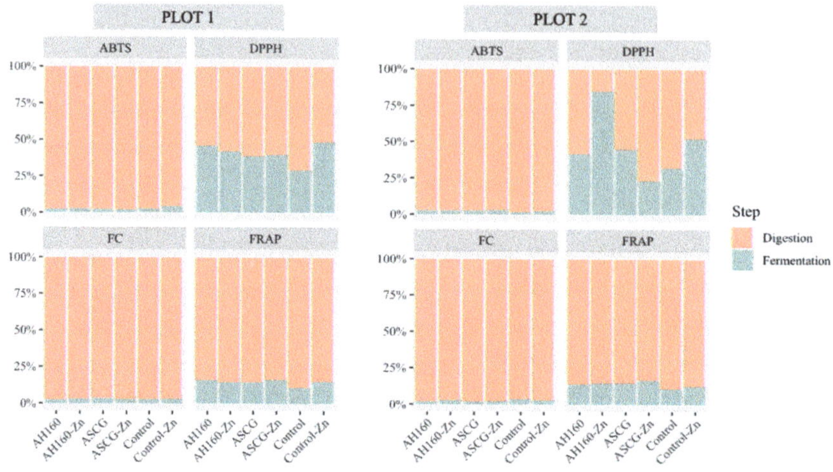

Figure 3. Contribution to the total antioxidant capacity (ABTS, DPPH, FC and FRAP) of each fraction in every group of cucumbers for both plots.

3.3. *Short-Chain Fatty Acids Produced after In Vitro Fermentation*

The production of SCFAs after in vitro fermentation of cucumbers from all groups was measured. First, comparisons were made between the different cucumber groups and the control, and then comparisons were made between all groups. The results of plot 1 (Figure 4a) showed that, for acetic acid, all cucumber groups were statistically significantly different ($p < 0.05$) to the control group, except the ASCGs and control-Zn groups, where no differences were found. For butyric acid, higher levels were produced after fermentation of AH160, AH160-Zn and ASCG-Zn cucumbers ($p < 0.05$) compared to the control group. For lactic and propionic acids, significant differences ($p < 0.05$) were found in the production of these short-chain fatty acids in all groups of cucumbers with respect to the control group (except for lactic acid production after control-Zn group fermentation). The same was true for succinic acid, although here, the ASCG-Zn group did not show significant differences with the control group. Finally, regarding the production of total SCFAs, statistically significant differences ($p < 0.05$) were found for AH160 and ASCG-Zn cucumber groups with the control group. Similar results were obtained for plot 2 (Figure 4b). When multiple comparisons (ANOVA) were made between all groups (Table 3), numerous statistically significant differences were found. Most of the differences were found for succinic acid production than for the other SCFAs.

Table 3. Differences in short-chain fatty acids production between cucumber groups. Differences in the two plots were studied separately. Statistical analysis was performed using ANOVA and Tukey's test. Statistical labels: *: $p < 0.05$, ns: not significant.

Plot 1	Acetic Acid	Butyric Acid	Lactic Acid	Propionic Acid	Succinic Acid	Total SCFAs
AH160-Zn/AH160	*	ns	ns	*	ns	*
ASCGs/AH160	*	*	*	*	*	*
ASCG-Zn/AH160	ns	*	*	*	*	*
Control-Zn/AH160	*	*	*	*	*	*

Table 3. *Cont.*

ASCGs/AH160-Zn	*	*	*	*	ns	*
ASCG-Zn/AH160-Zn	*	*	*	*	*	*
Control-Zn/AH160-Zn	*	*	*	*	*	ns
ASCG-Zn/ASCGs	*	*	*	ns	*	*
Control-Zn/ASCGs	*	*	*	*	ns	*
Control-Zn/ASCG-Zn	*	*	*	*	*	*
Plot 2	**Acetic Acid**	**Butyric Acid**	**Lactic Acid**	**Propionic Acid**	**Succinic Acid**	**Total SCFAs**
AH160-Zn/AH160	*	*	*	ns	*	*
ASCGs/AH160	ns	*	*	*	*	ns
ASCG-Zn/AH160	ns	ns	*	ns	*	ns
Control-Zn/AH160	ns	*	*	*	*	ns
ASCGs/AH160-Zn	*	*	*	*	*	*
ASCG-Zn/AH160-Zn	*	*	*	ns	ns	*
Control-Zn/AH160-Zn	*	*	*	ns	ns	*
ASCG-Zn/ASCGs	*	*	*	*	*	ns
Control-Zn/ASCGs	ns	ns	*	*	*	ns
Control-Zn/ASCG-Zn	ns	*	*	ns	*	*

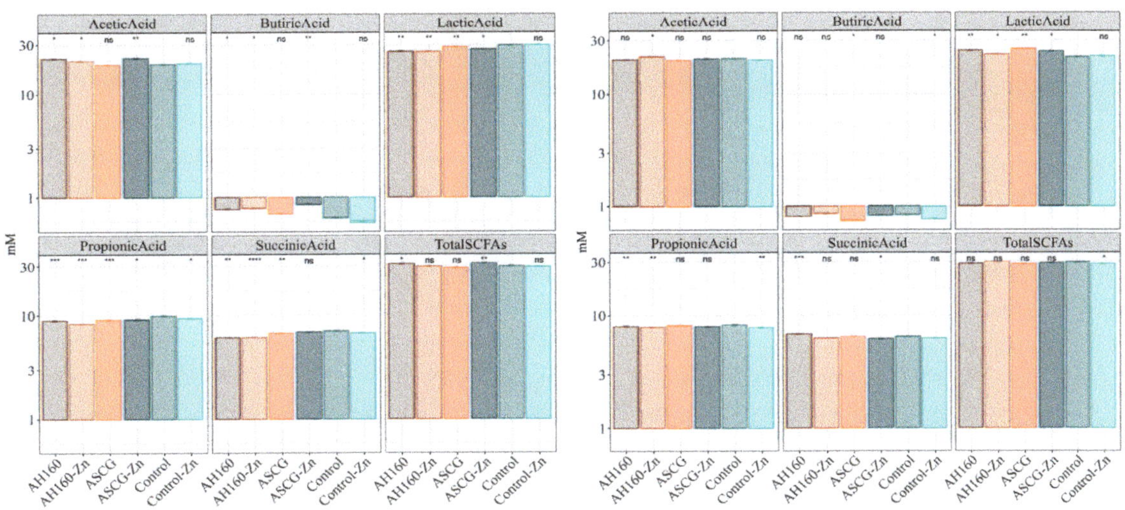

(a) (b)

Figure 4. SCFAs produced after in vitro fermentation of cucumbers. (**a**) Plot 1. (**b**) Plot 2. Results were log10 transformed to improve visualization. Statistical analysis performed using Student's t test (control group as reference). Statistical labels: *: $p < 0.05$, **: $p < 0.01$, ***: $p < 0.001$, ****: $p < 0.0001$, ns: not significant.

3.4. Influence of the Different Study Variables in Both Plots

In order to study the significance of the different variables studied in both plots, a principal component analysis was performed (Figure 5). Since the variables included in the analysis where very different (Zn content, antioxidant capacity and SCFAs) the variability explained for both plots was not very high (total 58.2%, 37.6 for PC1 and 20.6 for PC2). The samples from plot 1 showed higher contributions to the explained variability of lactic, succinic, propionic and acetic acids, total SCFAs, ABTS and FRAP. In the case of plot 2, DPPH, FC and butyric acid showed high contributions. It is well known that soil structure affects plant growth in many ways. The uptake of water and nutrients by plants can be limited by inadequate contact with the solid and liquid phases of the soil [28]. This could explain why such differences in soil could have a direct impact on the nutritional properties of cucumbers.

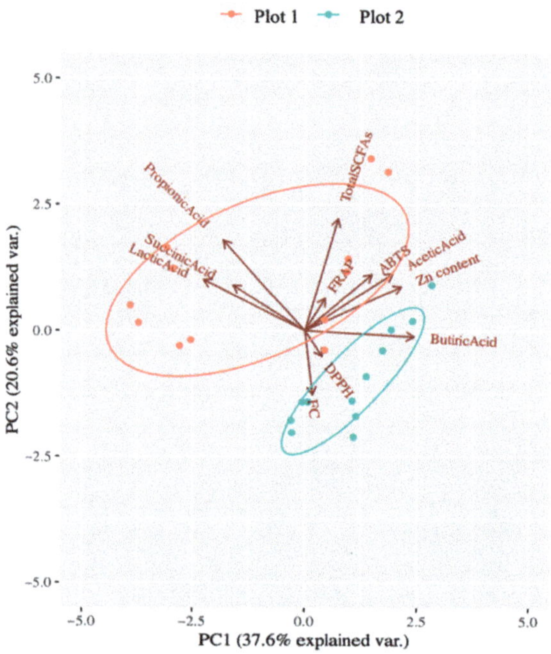

Figure 5. Principal component analysis (PCA) biplot chart. Total antioxidant capacity (FRAP, DPPH, ABTS and FC), short-chain fatty acids production and Zn content of the samples were taken into account. Arrows indicate in which plot the variable influence is higher.

4. Discussion

In this research it has been tested if spent coffee grounds, a bio-waste with a toxic potential for plants that also contributes to CO_2 release into the atmosphere [3,5], can be re-used after proper transformation to improve the nutritional value of vegetables by means of agronomic biofortification with Zn. In general, the bio-chelates that mostly increased the Zn content in cucumbers were the activated and functionalized hydrochar from SCGs in both plots (Figure 1a). To better understand the dynamics of Zn in this greenhouse experiment, the Zn utilization efficiency (UE) for cucumbers in relation to the total amount of added Zn was calculated. The following order was obtained in plot 1: AH160 (40.034 ± 0.132) > ASCG-Zn (0.174 ± 0.003) > AH160-Zn (0.024 ± 0.005) > control-Zn (−0.133 ± 0.008) > ASCGs (−162.857 ± 7.519). In the case of plot 2, the results for this parameter were as follows: ASCG-Zn (−0.081 ± 0.019) > AH160-Zn (−0.108 ± 0.003) > control-Fe (−0.169 ± 0.007) > AH160 (−129.716 ± 2.504) > ASCGs (−133.728 ± 0.140). If we compare these results with those in [3], which used activated and functionalized Zn hydrochar for growing lettuces, very similar trends were obtained. The activated and functionalized spent coffee grounds (ASCG-Zn) had a higher UE than the activated and functionalized hydrochar at 160 °C, both being higher than the commercial Zn-chelate. This is a positive outcome when comparing the bio-chelates with the control-Zn cucumbers. In fact, the Zn contained in the hydrochar particles [3] and activated and functionalized SCGs (ASCG-Zn) could be released into the medium over time and have a residual effect in subsequent cropping cycles [29]. On the other hand, the commercial chelate (control-Zn) had a lower utilization efficiency, which can be explained by the leaching of Zn over the cultivation period. The soil moisture was been maintained at a potential of −20 kPa (which is higher than field capacity, −33 kPa), indicating excess water and a continuous leaching process. According to Weil and Brady [30], Zn is a micronutrient that is less available under conditions of high soil leaching.

The aim of agronomic biofortification is to improve the nutritional value of selected cultivars; in our case, cucumbers enriched with Zn. Thus, once demonstrated that SCGs could be a source of bio-chelates with biofortification activity, the amount of Zn provided to the diet by each group of cucumbers was studied and compared with the European Zn reference intake for middle-aged men [27]. The AH160 and ASCG-Zn groups (Table 4) were the groups with the highest contribution of Zn per serving (AH160: 0.20 mg for plot 1 and 0.17 mg for plot 2; ASCG-Zn: 0.15 mg for plot 1 and 0.20 for plot 2) as well as when compared to the reference intakes. In contrast, the ASCGs group provided the lowest daily Zn levels to the diet per intake (and per serving) in plot 1, and the control group in plot 2, which is justified by the utilization efficiency data previously presented. The contribution of the Zn provided by cucumbers taking into account their daily intake in Spain was very low (from 0.06 to 0.09% of the daily needs). The percentages of contribution per serving were higher (from 1.39 to 2.25% of the daily Zn requirements) but still low, as most of the Zn in the diet comes from foods of animal origin [31].

Table 4. Contribution of cucumber consumption to daily antioxidant capacity and Zn intake in the Spanish diet.

Group	Analytical Assay	Zn/Daily Intake * (mg Zn/day)		Zn/Serving Intake † (mg Zn/Serving)		Mean Contribution to Daily Zn Intake Compared with Previous Data # (%)		Mean Contribution to Daily Zn per Serving Intake # (%)	
		Plot 1	Plot 2	Plot 1	Plot 2	Plot 1	Plot 2	Plot 1	Plot 2
Control	Zn content	0.01	0.01	0.17	0.14	0.07	0.06	1.87	1.60
Control-Zn	Zn content	0.01	0.01	0.16	0.16	0.07	0.07	1.77	1.81
ASCGs	Zn content	0.00	0.01	0.12	0.17	0.06	0.08	1.39	1.94
AH160	Zn content	0.01	0.01	0.20	0.17	0.09	0.08	2.25	1.92
ASCG-Zn	Zn content	0.01	0.01	0.15	0.20	0.07	0.09	1.68	2.18
AH160-Zn	Zn content	0.01	0.01	0.13	0.17	0.06	0.08	1.46	1.90
Group	**Analytical assay**	**Polyphenols/daily intake * (mg gallic acid/day)**		**Polyphenols/serving intake † (mg gallic acid/serving)**		**Mean contribution to daily polyphenols intake compared with previous data ★ (%)**		**Mean contribution to daily polyphenols intake per serving ★ (%)**	
		Plot 1	**Plot 2**	**Plot 1**	**Plot 2**	**Plot 1**	**Plot 2**	**Plot 1**	**Plot 2**
Control	FC assay	21.1	16.3	529	407	1.81	1.39	45.1	34.7
Control-Zn	FC assay	16.7	19.3	417	484	1.42	1.65	35.6	41.3
ASCGs	FC assay	16.8	29.9	421	748	1.44	2.56	35.9	63.9
AH160	FC assay	21.8	32.9	546	822	1.87	2.81	46.6	70.2
ASCG-Zn	FC assay	20.3	24.9	506	622	1.73	2.13	43.2	53.2
AH160-Zn	FC assay	16.7	19.3	418	482	1.43	1.65	35.7	41.1

* Considering daily consumption for the Spanish population. † Considering the intake of 1 serving. # Considering dietary reference intakes [27]. ★ Considering previous data [26].

Cucumbers are vegetable foods that can contribute to the daily intake of bioactive compounds such as polyphenols [31]. According to a previous study [26], the daily intake of phenolic compounds in Spain is 1171 mg gallic acid equivalents/person/day (assessed with the Folin–Ciocalteu assay). Of these, approximately 80% come from fruits, vegetables and cereals. In our study, Table 4 shows that the group of cucumbers with the lowest contribution to the daily intake of polyphenols was the control-Zn and AH160-Zn groups for plot 1 and the control group for plot 2. On the other hand, the cucumbers with the highest contribution to the daily intake of phenolic compounds were the AH160 group, reaching 546 (plot 1) and 822 (plot 2) mg/gallic acid/serving (Table 4). When the percentage of contribution to the daily intake of polyphenols was studied, it ranged from 1.39 to 2.13%

when the mean daily consumption of cucumbers in Spain was taken into account. However, if a more realistic approach is used (the intake per serving), then the contribution will reach up to a 70.2% (Table 4). Finally, in the case of other antioxidant capacity detection methods, the ASCG-Zn group stood out as the one with the highest antioxidant capacity, for example, with the ABTS method in plot 1 and the DPPH method in plot 2 (Figure 2b,c).

Another important fact to be mentioned is the higher antioxidant capacity released during digestion compared to that released after fermentation. This could be related with the high digestibility of cucumber, since most of the compounds with antioxidant capacity were released after in vitro digestion and were not available for metabolization by the colonic microbiota.

Other authors have published data on the antioxidant capacity of different vegetables. In particular, for cucumber the values were 0.7 and 0.4 mmol Trolox/Kg cucumber for $TEAC_{FRAP}$ and $TEAC_{ABTS}$, respectively [32]. If we compare these data with those presented in this article, our results are much higher, ranging from 33.8 to 85.8 mmol Trolox/Kg cucumber for $TEAC_{FRAP}$, and from 2.6 to 4.2 and mmol Trolox/Kg cucumber for $TEAC_{ABT}$. These differences come from the previous in vitro digestion and fermentation step applied to our samples, which increases the release of antioxidant species. Wojtunik-Kulesza et al. [33] argued that in vitro digestion can have a significant effect on the release of antioxidant compounds from foods because those antioxidant compounds present in foods undergo structural changes during digestion, affecting their antioxidant capacity. Other authors also argue [34] that during in vitro digestion there is an increase in the concentration of bioactive compounds, improving their bioavailability at target sites, thus enhancing their antioxidant properties.

Regarding SCFAs, these are health-promoting metabolites produced by gut microbes that feed on undigested nutrients such as dietary fiber [35]. The differences found in SCFAs production were numerous in both plots and are important to be discussed. Van Leeuwen and colleagues [36] studied the influence of soil on vine development and grape quality. They described how the soil had a great influence on plant production, nutritional status and grape composition, so that different soil types can provide unique conditions that affect both plant growth and vegetable and fruit production, as observed in our experiments with Dutch cucumbers.

5. Conclusions

In this work, it has been demonstrated that SCGs can be transformed into smart Zn biochelates. Therefore, SCGs can be re-used for the biofortification of Dutch cucumbers, also avoiding their contaminant properties when disposed in landfills. The cucumbers grown with spent coffee grounds functionalized with Zn (ASCG-Zn) had the highest Zn levels and contribution to the daily intake of Zn for plot 2, while in the case of plot 1 activated SCGs hydrochar (AH160) were those with the highest contribution to Zn intake; however, in general, the Zn contribution of cucumbers to the human diet was low. Cucumbers grown with SCGs hydrochar (AH160) showed the highest contribution to the daily intake of polyphenols; however, in terms of antioxidant capacity, statistically significant differences were hardly found. Diversity in SCFAs production was also observed between groups of cucumbers, showing that the groups had a different chemical composition. Both the Zn content and the chemical composition of cucumbers may vary significantly depending on the growing conditions, which in turn may affect the contribution of these cucumbers to the dietary intake of nutrients and antioxidants, which may have important implications for human health and nutrition.

Author Contributions: Conceptualization, S.P. and J.Á.R.-H.; methodology, A.C.-M., M.N.-A., B.N.-P., D.H.-N., M.N.-M. and A.D.-O.; software, A.F.-A., M.N.-A. and M.N.-M.; validation, M.N.-A. and S.P.; formal analysis, G.D., M.N.-A., M.N.-M. and S.P.; investigation, B.N.-P., D.H.-N., A.D.-O. and G.D.; resources, J.Á.R.-H.; data curation, B.N.-P., A.C.-M. and M.N.-A.; writing—original draft, B.N.-P., A.C.-M., M.N.-M. and J.Á.R.-H.; writing—review and editing, A.F.-A., G.D., M.N.-A. and J.Á.R.-H.; visualization, S.P. and J.Á.R.-H.; supervision, M.N.-A. and J.Á.R.-H.; project administration,

J.Á.R.-H.; funding acquisition, J.Á.R.-H. All authors have read and agreed to the published version of the manuscript.

Funding: This work was supported by research project P20_00585 from the Consejería de Economía, Conocimiento, Empresas y Universidad of the Andalusia Government; by the European Research Commission (Research Executive Agency) under the research project Stance4Health (Contract N° 816303); and by the Plan propio de Investigación y Transferencia of the University of Granada under the program "Intensificación de la Investigación, modalidad B".

Institutional Review Board Statement: Not applicable.

Informed Consent Statement: Not applicable.

Data Availability Statement: The original contributions presented in the study are included in the article, further inquiries can be directed to the corresponding author.

Acknowledgments: The authors thank the support of the Unit of Excellence 'UNETE' from the University of Granada (reference UCE-PP2017-05).

Conflicts of Interest: The authors declare no conflicts of interest.

Abbreviations

SCG: Spent Coffee Grounds, TEAC: Trolox Equivalent Antioxidant Capacity, GAE: Gallic Acid Equivalents, FC: Folin-Ciocalteu, SCFAs: Short Chain Fatty Acids.

References

1. Kamil, M.; Ramadan, K.M.; Awad, O.I.; Ibrahim, T.K.; Inayat, A.; Ma, X. Environmental impacts of biodiesel production from waste spent coffee grounds and its implementation in a compression ignition engine. *Sci. Total Environ.* **2019**, *675*, 13–30. [CrossRef] [PubMed]
2. Pérez-Burillo, S.; Cervera-Mata, A.; Fernández-Arteaga, A.; Pastoriza, S.; Rufián-Henares, J.Á.; Delgado, G. Why should we be concerned with the use of spent coffee grounds as an organic amendment of soils? A Narrative Review. *Agronomy* **2022**, *12*, 2771. [CrossRef]
3. Lara-Ramos, L.; Cervera-Mata, A.; Fernández-Bayo, J.; Navarro-Alarcón, M.; Delgado, G.; Fernández-Arteaga, A. hydrochars derived from spent coffee grounds as Zn bio-chelates for agronomic biofortification. *Sustainability* **2023**, *15*, 10700. [CrossRef]
4. Palansooriya, K.N.; Ok, Y.S.; Awad, Y.M.; Lee, S.S.; Sung, J.-K.; Koutsospyros, A.; Moon, D.H. Impacts of biochar application on upland agriculture: A review. *J. Environ. Manag.* **2019**, *234*, 52–64. [CrossRef] [PubMed]
5. Garbowski, T.; Bar-Michalczyk, D.; Charazińska, S.; Grabowska-Polanowska, B.; Kowalczyk, A.; Lochyński, P. An overview of natural soil amendments in agriculture. *Soil Till. Res.* **2023**, *225*, 105462. [CrossRef]
6. Larney, F.J.; Angers, D.A. The role of organic amendments in soil reclamation: A review. *Can. J. Soil Sci.* **2012**, *92*, 19–38. [CrossRef]
7. Szerement, J.; Szatanik-Kloc, A.; Mokrzycki, J.; Mierzwa-Hersztek, M. Agronomic biofortification with Se, Zn, and Fe: An effective strategy to enhance crop nutritional quality and stress defense—A review. *J. Soil Sci. Plant Nutr.* **2022**, *22*, 1129–1159. [CrossRef]
8. Buturi, C.V.; Mauro, R.P.; Fogliano, V.; Leonardi, C.; Giuffrida, F. Mineral biofortification of vegetables as a tool to improve human diet. *Foods* **2021**, *10*, 223. [CrossRef] [PubMed]
9. Di Gioia, F.; Petropoulos, S.A.; Ozores-Hampton, M.; Morgan, K.; Rosskopf, E.N. Zn and iron agronomic biofortification of brassicaceae microgreens. *Agronomy* **2019**, *9*, 667. [CrossRef]
10. Sharma, V.; Sharma, L.; Sandhu, K.S. Cucumber (*Cucumis sativus* L.). In *Antioxidants in Vegetables and Nuts—Properties and Health Benefits*; Nayik, G.A., Gull, A., Eds.; Springer: Singapore, 2020; pp. 333–340.
11. Thomson, C.; Garcia, A.L.; Edwards, C.A. Interactions between Dietary Fibre and the Gut Microbiota. *Proc. Nutr. Soc.* **2021**, *80*, 398–408. [CrossRef]
12. Campos-Perez, W.; Martinez-Lopez, E. Effects of Short Chain Fatty Acids on Metabolic and Inflammatory Processes in Human Health. *Biochim. Biophys. Acta Mol. Cell Biol. Lipids* **2021**, *1866*, 158900. [CrossRef] [PubMed]
13. Navajas-Porras, B.; Pérez-Burillo, S.; Valverde-Moya, Á.J.; Hinojosa-Nogueira, D.; Pastoriza, S.; Rufián-Henares, J.A. Effect of Cooking Methods on the Antioxidant Capacity of Plant Foods Submitted to In Vitro Digestion–Fermentation. *Antioxidants* **2021**, *9*, 1312. [CrossRef] [PubMed]
14. Ofori, K.F.; Antoniello, S.; English, M.M.; Aryee, A.N.A. Improving nutrition through biofortification-A systematic review. *Front. Nutr.* **2022**, *9*, 1043655. [CrossRef] [PubMed]
15. Zhao, A.; Yang, S.; Wang, B.; Tian, X. Effects of ZnSO4 and Zn-EDTA applied by broadcasting or by banding on soil Zn fractions and Zn uptake by wheat (*Triticum aestivum L.*) under greenhouse conditions. *J. Plant Nutr. Soil Sci.* **2019**, *182*, 307–317. [CrossRef]

16. Pérez-Burillo, S.; Molino, S.; Navajas-Porras, B.; Valverde-Moya, A.J.; Hinojosa-Nogueira, D.; López-Maldonado, A.; Pastoriza, S.; Rufián-Henares, J.A. An in vitro batch fermentation protocol for studying the contribution of food to gut microbiota composition and functionality. *Nat. Prot.* **2021**, *16*, 3186–3209. [CrossRef] [PubMed]
17. Dello Russo, M.; Russo, P.; Rufián-Henares, J.Á.; Hinojosa-Nogueira, D.; Pérez-Burillo, S.; de la Cueva, S.P.; Rohn, S.; Fatouros, A.; Douros, K.; González-Vigil, V.; et al. The Stance4Health Project: Evaluating a Smart Personalised Nutrition Service for Gut Microbiota Modulation in Normal- and Overweight Adults and Children with Obesity, Gluten-Related Disorders or Allergy/Intolerance to Cow's Milk. *Foods* **2022**, *11*, 1480. [CrossRef] [PubMed]
18. Pastoriza, S.; Delgado-Andrade, C.; Haro, A.; Rufián-Henares, J.A. A physiologic approach to test the global antioxidant response of foods. The GAR method. *Food Chem.* **2011**, *129*, 1926–1932. [CrossRef]
19. Benzie, I.F.; Strain, J.J. The ferric reducing ability of plasma (FRAP) as a measure of «antioxidant power»: The FRAP assay. *Anal. Biochem.* **1996**, *239*, 70–76. [CrossRef] [PubMed]
20. Yen, G.-C.; Chen, H.-Y. Antioxidant Activity of Various Tea Extracts in Relation to Their Antimutagenicity. *J. Agric. Food Chem.* **1995**, *43*, 27–32. [CrossRef]
21. Ozgen, M.; Reese, R.N.; Tulio, A.Z.; Scheerens, J.C.; Miller, A.R. Modified 2,2-Azino-bis-3-ethylbenzothiazoline-6-sulfonic Acid (ABTS) Method to Measure Antioxidant Capacity of Selected Small Fruits and Comparison to Ferric Reducing Antioxidant Power (FRAP) and 2,2′-Diphenyl-1-picrylhydrazyl (DPPH) Methods. *J. Agric. Food Chem.* **2006**, *54*, 1151–1157. [CrossRef]
22. Singleton, V.L.; Rossi, J.A. Colorimetry of total phenolics with phosphomolybdic-phosphotungstic acid reagents. *Am. J. Enol. Vitic.* **1965**, *16*, 144–158. [CrossRef]
23. Panzella, L.; Pérez-Burillo, S.; Pastoriza, S.; Martín, M.A.; Cerruti, P.; Goya, L.; Ramos, S.; Rufián-Henares, J.A.; Napolitano, A.; d'Ischia, M. High antioxidant action and prebiotic activity of hydrolyzed spent coffee grounds (hscg) in a simulated digestion–fermentation model: Toward the development of a novel food supplement. *J. Agric. Food Chem.* **2017**, *65*, 6452–6459. [CrossRef] [PubMed]
24. Alimentación en España. Consulted on November 16th, 2023. Available online: https://www.mercasa.es/publicaciones/alimentacion-en-espana/ (accessed on 15 December 2023).
25. Salas-Salvadó, J.; Sanjaume, A.B.; Casañas, R.T.; Solà, M.E.S.; Peláez, R.B. (Eds.) *Nutrición y Dietética Clínica*, 4th ed.; Elsevier Health Sciences: Barcelona, España, 2019.
26. Saura-Calixto, F.; Goñi, I. Antioxidant capacity of the Spanish Mediterranean diet. *Food Chem.* **2006**, *94*, 442–447. [CrossRef]
27. Dietary Reference Values. EFSA. Consulted on November 16th, 2023. Available online: https://multimedia.efsa.europa.eu/drvs/index.htm (accessed on 15 December 2023).
28. Passioura, J.B. Soil structure and plant growth. *Soil Res.* **1991**, *29*, 717–728. [CrossRef]
29. Maqueda, C.; Morillo, E.; Lopez, R.; Undabeytia, T.; Cabrera, F. Influence of organic amendments on Fe, Cu, Mn, and Zn availability and clay minerals of different soils. *Arch. Agron. Soil Sci.* **2015**, *61*, 599–613. [CrossRef]
30. Weil, R.R.; Brady, N.C. Calcium, magnesium, silicon and trace microelements. In *The Nature and Properties of Soils*, 15th ed.; Weil, R.R., Brady, N.C., Eds.; Pearson: New York, NY, USA, 2017; pp. 696–744.
31. Dave, L.A.; Hodgkinson, S.M.; Roy, N.C.; Smith, N.W.; McNabb, W.C. The role of holistic nutritional properties of diets in the assessment of food system and dietary sustainability. *Crit. Rev. Food Sci. Nutr.* **2023**, *63*, 5117–5137. [CrossRef] [PubMed]
32. Gong, Z.; Han, X. The Determination of Grape Wine Quality in China. *J. Food Sci.* **2007**, *72*, 429–434.
33. Wojtunik-Kulesza, K.; Oniszczuk, A.; Oniszczuk, T.; Combrzyński, M.; Nowakowska, D.; Matwijczuk, A. Influence of In Vitro Digestion on Composition, Bioaccessibility and Antioxidant Activity of Food Polyphenols—A Non-Systematic Review. *Nutrients* **2020**, *12*, 1401. [CrossRef]
34. Ketnawa, S.; Reginio, F.C., Jr.; Thuengtung, S.; Ogawa, Y. Changes in bioactive compounds and antioxidant activity of plant-based foods by gastrointestinal digestion: A review. *Crit. Rev. Food Sci. Nutr.* **2022**, *62*, 4684–4705. [CrossRef]
35. Kolodziejczyk, A.A.; Zheng, D.; Elinav, E. Diet-microbiota interactions and personalized nutrition. *Nat. Rev. Microbiol.* **2019**, *17*, 742–753. [CrossRef]
36. van Leeuwen, C.; Friant, P.; Choné, X.; Tregoat, O.; Koundouras, S.; Dubourdieu, D. Influence of climate, soil, and cultivar on terroir. *Am. J. Enol. Vitic.* **2004**, *55*, 207–217. [CrossRef]

foods

MDPI

Article

Ultrasonic-Assisted Extraction of *Dictyophora rubrovolvata* Volva Proteins: Process Optimization, Structural Characterization, Intermolecular Forces, and Functional Properties

Yongqing Zhang [1,2], Shinan Wei [1], Qinqin Xiong [1], Lingshuai Meng [2], Ying Li [2], Yonghui Ge [2], Ming Guo [3], Heng Luo [1,*] and Dong Lin [2,*]

1 State Key Laboratory of Functions and Applications of Medicinal Plants, Guizhou Medical University, Guiyang 550014, China; gyuzhyq@163.com (Y.Z.); 15761632673@163.com (S.W.); 18208490421@163.com (Q.X.)
2 Guizhou Higher Education Key Laboratory of Functional Food, Guizhou Engineering Research Center for Fruit Processing, College of Food Science and Engineering, Guiyang University, Guiyang 550005, China; 15040260380@163.com (L.M.); li.ying.1990@163.com (Y.L.); skyge@163.com (Y.G.)
3 Guizhou Jin Chan Da Shan Biotechnology Company Limited, Bijie 553300, China; guoming@163.com
* Correspondence: luo_heng@gmc.edu.cn (H.L.); gyulindong@gyu.edu.cn (D.L.)

Abstract: *Dictyophora rubrovolvata* volva, an agricultural by-product, is often directly discarded resulting in environmental pollution and waste of the proteins' resources. In this study, *D. rubrovolvata* volva proteins (DRVPs) were recovered using the ultrasound-assisted extraction (UAE) method. Based on one-way tests, orthogonal tests were conducted to identify the effects of the material–liquid ratio, pH, extraction time, and ultrasonic power on the extraction rate of DRVPs. Moreover, the impact of UAE on the physicochemical properties, structure characteristics, intermolecular forces, and functional attributes of DRVPs were also examined. The maximum protein extraction rate was achieved at 43.34% under the best extraction conditions of UAE (1:20 g/mL, pH 11, 25 min, and 550 W). UAE significantly altered proteins' morphology and molecular size compared to the conventional alkaline method. Furthermore, while UAE did not affect the primary structure, it dramatically changed the secondary and tertiary structure of DRVPs. Approximately 13.42% of the compact secondary structures (α-helices and β-sheets) underwent a transition to looser structures (β-turns and random coils), resulting in the exposure of hydrophobic groups previously concealed within the molecule's core. In addition, the driving forces maintaining and stabilizing the sonicated protein aggregates mainly involved hydrophobic forces, disulfide bonding, and hydrogen bonding interactions. Under specific pH and temperature conditions, the water holding capacity, oil holding capacity, foaming capacity and stability, emulsion activity, and stability of UAE increased significantly from 2.01 g/g to 2.52 g/g, 3.90 g/g to 5.53 g/g, 92.56% to 111.90%, 58.97% to 89.36%, 13.85% to 15.37%, and 100.22% to 136.53%, respectively, compared to conventional alkali extraction. The findings contributed to a new approach for the high-value utilization of agricultural waste from *D. rubrovolvata*.

Keywords: *Dictyophora rubrovolvata*; edible mushroom proteins; ultrasonic extraction; functional properties; intermolecular forces

1. Introduction

The next 30 years are expected to witness a significant increase in the demand for high-quality proteins due to the fast expansion of the world's population and growing consciousness about nutrition. Nowadays, food proteins are mainly derived from animals and plants. However, the high cost of animal breeding, along with greenhouse gas production and environmental pollution, as well as the absence of one or more essential amino acids in plant proteins, have limited their application [1]. Given this, new alternative sources of proteins are being explored. Edible mushrooms are preferred by consumers all over the world owing to their delicious taste and rich nutrition (e.g., proteins, dietary fibers, micronutrients). At present, benefiting from the well-established artificial cultivation technology, the edible fungi industry has been flourishing across the world, and the annual production exceeds 40 million tons merely in China. In comparison with animal raising and crop planting, the cultivation of edible mushrooms is low-cost and eco-friendly with a shorter growth cycle. Edible mushrooms are considered promising food protein sources due to their high protein content (approximately 19–35% in dry weight) and well-balanced ratio of amino acids, effectively meeting the nutritional requirements of essential amino acids for the human body [2].

Dictyophora rubrovolvata, a type of saprophytic fungus, is classified as a member of the *Phallus* species [3]. Previous research has indicated that *D. rubrovolvata* is abundant in nutrients such as proteins, polysaccharides, flavonoids, and terpenoids, which possess physiological effects including immunomodulation [4], antioxidant properties [5], anti-tumor effects [6], and antimicrobial properties [7]. Its appealing appearance, delightful taste, and exceptional nutritional content have made *D. rubrovolvata* highly sought after in China, Japan, Korea, and various other Asian nations. Due to its high market demand and significant commercial worth, *D. rubrovolvata* has emerged as a prominent and thriving species of edible fungus in the southwestern provinces of China, namely Guizhou, Yunnan, and Sichuan. In Guizhou province alone, the annual production reaches approximately 10,000 tons. About 35% of its overall weight is made up of the stipe and the veil, which are the edible parts [2]. However, the volva of *D. rubrovolvata*, which makes up approximately 40% of the entire fruiting body, is discarded after harvesting, leading to significant environmental pollution and wastage of bio-resources [3]. Zhuang and Sun [2] revealed that the crude proteins content in *D. rubrovolvata* volva was 26.74%, implicating its potential as a source of high-quality protein.

The use of ultrasound-assisted extraction (UAE) has become prevalent in protein recovery due to its benefits, including efficient extraction, time and solvent conservation, ease of operation, and environmental friendliness [8]. Ultrasonic waves propagate through the fluid medium as pressure oscillations, generating numerous rapidly expanding cavitation cavities. The cavitation region experiences transient high temperatures (5000 k), high pressures (1000 atm), shock waves, turbulence, and shear due to the expansion and rupture of cavitation bubbles [8]. On one hand, cavitation causes the bio-matrix to perforate and fragment, enhancing solvent accessibility and improving protein extraction efficiency [9]. On the other hand, the cavitation effect partially unfolds the proteins and alters their advanced structure (secondary, tertiary, and quaternary structures), disrupting the non-covalent interactions among proteins molecules [10]. These ultrasonic structural modifications can greatly enhance the functional properties of proteins for various processing purposes, including solubility, foaming, emulsification, and gelation. Reports indicated that sonication is effective in modifying the properties of proteins in soy [10], cod [11], and peanuts [12]. Therefore, it is hypothesized that UAE applied to *D. rubrovolvata* volva might substantially improve its structural and functional properties.

Currently, there are few studies about the effects of UAE on *D. rubrovolvata* volva proteins (DRVPs). Therefore, the purpose of this work was to study the DRVP modification using UAE in terms of physicochemical and structural characterization, intermolecular forces, and functional properties. The results will establish the groundwork for the advancement of DRVPs as a novel food protein source and its utilization in the food sector. It could provide a significant reference for realizing the sustainable development goal of the *D. rubrovolvata* industry.

2. Materials and Methods

2.1. Materials

D. rubrovolvata volva was kindly provided by Guizhou Jin Chan Da Shan Biotechnology Co., Ltd. (Bijie, China). Tris (hydroxymethyl) aminomethane, sodium dodecyl sulfate (SDS), β-mercaptoethanol (β-ME), and bromophenol blue were purchased from Sangon Bioengineering Co., Ltd. (Shanghai, China). Macklin Biochemical Technology Co., Ltd. (Shanghai, China) provided 5,5′-Dithio bis-(2-nitrobenzoic acid) (DTNB) and 8-Anilino-1-naphthalenesulfonic acid (ANS). All other analytical-grade chemicals and reagents, unless otherwise specified, were obtained from Sinopharm Chemical Reagent Co., Ltd. (Shanghai, China).

2.2. Optimization of UAE Process of DPVP

2.2.1. Extraction Method and Determination of Proteins Extraction Rate

Fresh *D. rubrovolvata* volvas were washed, freeze-dried, pulverized, sieved (60 mesh), and then degreased with petroleum ether (1:2, *w*/*v*) for 24 h. We weighed accurately 10.00 g of defatted powder (proteins content m_1, determined using the Kjeldahl method) into a beaker and mixed it with a specific volume of distilled water. Afterwards, the mixtures were adjusted to the assigned pH with 1 M HCl or NaOH. The extraction was carried out for a certain period at a specified power in a water bath ultrasonic reactor (KQ 3200DB, 40 kHz, Kunshan Ultrasonic Instrument Co., Ltd., Kunshan, China). Upon completion of the extraction, the supernatant (proteins content m_2, determined using the Coomassie brilliant blue method) was collected using centrifugation at 5000 r/min for 20 min. After the pH adjustment to the proteins isoelectric point (pH 2.5) with 1 M HCl, the sediment was collected using centrifugation again (5000 r/min, 20 min). Then, the precipitates were lyophilized in a vacuum freeze dryer (LC-12N-50A, Lichen Instrument Technology Co., Ltd., Shaoxing, China) for 24 h and stored in a refrigerator at −20 °C for further analysis. The protein extraction rate was calculated using the following equation: protein extraction rate (%) = $m_2/m_1 \times 100$ [13].

2.2.2. Single-Factor and Orthogonal Experiments Design

The effects of four factors (pH, ultrasonic time, ultrasonic power, and material–liquid ratio) on the extraction rate of DRVPs were investigated in the one-factor experiments, respectively [14]. The reference conditions were set at pH 11, ultrasound power 550 W, ultrasound time 25 min, and material–liquid ratio 1:10. Only one factor was varied in each trial, and the other factor levels were kept constant. The ranges of levels for each factor are as follows: pH (9.0, 9.5, 10.0, 10.5, 11.0, 11.5), ultrasound time (10, 15, 20, 25, 30 min), ultrasound power (330, 385, 440, 495, 550 W), and material–liquid ratio (1:10, 1:15, 1:20, 1:25, 1:30). Based on the results of the one-factor test, a four-factor three-level orthogonal test was conducted with protein extraction rate as the response variable. The factors and levels of the orthogonal test were listed in Table S1. The optimal UAE process of DRVPs was obtained and validated by integrating it with the range analysis.

2.3. Determinations of the Physicochemical Properties

2.3.1. Microstructures

The scanning electron microscope (SEM) (Gemini500, Carl Zeiss AG, Oberkochen, Germany), magnified at 30,000 times, was used to observe the microstructures of the samples. After applying conductive adhesive, the lyophilized proteins powders were attached to the sample stage. Subsequently, the sample micromorphology was observed and photographed at an accelerating voltage of 15 kV following the gold sputtering process [15].

2.3.2. Turbidity

The samples were prepared as a 5 mg/mL solution in water, and their absorbance at 660 nm was measured at room temperature using a UV–Vis spectrophotometer (Cary 60, Agilent Technologies, Santa Clara, CA, USA) [16].

2.3.3. Particle Size Distribution and Zeta Potential

Minor modifications were made to Zhong's and Xiong's [16] method for analyzing particle size and zeta potential. The 1 mg/mL of proteins was dissolved in phosphate-buffered saline (0.01 M, pH 7.0). At 25 °C, the Malvern Company's laser particle size analyzer (Nano ZS90, London, UK) was employed to ascertain the distribution of particle sizes and zeta potential.

2.3.4. Thermal Properties

The thermal characteristics of the proteins samples were assessed employing a DSC4000 differential scanning calorimeter (PerkinElmer Ltd., Waltham, MA, USA). The approach described by Wang et al. [9] was followed. Nitrogen was passed through at a flow rate of 40 mL/min, and an aluminum pan was used to enclose 5 mg of the sample. Using an aluminum pan devoid of any contents as a control, the temperature was gradually increased at a speed of 5 °C/min across the temperature range of 20 to 180 °C. The thermal profiles were scanned and graphed during this process.

2.4. Characterization of the Structures

2.4.1. Amino Acid Profile

The samples were analyzed for their amino acid makeup following the procedure outlined by Eze, Chatzifragkou, and Charalampopoulos [17]. More specifically, 80 mg of proteins powder was combined with HCl (6 M, 10 mL) in a hydrolysis tube and subjected to hydrolysis at 110 °C for 24 h under a nitrogen atmosphere. After passing through a 0.45 μm membrane, the hydrolysis products underwent filtration. Subsequently, amino acid content was monitored with an automated amino acid analyzer (L-8900, Hitachi Co., Tokyo, Japan).

2.4.2. Molecular Weight Distribution

The sodium dodecyl sulfate–polyacrylamide gel electrophoresis (SDS-PAGE) was employed to visualize the molecular weight distribution of DRVPs. The SDS-PAGE was carried out following the procedure described by Zhong and Xiong [16]. There were two types of gels used, 12% separating gel and 5% concentrating gel. The proteins solution, with a concentration of 5 mg/mL, was mixed with buffer containing 0.05 M Tris-HCl buffer, 5% glycerol, 1% SDS, 2.5% β-ME, and 0.02% bromophenol blue in an equal volume. The mixture was then heated at 95 °C for 5 min to denature. The sample (20 μL) was placed into the gel and the electrophoresis was conducted at a steady voltage of 100 V. The gel was dyed using BeyoBlue Ultrafast Staining Solution (P0017F, Beyotime, Nantong, China) while being shaken for 1 h, then washed in distilled water for 2 h to remove the color. To determine the molecular weight of the proteins samples, a non-prestained protein marker (P0060M, Beyotime, Nantong, China) with a range of 10 to 150 kDa was utilized.

2.4.3. Secondary Structure

The fourier transform infrared (FTIR) spectrometer (Spectrum Two, Perkin Elmer Ltd., Waltham, MA, USA) was utilized to measure proteins samples, scanning 32 times at a resolution of 4 cm^{-1} within a wavelength range of 500 to 4000 cm^{-1}. The absorption peaks in the amide I band (1600–1700 cm^{-1}) were subjected to Gaussian deconvolution using Peakfit software (v4.12, Thermo Electron Co., Waltham, MA, USA). The area of each subpeak was used to calculate the percentage of each secondary structure [18].

2.4.4. Surface Hydrophobicity (H_0)

A hydrophobic fluorescent probe created using the ANS was employed to determine the surface hydrophobicity of the proteins sample, following the method described by Li et al. [19]. In short, proteins dispersion (1 mg/mL) was produced using phosphate buffer (0.01 M, pH 7.0). Then, 4 mL proteins solution was mixed with 20 μL freshly prepared ANS solution (8.0 M). The blend underwent a reaction for 20 min at ambient temperature while being shielded from light. The fluorescence intensity in the emission wavelength range of 470–520 nm was recorded with 370 nm as the excitation wavelength using a fluorescence spectrophotometer (F-320, Tianjin Gangdong Technology Co., Tianjin, China).

2.4.5. Free Sulfhydryl (SH) Content

With slight modifications, Ellman's reagent was used to determine the SH content [19]. The Tris-glycine buffer (pH 8.0) was configured as follows: 0.086 M Tris, 0.09 M glycine, 4 mM EDTA-Na2, 0.5% SDS and 8 M urea. Ellman's reagent was made by combining 20 mg of DTNB with 5 mL of the buffer above. To 4 mL of proteins solution (2 mg/mL), Ellman's reagent was added in the amount of 40 μL. Following vortex, the reaction was conducted at 25 °C for 15 min in a light-free environment. The absorbance value at 412 nm was recorded using a Cary 60 UV spectrophotometer (Agilent Technologies, Santa Clara, CA, USA). The precise SH content was determined using the following calculation.

$$\text{SH content } (\mu\text{mol/g}) = \frac{73.53 \times A_{412} \times D}{C} \quad (1)$$

where A_{412} represents the absorbance value of the sample at 412 nm, D represents the sample dilution factor, and C represents the mass concentration of the proteins (mg/mL).

2.5. Proteins Aggregates Dissociation Test

The method was modified slightly from Wang et al. [20]. Specifically, the proteins solution of 2 mg/mL was combined with various concentrations of bond dissociation agents (HCl/NaOH, SDS, urea, and β-ME) at a volume ratio of 1:9 and allowed to stand overnight. The size distribution of the sample was then recorded with a laser nanoparticle sizer (Nano ZS90, Malvern Company, Malvern, UK). A proteins dispersion without bond dissociation reagents was used as a control.

2.6. Measurements of Functional Properties

2.6.1. Solubility

Solubility was determined according to Alavi et al. [21] with slight modifications. The proteins samples were 1 mg/mL in distilled water and then pH values of 2, 4, 6, 8, and 10 were adjusted. To collect the supernatant, the mixture was centrifuged at 5000 r/min for 20 min following 10 min vortex. The Coomassie Brilliant Blue method was employed to determine the proteins amount in the supernatant. The protein solubility was calculated using the provided Equation (2):

$$\text{Protein solubility } (\%) = \frac{C_1}{C_0} \times 100 \quad (2)$$

where C_1 is the proteins. concentration in the supernatant (mg/mL), and C_0 is the total proteins concentration (mg/mL).

2.6.2. Water Holding Capacity (WHC) and Oil Holding Capacity (OHC)

After making slight adjustments, the evaluation of WHC and OHC was conducted using Huang et al.'s [15] method. Mix 10 mL of distilled water or soybean oil in a centrifuge tube containing 1 g of proteins sample. Then, the tubes were placed in the water baths at various temperatures (30, 40, 50, 60, 70 °C) for 30 min. Subsequently, they were centrifuged at a speed of 5000 r/min for 20 min, and the resulting supernatant was removed. WHC and OHC were calculated using Equation (3):

$$\text{WHC/OHC } (\text{g/g}) = \frac{W_3 - W_2}{W_1} \quad (3)$$

where W_3 is the weight of the empty tube plus the absorbed water/oil proteins sample (g), W_2 is the weight of the empty tube plus the proteins sample (g), and W_1 is the weight of the proteins sample (g).

2.6.3. Emulsion Activity Index (EAI) and Emulsion Stability Index (ESI)

EAI and ESI were assessed as described by Li et al. [19] with minor modifications. To create an emulsion, 2 mL of soybean oil was added to 8 mL of proteins dispersion. The pH of the emulsion was then modified to various levels (2, 4, 6, 8, and 10), and homogenization was performed for 2 min at 10,000 r/min using a homogenizer from IKA, Germany (T25 digital ULTRA-TURRAX). A 50 μL aliquot was aspirated from the lower part of the tube (0 and 10 min) and diluted 100 times with SDS solution (*w*/*v*, 0.1%). Afterward, it was mixed by vortexing for 10 s. A spectrophotometer was used to measure the absorbance of the diluted emulsion at 500 nm. The calculation of EAI and ESI was performed using Equations (4) and (5), respectively:

$$\text{EAI}\ (\text{m}^2/\text{g}) = \frac{2 \times 2.303 \times \text{N} \times \text{A}_0}{\text{C} \times \text{L} \times \varphi \times 10000} \tag{4}$$

$$\text{ESI}\ (\text{min}) = \frac{\text{A}_0}{\text{A}_0 - \text{A}_{10}} \times 10 \tag{5}$$

where N represents the dilution coefficient (100), C is the proteins concentration (g/mL), L represents the optical path (1 cm), φ represents a fraction of the oil phase (0.2), and A_0, A_{10} denote absorbance at 0 and 10 min.

2.6.4. Foaming Capacity (FC) and Foaming Stability (FS)

FC and FS were determined according to the method outlined by Li et al. [22]. After making slight adjustments, DRVP solutions (14 mL, 5 mg/mL) with varying pH levels (2, 4, 6, 8, and 10) were agitated for 2 min at a speed of 10,000 r/min using a T25 digital homogenizer (ULTRA-TURRAX, IKA, Staufen, Germany). Subsequently, the foam volume (V_0) was promptly measured using a graduated cylinder. After 30 min of settling, the foam volume (V_{30}) was checked again. The equations below were used to calculate FC and FS.

$$\text{FC}(\%) = \frac{\text{V}_1}{\text{V}_0} \times 100 \tag{6}$$

$$\text{FS}(\%) = \frac{\text{V}_{30}}{\text{V}_1} \times 100 \tag{7}$$

where V_0 is the volume (mL) before homogenization, V_1 is the volume (mL) after homogenization, and V_{30} is the volume (mL) after 30 min of standing.

2.7. Statistical Analysis

The trials were performed thrice, and the outcomes were displayed as the means ± standard deviation. Origin 2018 software from Origin Lab Inc. (Northampton, MA, USA) was utilized to arrange and graph the data. Statistical analysis was conducted using SPSS software (version 18.0, SPSS Inc., Chicago, IL, USA). The means of two groups were analyzed using the Student's *t*-test. For multiple comparisons, one-way analysis of variance (ANOVA) followed by Duncan's multiple range test was performed. The results were deemed significantly distinct with a *p*-value less than 0.05.

3. Results and Discussion

3.1. Optimization of the Extraction Process of DRVPs

Different extraction conditions have varying effects on the extractability of proteins. To improve the efficiency of proteins preparation, it was evaluated how pH, time (min), ultrasonic power (W), and material–liquid ratio (g/mL) affected the DRVP extraction rate. As shown in Figure 1A, the protein extraction rate increased gradually with the rise in pH. Proteins are negatively charged in an alkaline environment and repel each other, which helps them to increase their solubility [14]. The highest protein extraction rate of 34.60% was achieved at pH = 11. With the prolongation of ultrasonication time, the protein extraction rate was first increased and then decreased. The highest protein extraction rate of 27.97% was reached at 25 min (Figure 1B). The shock wave and shear force arising from ultrasonic cavitation broke the cell wall and promoted the release of proteins. However, too-long ultrasonication time can cause protein aggregation and precipitation, thereby resulting in a decline in the protein extraction rate [23]. With the elevation of power, the cavitation effect also increased, significantly enhancing the mass transfer efficiency. Consequently, as the ultrasonic power rose, the protein extraction rate increased accordingly. At 495 W, the extraction rate reached the maximum value of 30.45% (Figure 1C). However, when the ultrasonic power was increased further, the extraction rate decreased instead (Figure 1C). This is because the local thermal effect caused by high-intensity cavitation can lead to protein degradation, thus reducing the extraction rate [13]. Figure 1D describes the effect of the material–liquid ratio on protein extraction rate. With the increase in solvent amount, the powders were in fuller contact with the solvent, which favored the protein extraction. Up to a material–liquid ratio of 1:20, a maximum value of 28.61% was reached. Continuing to increase the solvent volume, the protein extraction rate no longer increased, owing to the progressive dilution of the ultrasound energy density.

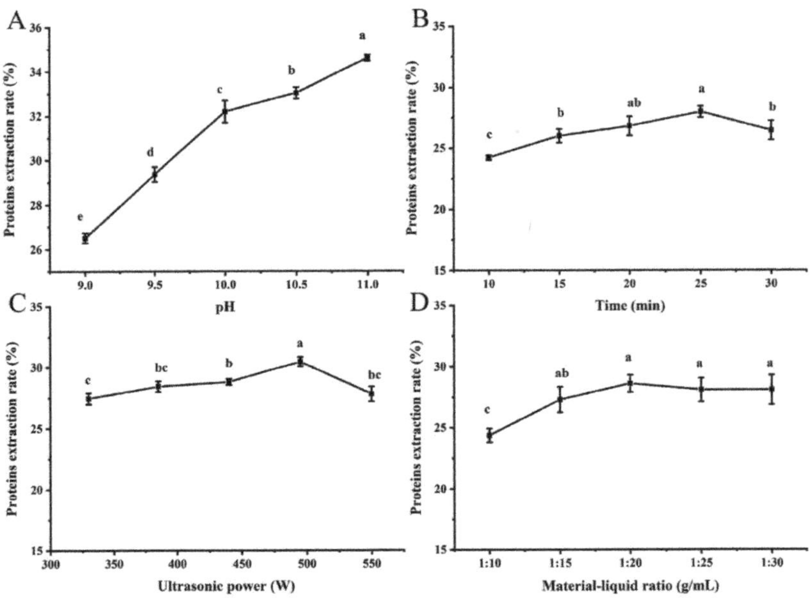

Figure 1. Effects of different process conditions on protein extraction rate of DRVPs. (**A**) pH, (**B**) time, (**C**) ultrasonic power, and (**D**) material–liquid ratio. Among different levels of the same single factor, the different letters indicate a significant difference ($p < 0.05$).

Concerning the results of the single-factor tests, the $L_9(3^4)$ orthogonal table was designed to optimize the extraction process of DRVPs. The experimental design and results are listed in Table S2. Based on the relative magnitude of the R value and the sum of squares, the order of the factors affecting the protein extraction rate was obtained as follows: pH > ultrasonic power > material–liquid ratio > ultrasonic time (Tables S2 and S3). The variance analysis revealed that pH ($p < 0.01$) and ultrasonic power ($p < 0.01$) had a significant effect on protein yield, while ultrasonic time ($p > 0.05$) and material-liquid ratio ($p > 0.05$) did not show any significant effect (Table S3). This suggested that the increase in protein extraction rate was mainly dependent on pH and ultrasonic power. A prolonged sonication time and an increase in the liquid-liquid ratio had a limited effect on the extractability. According to the k-value analysis, the optimization condition for the extraction of DRVP was $A_3B_2C_3D_2$ (Table S2), namely: pH = 11, ultrasonic power of 550 W, ultrasonic time of 25 min, and material-liquid ratio of 1:20, which resulted in the highest protein extraction rate of 43.37%. Proteins extracted in the above condition (designated as U-DRVP) were used for subsequent structural characterization and functional property evaluation, while those extracted by conventional alkaline extraction without UAE (designated as C-DRVP, 32.80% protein extraction rate) were employed as control.

3.2. Physical and Chemical Properties of U-DRVP

3.2.1. SEM Analysis

The changes in the micromorphology of the proteins surface after UAE were examined with SEM. As shown in Figure 2A,B, UAE noticeably altered the apparent morphology of DRVP. C-DRVP showed a highly cross-linked, porous, and rough surface (Figure 2A), whereas the surface of U-DRVP was more dense, smooth, and flat with better continuity (Figure 2B). On the one hand, ultrasonic cavitation induced changes in the structure of DRVP, exposing the functional groups buried inside the molecules and destroying the original aggregation pattern between proteins molecules [24]. On the other hand, the external energy afforded by ultrasound increased the collision frequency between proteins molecules [25], approaching each other via new molecular interactions (non-covalent or/and covalent interactions) [15], accomplishing the reconfiguration of the protein aggregates, and eventually presenting the microstructure shown in Figure 2B. Lv et al. [26] also reported similar findings.

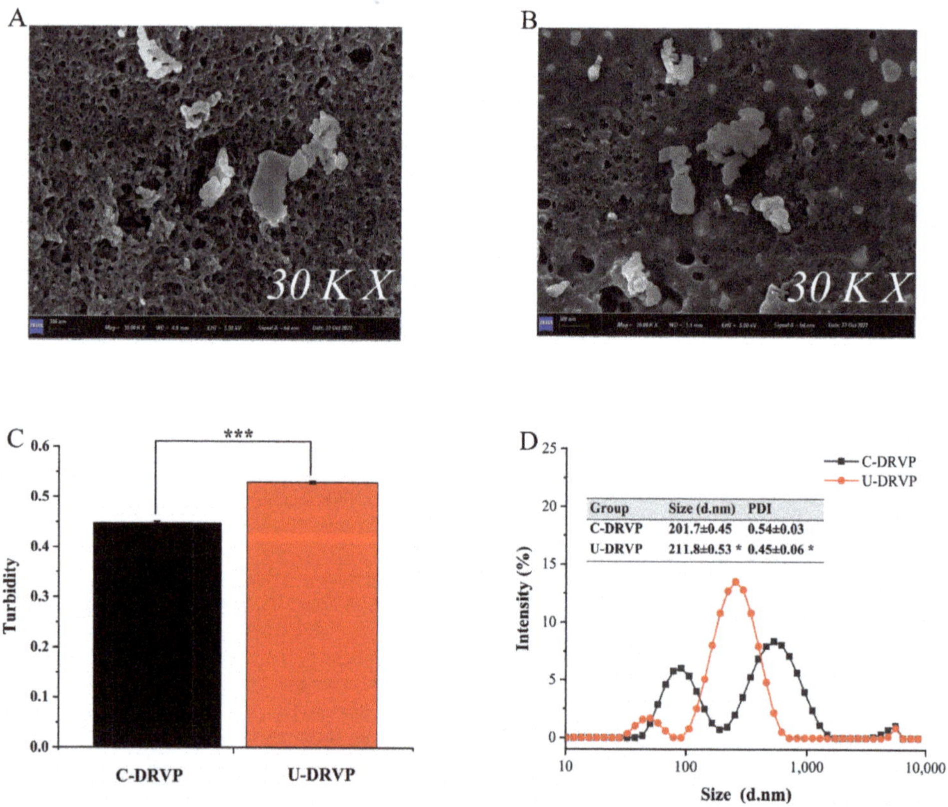

Figure 2. *Cont.*

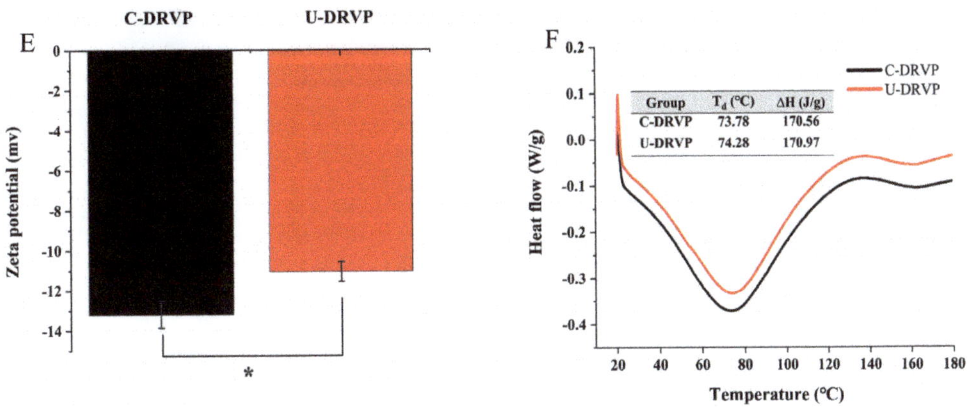

Figure 2. Physicochemical properties of C-DRVP and U-DRVP. (**A**,**B**) SEM images, (**C**) turbidity, (**D**) particle size distribution, (**E**) zeta potential, and (**F**) DSC. Note: The statistical method is Student's *t*-test. The symbols indicate statistical significance between the means of two groups: *: $p < 0.05$, ***: $p < 0.001$. C-DRVP, *D. rubrovolvata* volva proteins extracted using conventional alkaline extraction without UAE; U-DRVP, *D. rubrovolvata* volva proteins extracted using UAE.

3.2.2. Turbidity, Particle Size Distribution, and Zeta-Potential Analysis

Larger-sized colloidal particles usually have a greater light scattering capacity and proportionally higher turbidity values [16]. As shown in Figure 2C, the turbidity values for C-DRVP and U-DRVP were 0.45 and 0.53, respectively. Higher turbidity values for U-DRVP suggested larger molecular sizes. Consistent with the turbidity measurements, the average particle size of U-DRVP was significantly elevated from 201.7 nm to 211.8 nm compared to that of C-DRVP (Figure 2D). The particle size distribution of U-DRVP appeared as a single peak, while C-DRVP as a double peak. Correspondingly, the PDI values were 0.45 and 0.54, respectively, indicating that U-DRVP had a more homogeneous particle size distribution. The results above can be attributed to two points: for one thing, the ultrasound-induced shear and microjet loosened the proteins structure, exposing internal active groups that favor molecular aggregation [27]; for another, the thermal effect produced with ultrasound accelerated the collision and aggregation rate of DRVP particles, which drove them to form more homogeneous and larger aggregates by self-assembly [16].

The zeta potential could reflect the charged properties of the particle surface [27]. As illustrated in Figure 2E, the potential of U-DRVP rose from −13.20 mV to −11.07 mV in comparison with C-DRVP. In general, the higher potential of negatively charged particles implies less intermolecular repulsion and allows for easier formation of larger particles, which is confirmed by the results of turbidity and average particle size (Figure 2C,D). However, the opposite finding was observed by Huang et al. [15], who claimed that sonication caused the fragmentation of the proteins into smaller pieces, exposing more negatively charged groups inside the molecule, consequently resulting in lower potential values. The contradictory results mentioned above may be because there were differences in proteins sources, power, and duration of sonication in each study.

3.2.3. Differential Scanning Calorimetry (DSC) Analysis

DSC is widely deployed to analyze the thermal behavior of proteins [9]. As depicted in Figure 2F, the denaturation temperature (T_d) of U-DRVP was slightly elevated compared with C-DRVP, moving from 73.78 °C to 74.28 °C. The denaturing enthalpy change (ΔH) followed a similar trend, rising from 170.56 J/g to 170.97 J/g. T_d usually reflects the thermal stability of the polymer, and ΔH indicates the overall structural order. Ultrasonic cavitation enhanced intermolecular forces and promoted the reorganization of protein particles into more homogeneous aggregates (Figure 2D), thus requiring higher T_d and ΔH to induce protein thermal denaturation [9].

3.3. Changes in the Structure of U-DRVPs

3.3.1. Primary Structure

The nutritional value and functional properties of proteins greatly depend on the composition of amino acids [17]. Hence, the amino acid profiles of U-DRVPs were measured. As indicated in Table 1, U-DRVPs were abundant in aspartic acid, glutamic acid, leucine, threonine, and phenylalanine, leaving cysteine as the lowest ingredient. The proportion of negatively charged amino acids (23.89%) was noticeably higher than that of positively charged amino acids (10.09%), partly accounting for the negative potential of U-DRVPs (Figure 2E). Furthermore, U-DRVPs have great nutritional value due to their richness in essential amino acids (316.53 mg/g, 42.09%), surpassing the percentage recommended by WHO. The high percentage of hydrophobic amino acids (356.41 mg/g, 49.05%) was important for U-DRVPs to exhibit excellent processing functional properties. Compared to C-DRVPs, UAE significantly improved the absolute levels of total amino acids of DRVPs, increasing from 683.90 mg/g to 751.89 mg/g. However, UAE had little change in the proportion of amino acid composition, in agreement with the analyses for okara proteins by Eze, Chatzifragkou, and Charalampopoulos [17].

Table 1. Amino acid compositions of C-DRVPs and U-DRVPs.

Amino Acids	Absolute Quantities (mg/g)		Relative Quantities (%)	
	C-DRVPs	U-DRVPs	C-DRVPs	U-DRVPs
Histidine (His)	16.98 ^{b}	18.79 ^{a}	2.48 ^{A}	2.50 ^{A}
Isoleucine (Ile)	39.04 ^{b}	43.01 ^{a}	5.71 ^{A}	5.72 ^{A}
Leucine (Leu)	53.96 ^{b}	60.54 ^{a}	7.89 ^{B}	8.05 ^{A}
Lysine (Lys)	24.16 ^{b}	27.65 ^{a}	3.53 ^{B}	3.68 ^{A}
Methionine (Met)	9.06 ^{b}	9.89 ^{a}	1.33 ^{A}	1.31 ^{A}
Phenylalanine (Phe)	47.28 ^{b}	51.68 ^{a}	6.91 ^{A}	6.87 ^{A}
Threonine (Thr)	52.64 ^{b}	56.99 ^{a}	7.70 ^{A}	7.58 ^{B}
Valine (Val)	43.74 ^{b}	47.98 ^{a}	6.39 ^{A}	6.38 ^{A}
Aspartic acid (Asp)	85.52 ^{b}	93.15 ^{a}	12.51 ^{A}	12.39 ^{B}
Glutamic acid (Glu)	77.70 ^{b}	86.48 ^{a}	11.36 ^{A}	11.50 ^{A}
Serine (Ser)	46.66 ^{b}	50.81 ^{a}	6.82 ^{A}	6.76 ^{B}
Cystine (Cys)	2.26 ^{b}	2.44 ^{a}	0.33 ^{A}	0.33 ^{A}
Glycine (Gly)	40.54 ^{b}	44.51 ^{a}	5.93 ^{A}	5.92 ^{A}
Tyrosine (Tyr)	27.02 ^{b}	29.76 ^{a}	3.95 ^{A}	3.96 ^{A}
Arginine (Arg)	26.24 ^{b}	29.42 ^{a}	3.84 ^{B}	3.91 ^{A}
Alanine (Ala)	44.86 ^{b}	49.66 ^{a}	6.56 ^{A}	6.60 ^{A}
Proline (Pro)	46.24 ^{b}	49.15 ^{a}	6.76 ^{A}	6.54 ^{B}
HAA	324.72 ^{b}	356.41 ^{a}	49.25 ^{A}	49.05 ^{B}
EAA	286.86 ^{b}	316.53 ^{a}	41.94 ^{B}	42.09 ^{A}
NCAA	163.22 ^{b}	179.63 ^{a}	23.87 ^{A}	23.89 ^{A}
PCAA	67.38 ^{b}	75.85 ^{a}	9.85 ^{B}	10.09 ^{A}
Total	683.90 ^{b}	751.89 ^{a}	100.00 ^{A}	100.00 ^{A}

Note: HAA, hydrophobic amino acids (Thr, Ala, Val, Met, Leu, Ile, Phe, Pro); EAA, essential amino acids (Thr, Val, Met, Ile, Leu, Phe, Lys, His); NCAA, negatively charged amino acids (Asp, Glu); and PCAA, positively charged amino acids (Arg, His, Lys). Means with different superscript lowercase (uppercase) letters within the same line are significantly different ($p < 0.05$). C-DRVP, *D. rubrovolvata* volva proteins extracted using conventional alkaline extraction without UAE; U-DRVPs, *D. rubrovolvata* volva proteins extracted using UAE.

The SDS-PAGE was employed to visualize the molecular weight distribution of DRVPs [16]. As indicated in Figure 3A, the protein bands displayed identical patterns regardless of the extraction method. Both protein samples consisted of three major subunit bands with molecular weights located at 23, 63, and 72 kDa, respectively. The ultrasound energy was not strong enough to cleave the peptide bonds, yielding protein fragments with lower molecular weights. Consequently, the primary structure of U-DRVPs remained unchanged. Previous studies also found that the electrophoresis profiles of sonicated amaranth [27] and pea [28] proteins remained unmodified.

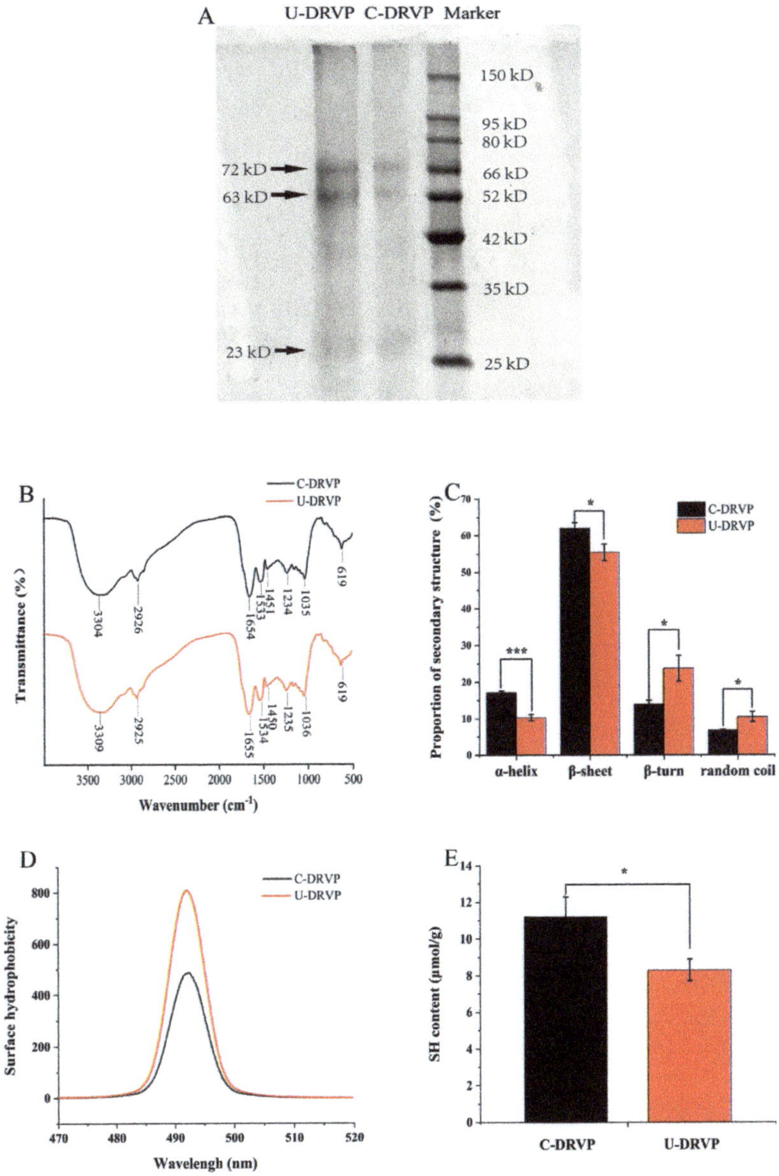

Figure 3. Structural characterization of C-DRVPs and U-DRVPs. (**A**) SDS-PAGE pattern, (**B**) FTIR spectra, (**C**) proportion of secondary structures, (**D**) surface hydrophobicity, (**E**) SH content. Note: The statistical method is Student's *t*-test. The symbols indicate statistical significance between the means of two groups: *: $p < 0.05$, ***: $p < 0.001$. C-DRVPs, *D. rubrovolvata* volva proteins extracted using conventional alkaline extraction without UAE; U-DRVPs, *D. rubrovolvata* volva proteins extracted using UAE.

3.3.2. Secondary Structure

FTIR has been commonly used for the detection of characteristic functional groups, which can be applied to quantitatively measure the proportions of secondary structure in proteins [18]. Generally, the amide bond (peptide bond) in the proteins backbone exhibits five characteristic absorption bands in FTIR: amide A, amide B, amide I, amide II, and amide III. Upon UAE, the IR absorption at 3304 cm^{-1} (amide A) was blueshifted to 3309 cm^{-1} (Figure 3B). It is inferred that the inter- or intramolecular hydrogen bonding rearrangements induced by sonocavitation affected the N-H stretching vibration. Furthermore, after UAE, the IR absorptions at 1654 (amide I), 1533 (amide II), and 1234 cm^{-1} (amide III) were blue-shifted to 1655, 1534, and 1235 cm^{-1}, respectively, suggesting that the stretching and bending vibration profiles of the C=O, C-N, and N-H bonds changed. Anyway, changes in the vibration of the functional group will affect the intramolecular hydrogen bonding and thus modify the percentage of secondary structures [9].

Among the characteristic absorptions of the five amides, the amide I band is the most responsive to alterations in the secondary structure of proteins. Hence, it was chosen for secondary structure fitting analysis [18]. As noted in Figure 3C, β-sheet was the dominant structure in four secondary structures for both proteins. In comparison to C-DRVPs, α-helix and β-sheet were reduced by 6.85% and 6.56%, corresponding to an increase in β-turn and random coil by 9.70% and 3.61% in U-DRVPs. The α-helix and β-sheet were more compact than the β-turn and random coil. As a result, the structure of U-DRVPs was made looser and more flexible, which was significant for it to realize desirable techno-functional properties. The above outcomes may be because the shock waves, turbulence, microjets, and shear generated with ultrasound disrupt the structure of a DRVP, exposing the groups within the molecule, remodeling the hydrogen bonding, and ultimately altering the ratio of the secondary structure [27].

3.3.3. Tertiary Structure

H_0 can indirectly reflect the degree of tertiary structure modification and denaturation by measuring the amount of hydrophobic clusters exposed to the proteins surface [19]. As revealed in Figure 3D, U-DRVPs had higher ANS fluorescence intensity (H_0) than C-DRVPs. It is because the ultrasonic cavitation broke down the interaction forces within DRVP, making the structure became more relaxed and stretched. The hydrophobic groups originally situated within the molecule relocated to the surface. The more fully the hydrophobic groups were exposed, then the more ANS probes were bound, and hence higher fluorescence values were observed [29]. Similar results regarding H_0 were also reported by Li et al. [19].

The variation in SH content reflects dynamic rearrangements in the spatial conformation of proteins [16]. Similarly to H_0, SH content also impacts proteins' functional properties, including solubility and emulsification. Thus, the SH content of DRVP was assayed in Figure 3E. Compared to C-DRVPs, the content of SH in U-DRVPs dropped by 2.91 μmol/g. Kang et al. [30] identified that the SH content of chickpea proteins initially rose and then fell with the increase in sonication time, and that the sonication time had a pronounced effect on the SH content. Accordingly, it was speculated that the internal SH was gradually transferred to the surface of DRVPs at the beginning of UAE, causing an increase in SH content. Instead, SH content continued to decrease rapidly after a critical time point. This may be because the energy delivered by ultrasonic cavitation can split water molecules into hydrogen atoms and high-reactivity hydroxyl radicals. Upon oxidation by hydroxyl radicals, cross-linking between DRVP occurred by exchanging thiols for disulfides. This resulted in the formation of soluble protein aggregates (Figure 2C), thereby decreasing the amount of SH [22].

3.4. Intermolecular Forces Analysis

Compared with C-DRVPs, UAE was accompanied by a pronounced expansion of U-DRVP aggregates (Figure 2D), which suggested their different aggregation modes among the proteins particles. To further understand the intermolecular forces involved in the formation and maintenance of DRVP aggregates, the influence of different dissociating agents (HCl/NaOH, SDS, urea, and β-ME) on the particle size of the aggregates were determined (Figures S1 and S2).

3.4.1. Electrostatic Interactions

Modifying pH can change the system charge and subsequently affect the electrostatic interactions between protein particles [31]. Therefore, the role of ionic bonding in DRVP aggregates can be ascertained by varying the pH value. The average particle size of U-DRVPs was slightly reduced in both acidic (pH = 5) and basic (pH = 9) conditions compared to the neutral condition (pH = 7) as illustrated in Figure 4A, whereas in the case of C-DRVPs, the average particle size was reduced under acidic (pH = 5) conditions and slightly increased under alkaline (pH = 9) conditions. These results revealed that the stability of U-DRVP aggregates was affected less by electrostatic interactions (ionic bonding) than that of C-DRVP.

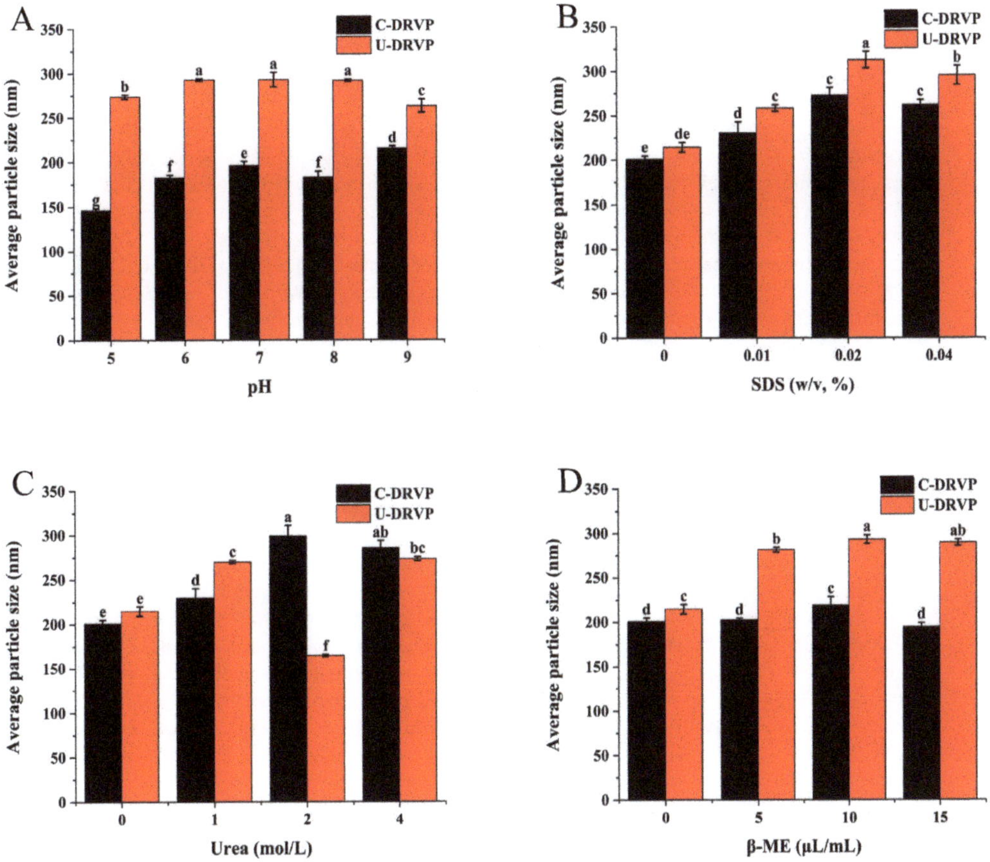

Figure 4. Effects of different dissociation agents on average particle size of C-DRVPs and U-DRVPs. (**A**) pH, (**B**) SDS, (**C**) urea, and (**D**) β-ME. Note: Different letters indicate a significant difference between the means of any two groups ($p < 0.05$). C-DRVPs, *D. rubrovolvata* volva proteins extracted using conventional alkaline extraction without UAE; U-DRVPs, *D. rubrovolvata* volva proteins extracted using UAE.

3.4.2. Hydrophobic Interaction

As a surfactant, SDS can interact with nonpolar groups in protein side chains, disrupting hydrophobic forces and causing protein denaturation [20]. Therefore, the impact of SDS on particle size was evaluated to explore the contribution of hydrophobic interaction in maintaining DRVP stability, as depicted in Figure 4B. The results signified that the aggregation of U-DRVPs was related to the concentration of SDS. To be specific, concentration raised from 10 mg/mL to 40 mg/mL, resulting in an increment in average particle size, forming larger aggregates. The hydrophobic forces between U-DRVPs were disrupted by raising the SDS concentration, exposing the hydrophobic regions hidden inside. Subsequently, the long-chain hydrophobic tail of SDS is bound to the exposed region, and the negatively charged head is bound to another positively charged protein via ionic bonding, ultimately forming larger aggregates [20]. A similar effect was also observed in C-DRVP (Figure 4B). Overall, hydrophobic interaction plays an important role in aggregate formation, regardless of sonication.

3.4.3. Hydrogen Bonding

Urea binds competitively to the amide groups in proteins, forming new hydrogen bonds and disrupting the pre-existing hydrogen bonds in the aggregates [20]. Therefore, urea was added to the aggregates dispersion to probe the role of hydrogen bonding in maintaining DRVP conformation. As seen in Figure 4C, the average particle size of the U-DRVP aggregates followed an increasing–decreasing–increasing trend with urea concentration from 1 M to 4 M. Intramolecular hydrogen bonds were separated by 1 M urea, resulting in swelling of the molecules; 2 M urea disrupted the hydrogen bonds, fragmenting the protein aggregates into smaller particulate; and when the concentration reached 4 M, the urea facilitated intermolecular hydrophobic forces by depriving surfaces moisture, ultimately leading to the formation of larger aggregates [20]. Regarding C-DRVP, its particle size gradually enlarged with increasing urea concentration (Figure 4C). In conclusion, the results confirmed the involvement of hydrogen bonding in the assembly of DRVP aggregates.

3.4.4. Disulfide Bonding

β-ME cleaves disulfide bonds in proteins, releasing more free sulfhydryl groups [20]. Thus, the involvement of disulfide bonds can be examined by adding β-ME. As shown in Figure 4D, simply 5 μL/mL β-ME dramatically raised the particle size of U-DRVP aggregates. This may be due to two reasons: On the one hand, β-ME cleaves disulfide bonds and releases a variety of reactive sites facilitating the action of other intermolecular forces and thus particle size enlargement; on the other hand, β-ME, as a linker, promotes the connection between U-DRVPs containing reactive functional groups [20]. Interestingly, β-ME had a negligible effect on C-DRVP size. The disulfide bonds were essential for U-DRVP aggregates formation, but not for C-DRVPs, which was supported by the data in Figure 4D. Collectively, the forces that maintain the two DRVP aggregates were different. It was mainly hydrophobic forces, disulfide bonding, and hydrogen bonding for U-DRVPs, whereas it was mainly hydrogen bonding, hydrophobic forces, and ionic bonding for C-DRVPs.

3.5. *Functional Properties of DRVPs*

3.5.1. Solubility

Protein solubility is the most crucial functional characteristic, influencing properties like emulsification, foaming, and gelation [21]. These properties determine the practical uses of proteins in various food systems. In this study, solubility was measured at different pH levels (ranging from 2 to 10) to simulate acidic, neutral, and alkaline conditions in food matrices (Figure 5A). Due to their close proximity to the isoelectric point, both proteins were least soluble at pH 2. As the pH deviated from the isoelectric point, the solubility increased dramatically, which was related to the homo-charge repulsion [21]. Interestingly, compared to C-DRVPs, U-DRVPs exhibited better solubility under acidic conditions (pH < 7), while the results were opposite under alkaline conditions. Several findings found that the higher solubility of sonicated proteins was dependent on size reduction [9]. However, the present observations suggested that the relationship between particle size and solubility may not be so straightforward [32]. Through sonocavitation, U-DRVPs partially unfolded, exposing its internal hydrophilic groups. As a result, U-DRVPs were able to interact more strongly with the surrounding water molecules [9].

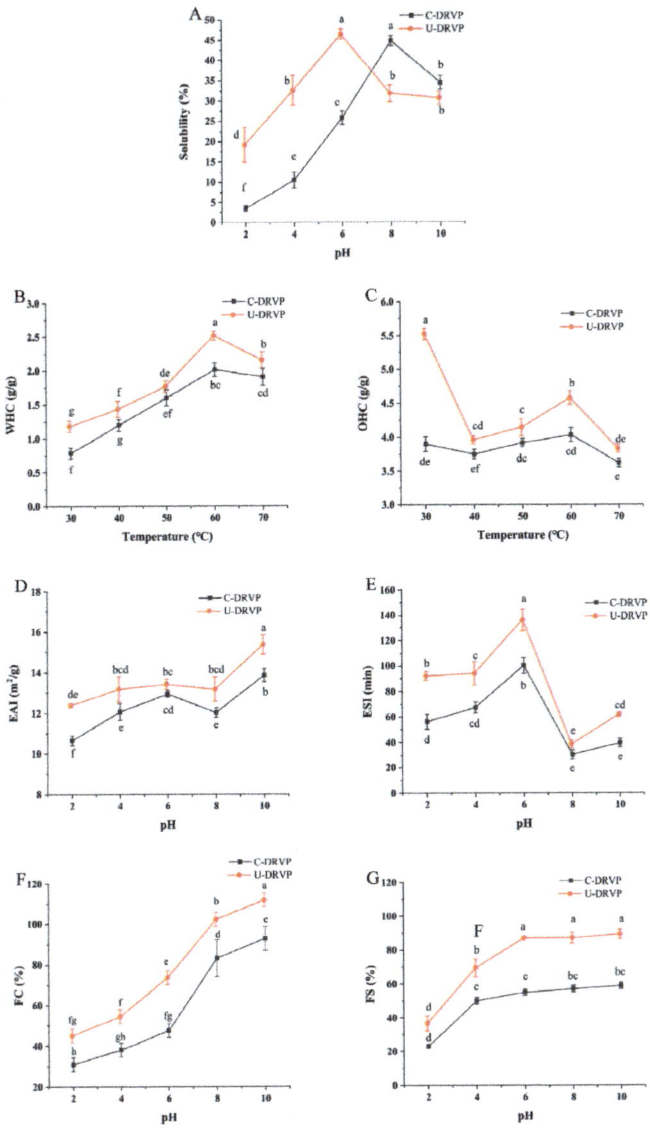

Figure 5. Functional properties of C-DRVP and U-DRVP. (**A**) Solubility, (**B**) WHC, (**C**) OHC, (**D**) EAI, (**E**) ESI, (**F**) FC, and (**G**) FS. Note: Different letters indicate a significant different between the means of any two groups ($p < 0.05$). C-DRVPs, *D. rubrovolvata* volva proteins extracted using conventional alkaline extraction without UAE; U-DRVPs, *D. rubrovolvata* volva proteins extracted using UAE.

3.5.2. WHC and OHC

WHC and OHC are very important as they severely affect the taste and texture of diverse food types [33]. As illustrated in Figure 5B, the WHC of DRVP increased with rising temperature regardless of sonication, reaching the maximum value at 60 °C for both U-DRVP (2.52 g/g) and C-DRVP (2.01 g/g). Continuing to increase the temperature, WHC gradually declined. U-DRVP presented higher WHC than C-DRVP throughout the temperature range tested. This may be due to the transient high temperature and pressure created by the cavitation bubble burst, which loosened the spatial structure of U-DRVP, increasing the contact area of water molecules with the hydrophilic groups in U-DRVP, and subsequently improving WHC [33]. Additionally, the increase in temperature further encouraged the stretching of DRVP, which contributed to the further enhancement of WHC. However, excessive temperature severely damaged the protein structures, resulting in a significant reduction in WHC.

Likewise, U-DRVP had higher OHC than C-DRVP. In general, lipids bind to proteins via hydrophobic forces between fatty chains and nonpolar amino acid side chains. Thus, it was hypothesized that U-DRVP with higher hydrophobicity would have an improved lipid-binding capacity [34,35]. OHC of U-DRVP and C-DRVP peaked at the initial temperature (30 °C) at 5.53 g/g and 3.90 g/g (Figure 5C), respectively. Keeping the temperature elevated, OHC showed a trend of decreasing–increasing–decreasing, but always lower than that at the starting temperature. This suggests that high temperature was unfavorable for oil uptake by DRVPs. Changes in physicochemical properties and spatial structure may be responsible for the complex temperature effects on OHC.

3.5.3. Emulsifying Properties

Emulsification characterized by EAI and ESI is an important quality index in the food manufacturing process. EAI refers to the ability of proteins to promote emulsion formation, whilst ESI reflects the proteins property to keep the emulsion stable over time [36]. As displayed in Figure 5D, both pH and UAE had significant effects on the EAI of DRVP. The EAI of both DRVPs improved with pH increasing, and U-DRVPs showed a higher EAI over the entire pH range. It was worth noting that the poorer solubility (Figure 5A) and larger particle size (Figure 2D) of U-DRVPs did not prevent them from exhibiting better EAI in alkaline conditions as compared to C-DRVPs. Thus, in this case, the EAI may be influenced by the combination of several factors, such as solubility, particle size, spatial conformation, hydrophilicity, and lipophilicity [37–39]. Under the effect of sonication cavitation and shear force, the spatial structure of U-DRVP was more stretched, exposing more hydrophilic and hydrophobic groups [36]. As a result, the rapid migration and absorption of U-DRVPs at the oil–water interface lowered the interfacial tension and ultimately improved the emulsification [35]. Consistent with the results of EAI, U-DRVPs also had higher ESI values than C-DRVPs over the given pH range, with a maximum value of 136.53 min at pH = 6. In conclusion, U-DRVPs with high EAI and ESI values possess improved emulsification properties. Sauces, dressings, baked foods, and meat analogs can all make use of it as an emulsifier.

3.5.4. Foaming Properties

The foaming properties of proteins are essential for the production of foaming foods such as ice cream, cakes, and beer [22]. Two indicators serve to evaluate the foaming properties of proteins, FC and FS. The FC of both U-DRVPs and C-DRVPs grew with increasing pH. Their maximum values were 111.90% and 92.85% at pH 10, respectively (Figure 5F). UAE was effectively used to improve FC in the assayed pH range. The possible explanation involved a decrease in the proportion of α-helix, the densest secondary structure (Figure 3C). As a result, the flexibility of the proteins increased and the buried hydrophobic groups were uncovered [40]. U-DRVPs were then rapidly dispersed at the air/water interface, entrapping the air bubbles and improving foaming performance [40]. Analogously, UAE substantially increased the FS of DRVPs (Figure 5G). Under alkaline conditions, despite being less soluble than C-DRVPs, U-DRVPs retained a higher FS. A possible explanation for this contradictory result might be that U-DRVPs were more uniform in size compared to C-DRVPs. This contributed to the reduction in intermolecular forces and interfacial surface tension, making it easier to build a firm, thick, cohesive, elastic and continuous film over the dispersed bubbles, thus stabilizing them [22].

4. Conclusions

The present study demonstrated that UAE can significantly improve the protein extraction rate and functional properties of DRVPs. Under the optimal process conditions, the extraction rate of DRVPs can reach 43.34%. Meanwhile, the results of SDS-PAGE patterns, FTIR, H_0, and SH contents indicated that the primary structure of U-DRVPs remained unchanged, whereas the secondary and tertiary structures varied obviously. Turbidity and particle size results suggested that ultrasound promoted the reorganization of U-DRVPs to form aggregates with increased particle size, which were maintained mainly by hydrophobic forces, disulfide bonding, and hydrogen bonding. Moreover, water/oil-holding ability, as well as the emulsifying and foaming properties of DRVPs were significantly enhanced using UAE. It was ascribed to the comprehensive effects induced by ultrasonic cavitation. Specifically, it referred to the changes in the advanced structure, molecular interactions, and physicochemical properties of DRVPs, rather than the reduction in particle size commonly assumed. In conclusion, these findings provided valuable information on *D. rubrovolvata* volva as a potential protein source. Further research should focus on the scale-up of DRVP manufacturing and the development of corresponding ultrasound equipment for its future application in the food industry.

Supplementary Materials: The following supporting information can be downloaded at: https://www.mdpi.com/article/10.3390/foods13081265/s1, Figure S1: Effects of different dissociation agents on particle size distribution of U-DRVP; Figure S2: Effects of different dissociation agents on particle size distribution of C-DRVP; Table S1: The factors and levels of the orthogonal test; Table S2: Orthogonal array design with experimental results; Table S3: Variance analysis of orthogonal test.

Author Contributions: Conceptualization: D.L. and H.L.; data curation: Y.Z.; formal analysis: Y.Z.; funding acquisition: D.L., Y.L. and Y.Z.; investigation: Y.Z., S.W. and Q.X.; methodology: Y.Z., S.W. and Q.X.; project administration: D.L., H.L. and M.G.; resources: D.L., H.L., Y.G. and M.G.; supervision, D.L. and H.L.; validation: L.M. and Y.L.; visualization: Y.Z.; writing—original draft: Y.Z.; writing—review and editing: D.L. and H.L. All authors have read and agreed to the published version of the manuscript.

Funding: This work was supported by the Scientific and Technological Plan Project of Guizhou province (QKHJC-ZK [2021]177, QKHPTRC-CXTD [2022]002), Discipline and Master's Site Construction Project of Guiyang University by Guiyang City Financial Support Guiyang University (2021-xk15), Scientific Research Project of Higher Education Department of Guizhou province (QJJ [2022]301, QJJ [2023]042), and the Graduate Innovation Program of Guiyang University in 2021 (GYU-YJS [2021]-37).

Institutional Review Board Statement: Not applicable.

Informed Consent Statement: Not applicable.

Data Availability Statement: The original contributions presented in the study are included in the article/Supplementary Material, further inquiries can be directed to the corresponding authors.

Conflicts of Interest: Author Ming Guo was employed by the company Guizhou Jin Chan Da Shan Biotechnology Company Limited. The remaining authors declare that the research was conducted in the absence of any commercial or financial relationships that could be construed as a potential conflict of interest.

References

1. Colgrave, M.L.; Dominik, S.; Tobin, A.B.; Stockmann, R.; Simon, C.; Howitt, C.A.; Belobrajdic, D.P.; Paull, C.; Vanhercke, T. Perspectives on future protein production. *J. Agric. Food Chem.* **2021**, *69*, 15076–15083. [CrossRef]
2. Zhuang, Y.; Sun, L. Nutritional characteristics of proteins from the volva and pileus in cultivated mushroom *Dictyophora rubrovolvata*. *Int. J. Food Sci. Nutr.* **2011**, *62*, 392–396. [CrossRef]
3. Bao, K.; Song, M.; Wang, S.; Li, T.; Wang, J.; Cheng, X.; Wang, L.; Wang, S.; Wen, T.; Zhu, Z. Isolation, purification, characterization and immunomodulatory effects of polysaccharides from *Dictyophora rubrovalvata* waste. *Ind. Crops Prod.* **2023**, *206*, 117754. [CrossRef]
4. Zhang, Y.; Xun, H.; Gao, Q.; Qi, F.; Sun, J.; Tang, F. Chemical constituents of the mushroom *Dictyophora indusiata* and their anti-inflammatory activities. *Molecules* **2023**, *28*, 2760. [CrossRef] [PubMed]
5. Zhang, Y.; Lei, Y.; Qi, S.; Fan, M.; Zheng, S.; Huang, Q.; Lu, X. Ultrasonic-microwave-assisted extraction for enhancing antioxidant activity of *Dictyophora indusiata* polysaccharides: The difference mechanisms between single and combined assisted extraction. *Ultrason. Sonochem.* **2023**, *95*, 106356. [CrossRef]
6. Guo, Q.; Liang, S.; Xiao, Z.; Ge, C. Research progress on extraction technology and biological activity of polysaccharides from edible fungi: A review. *Food Rev. Int.* **2023**, *39*, 4909–4940. [CrossRef]
7. Riad, N.; Zahi, M.R.; Trovato, E.; Bouzidi, N.; Daghbouche, Y.; Utczás, M.; Mondello, L.; El Hattab, M. Chemical screening and antibacterial activity of essential oil and volatile fraction of *Dictyopteris polypodioides*. *Microchem. J.* **2020**, *152*, 104415. [CrossRef]
8. Chen, J.; Chen, X.; Zhou, G.; Xu, X. Ultrasound: A reliable method for regulating food component interactions in protein-based food matrices. *Trends Food Sci. Technol.* **2022**, *128*, 316–330. [CrossRef]
9. Wang, Q.; Wang, Y.; Huang, M.; Hayat, K.; Kurtz, N.C.; Wu, X.; Ahmad, M.; Zheng, F. Ultrasound-assisted alkaline proteinase extraction enhances the yield of pecan protein and modifies its functional properties. *Ultrason. Sonochem.* **2021**, *80*, 105789. [CrossRef]
10. Chen, L.; Chen, J.; Ren, J.; Zhao, M. Effects of ultrasound pretreatment on the enzymatic hydrolysis of soy protein isolates and on the emulsifying properties of hydrolysates. *J. Agric. Food Chem.* **2011**, *59*, 2600–2609. [CrossRef]
11. Xie, Y.; Yang, F.; Zhao, K.; Zhang, W.; Liu, Q.; Yuan, Y. Regulation of protein flexibility and promoting the cod protein gel formation using ultrasound treatment. *J. Agric. Food Chem.* **2023**, *71*, 18601–18612. [CrossRef] [PubMed]
12. Chen, L.; Chen, J.; Wu, K.; Yu, L. Improved low pH emulsification properties of glycated peanut protein isolate by ultrasound Maillard reaction. *J. Agric. Food Chem.* **2016**, *64*, 5531–5538. [CrossRef] [PubMed]
13. Gao, J.; Wang, Y.; Yan, Y.; Li, Z. Ultrasonic-alkali method for synergistic breakdown of excess sludge for protein extraction. *J. Clean. Prod.* **2021**, *295*, 126288. [CrossRef]
14. Cui, Q.; Ni, X.; Zeng, L.; Tu, Z.; Li, J.; Sun, K.; Chen, X.; Li, X. Optimization of protein extraction and decoloration conditions for tea residues. *Hortic. Plant J.* **2017**, *3*, 172–176. [CrossRef]
15. Huang, D.; Li, W.; Li, G.; Zhang, W.; Chen, H.; Jiang, Y.; Li, D. Effect of high-intensity ultrasound on the physicochemical properties of Tenebrio Molitor protein. *Food Hydrocoll.* **2023**, *134*, 108056. [CrossRef]
16. Zhong, Z.; Xiong, Y.L. Thermosonication-induced structural changes and solution properties of mung bean protein. *Ultrason. Sonochem.* **2020**, *62*, 104908. [CrossRef] [PubMed]
17. Eze, O.F.; Chatzifragkou, A.; Charalampopoulos, D. Properties of protein isolates extracted by ultrasonication from soybean residue (okara). *Food Chem.* **2022**, *368*, 130837. [CrossRef]
18. Li, K.; Fu, L.; Zhao, Y.Y.; Xue, S.W.; Wang, P.; Xu, X.L.; Bai, Y.H. Use of high-intensity ultrasound to improve emulsifying properties of chicken myofibrillar protein and enhance the rheological properties and stability of the emulsion. *Food Hydrocoll.* **2020**, *98*, 105275. [CrossRef]
19. Li, X.; Luo, T.; Wang, L.; Song, H.; Wang, F.; Weng, Z.; Zhou, J.; Xiang, X.; Xiong, L.; Shen, X. Emulsifying properties of wheat germ protein: Effect of different ultrasonic treatment. *Ultrason. Sonochem.* **2023**, *98*, 106479. [CrossRef]
20. Wang, R.; Zhang, L.; Chi, Y.; Chi, Y. Forces involved in freeze-induced egg yolk gelation: Effects of various bond dissociation reagents on gel properties and protein structure changes. *Food Chem.* **2022**, *371*, 131190. [CrossRef]
21. Alavi, F.; Chen, L.; Emam-Djomeh, Z. Effect of ultrasound-assisted alkaline treatment on functional property modifications of faba bean protein. *Food Chem.* **2021**, *354*, 129494. [CrossRef]
22. Li, W.; Yang, H.; Coldea, T.E.; Zhao, H. Modification of structural and functional characteristics of brewer's spent grain protein by ultrasound assisted extraction. *LWT-Food Sci. Technol.* **2021**, *139*, 110582. [CrossRef]
23. Gordalina, M.; Pinheiro, H.M.; Mateus, M.; da Fonseca, M.M.R.; Cesário, M.T. Macroalgae as protein sources—A review on protein bioactivity, extraction, purification and characterization. *Appl. Sci.* **2021**, *11*, 7969. [CrossRef]
24. Zhao, F.; Liu, X.; Ding, X.; Dong, H.; Wang, W. Effects of high-intensity ultrasound pretreatment on structure, properties, and enzymolysis of soy protein isolate. *Molecules* **2019**, *24*, 3637. [CrossRef] [PubMed]
25. Jiang, L.; Wang, J.; Li, Y.; Wang, Z.; Liang, J.; Wang, R.; Chen, Y.; Ma, W.; Qi, B.; Zhang, M. Effects of ultrasound on the structure and physical properties of black bean protein isolates. *Food Res. Int.* **2014**, *62*, 595–601. [CrossRef]
26. Lv, S.; Taha, A.; Hu, H.; Lu, Q.; Pan, S. Effects of ultrasonic-assisted extraction on the physicochemical properties of different walnut proteins. *Molecules* **2019**, *24*, 4260. [CrossRef] [PubMed]

27. Figueroa-González, J.J.; Lobato-Calleros, C.; Vernon-Carter, E.J.; Aguirre-Mandujano, E.; Alvarez-Ramirez, J.; Martínez-Velasco, A. Modifying the structure, physicochemical properties, and foaming ability of amaranth protein by dual pH-shifting and ultrasound treatments. *LWT-Food Sci. Technol.* **2022**, *153*, 112561. [CrossRef]
28. Wang, F.; Zhang, Y.; Xu, L.; Ma, H. An efficient ultrasound-assisted extraction method of pea protein and its effect on protein functional properties and biological activities. *LWT-Food Sci. Technol.* **2020**, *127*, 109348. [CrossRef]
29. Arzeni, C.; Martínez, K.; Zema, P.; Arias, A.; Pérez, O.E.; Pilosof, A.M.R. Comparative study of high intensity ultrasound effects on food proteins functionality. *J. Food Eng.* **2012**, *108*, 463–472. [CrossRef]
30. Kang, S.; Zhang, J.; Guo, X.; Lei, Y.; Yang, M. Effects of ultrasonic treatment on the structure, functional properties of chickpea protein isolate and its digestibility in vitro. *Foods* **2022**, *11*, 880. [CrossRef]
31. Sun, N.; Wang, Y.; Bao, Z.; Cui, P.; Wang, S.; Lin, S. Calcium binding to herring egg phosphopeptides: Binding characteristics, conformational structure and intermolecular forces. *Food Chem.* **2020**, *310*, 125867. [CrossRef] [PubMed]
32. Cai, Y.; Li, Q.; Li, D.; Sun, C.; Bao, Y.; Li, F.; Jiang, S. Optimizing the extraction of protein from broken rice using response surface methodology and comparing the protein functional properties. *J. Cereal Sci.* **2023**, *113*, 130726. [CrossRef]
33. Loushigam, G.; Shanmugam, A. Modifications to functional and biological properties of proteins of cowpea pulse crop by ultrasound-assisted extraction. *Ultrason. Sonochem.* **2023**, *97*, 106448. [CrossRef] [PubMed]
34. Li, X.; Qi, B.; Zhang, S.; Li, Y. Effects of ultrasonic treatment on the structural and functional properties of cactus (*Opuntia ficus-indica*) seed protein. *Ultrason. Sonochem.* **2023**, *97*, 106465. [CrossRef] [PubMed]
35. Gani, A.; Ashraf, Z.; Noor, N.; Ahmed Wani, I. Ultrasonication as an innovative approach to tailor the apple seed proteins into nanosize: Effect on protein structural and functional properties. *Ultrason. Sonochem.* **2022**, *86*, 106010. [CrossRef] [PubMed]
36. Shi, R.; He, Y.; Wang, Q.; Cai, J.; Gantumur, M.; Jiang, Z. Insight into the physicochemical characteristics, functionalities and digestion behavior of protein isolate derived from *Lactarius volemus* (*L.volemus*): Impacts of microwave-assisted extraction. *Food Chem.* **2024**, *431*, 137070. [CrossRef] [PubMed]
37. Fang, B.; Chang, L.; Ohm, J.B.; Chen, B.; Rao, J. Structural, functional properties, and volatile profile of hemp protein isolate as affected by extraction method: Alkaline extraction–isoelectric precipitation vs salt extraction. *Food Chem.* **2023**, *405*, 135001. [CrossRef] [PubMed]
38. Görgüç, A.; Bircan, C.; Yılmaz, F.M. Sesame bran as an unexploited by-product: Effect of enzyme and ultrasound-assisted extraction on the recovery of protein and antioxidant compounds. *Food Chem.* **2019**, *283*, 637–645. [CrossRef] [PubMed]
39. Yang, J.S.; Dias, F.F.G.; Pham, T.T.K.; Barile, D.; de Moura Bell, J.M.L.N. A sequential fractionation approach to understanding the physicochemical and functional properties of aqueous and enzyme-assisted aqueous extracted black bean proteins. *Food Hydrocoll.* **2024**, *146*, 109250. [CrossRef]
40. Karabulut, G.; Yemiş, O. Modification of hemp seed protein isolate (*Cannabis sativa* L.) by high-intensity ultrasound treatment. Part 1: Functional properties. *Food Chem.* **2022**, *375*, 131843. [CrossRef]

foods

MDPI

Article

The Application of Protein Concentrate Obtained from Green Leaf Biomass in Structuring Nanofibers for Delivery of Vitamin B12

Bojana Balanč [1], Ana Salević-Jelić [2], Verica Đorđević [3,*], Branko Bugarski [3], Viktor Nedović [2], Predrag Petrović [1] and Zorica Knežević-Jugović [3]

[1] Innovation Centre of Faculty of Technology and Metallurgy, University of Belgrade, Karnegijeva 4, 11000 Belgrade, Serbia; bisailovic@tmf.bg.ac.rs (B.B.); ppetrovic@tmf.bg.ac.rs (P.P.)
[2] Faculty of Agriculture, University of Belgrade, Nemanjina 6, 11080 Beograd, Serbia; ana.salevic@agrif.bg.ac.rs (A.S.-J.); viktor.nedovic@mpn.gov.rs (V.N.)
[3] Faculty of Technology and Metallurgy, University of Belgrade, Karnegijeva 4, 11000 Belgrade, Serbia; branko@tmf.bg.ac.rs (B.B.); zknez@tmf.bg.ac.rs (Z.K.-J.)
* Correspondence: vmanojlovic@tmf.bg.ac.rs; Tel.: +381-11-3370472

Abstract: Nanofibers made of natural proteins have caught the increasing attention of food scientists because of their edibility, renewability, and possibility for various applications. The objective of this study was to prepare nanofibers based on pumpkin leaf protein concentrate (LPC) as a by-product from some crops and gelatin as carriers for vitamin B12 using the electrospinning technique. The starting mixtures were analyzed in terms of viscosity, density, surface tension, and electrical conductivity. Scanning electron micrographs of the obtained nanofibers showed a slight increase in fiber average diameter with the addition of LPC and vitamin B12 (~81 nm to 109 nm). Fourier transform infrared spectroscopy verified the physical blending of gelatin and LPC without phase separation. Thermal analysis showed the fibers had good thermal stability up to 220 °C, highlighting their potential for food applications, regardless of the thermal processing. Additionally, the newly developed fibers have good storage stability, as detected by low water activity values ranging from 0.336 to 0.376. Finally, the release study illustrates the promising sustained release of vitamin B12 from gelatin-LPC nanofibers, mainly governed by the Fickian diffusion mechanism. The obtained results implied the potential of these nanofibers in the development of functional food products with improved nutritional profiles.

Keywords: nanofibers; electrospinning; pumpkin leaves protein; gelatin; vitamin B12

1. Introduction

Nanofibers made of natural proteins have caught the increasing attention of food scientists in recent years because of their edibility, renewability, availability, and possibility for various applications, such as bioactive encapsulation, enzyme immobilization, food coating, food texture modification, the production of active packaging systems, or even wider [1–3]. Generally, the use of nanofibers allows for a large surface area, high porosity, and improved encapsulation efficiency. Electrospinning is an electro-hydrodynamic, simple, and cost-effective technique used for the fabrication of nanofibers using different raw materials [4,5]. In the context of food-grade applications, it is employed to create structures that can encapsulate and protect sensitive bioactive compounds, such as vitamins. There has been interest in nanofibers made of mixed biopolymers [6]. A prerequisite for obtaining nanofibers by electrospinning is to use at least one spinnable biopolymer. Gelatin is one of the few highly spinnable proteins with many favorable properties, such as biodegradability, biocompatibility, and low cost [7,8]. However, sole gelatin nanofibers do not exert bioactivity [7]. Therefore, gelatin was successfully blended with some globular proteins that cannot be spun by themselves [9]. In this way, the extra nutritional value of gelatin-based nanofibers was achieved by using whey protein isolate, ovalbumin, soy protein isolate, or sodium caseinate. In addition, protein nanofibers may interact with natural polyphenols present in protein isolates, which results in improved stability [2].

The need for cheap and high-quality protein is growing. Leaves available as by-products from some crops on a large scale could potentially be used as a major protein source for food applications. Most of the studies on leaf protein in the past decades focused on developing extraction protocols for soluble protein recovery in the form of leaf protein concentrate (LPC) [10,11], while its utilization as an ingredient in human food is still at the early stages [12]. Rubisco, the main compound of the soluble protein fraction, is found to have interesting functional properties (high foaming capacity, good solubility at food pH, and ability to form gels at low concentrations and low temperatures) [13,14]. However, the potential of this protein to be processed into a structurally appealing product has not been explored yet. To our knowledge, the functionality of rubisco when incorporated into fibrous mats has not yet been studied. This study aimed to investigate the spinnability of a mixed solution of gelatin and leafy vegetable protein concerning changes in the physicochemical properties of solutions. Pumpkin (*Cucurbita pepo*) will be used as a source of leaf protein. This annual green leafy vegetable is widely cultivated for its fruits and seeds, while leaves are mainly underutilized despite huge potential; only in the southern part of Nigeria do the leaves of pumpkin, *Telfairia occidentalis*, constitute an important vegetable with a protein content of ~30% dry matter [15,16]. According to Tanimowo Fadupin et al., the crude protein content in the leaf of *Cucurbita pepo* is 7.3% wet matter [17].

Electrospun gelatin nanofibers instantly dissolve in cold water and can be used for designing fast-dissolving systems aimed at supplementing food with nutraceuticals [8,18,19]. Higher stability of sensitive compounds was achieved by their incorporation in gelatin nanofibrous films [18,20]. In this study, we test the hypothesis that gelatin-leaf protein can serve as a carrier for vitamin B12. Vitamin B12, or cobalamin, is an essential vitamin present in milk, eggs, meat, and shellfish. It exhibits important biological functions such as red blood cell formation, DNA synthesis and regulation, cell division, and neurological functions [21,22]. However, it becomes degraded and loses its activity upon heating the food and during storage processes [23]. Thus, vitamin B deficiency becomes an increasing health issue, especially among the elderly population, vegetarians, and pregnant women [24,25]. Therefore, different structures with encapsulated vitamins have been developed in the last couple of years [26–28]. High hydrophilicity, high molecular weight (1355.38 g/mol), and sterically hindering of this compound may cause difficulties for encapsulation. The novel herein-developed protein matrices containing vitamin B12 may exhibit high potential to contribute to the food industry.

2. Materials and Methods

2.1. Materials

The pumpkin leaves used for protein extraction were collected on fields owned by the company JS&O (Novo Milosevo, Serbia). Beef-skin gelatin (Type B, Bloom 220) was purchased from Gelnex (Santa Catarina, Brazil), and Cyanocobalamin (vitamin B12) was provided from Sigma Chemical Co. (St. Louis, MO, USA). Glacial acetic acid (>99%) was purchased from Macron Fine Chemicals (Centre Valley, PA, USA). All other chemicals were of analytical grade, and the water used was double-distilled.

2.2. Preparation of Pumpkin Leaf Protein Concentrate (LPC)

Briefly, about 300 g of fresh pumpkin leaves were mechanically processed so that a protein-rich juice was squeezed out of the leaves. Then, the heat coagulation step was applied for the precipitation of green proteins from the obtained green juice (at 55 °C, 30 min). The final step implied the addition of acid to so-called "brown juice" to change the solubility of the proteins and provide their precipitation. Thus, the pH was adjusted to 4.5 using 2 M HCl, as the isoelectric point of Rubisco is between 4.4 and 4.7 [29]. After the centrifugation, the white protein fraction was separated and further subjected to the freeze-drying process. In this way, about 1 g of LPC was prepared.

2.3. Preparation of the Solutions

Gelatin (20% *w/v*) was dissolved in acetic acid (30% *v/v*) at room temperature for 24 h to ensure complete hydration. Then, the LPC was added to the gelatin solution at a concentration of 10 mg/mL, and again, the mixture was stirred overnight at room temperature. A vitamin B12 solution was added directly 30 min before the electrospinning procedure at a final concentration of 0.4 mg/mL. For viscosity, surface tension, density, and conductivity measurements, vitamin B12 was dissolved in acetic acid (30% *v/v*) to obtain the same final concentration.

2.4. Characterization of the Gelatin-LPC Solutions

Previously prepared solutions were characterized in terms of electrical conductivity using a digital Benchtop Conductivity Meter (HI 2315, HANNA Instruments, Ltd., Temse, Belgium).

The surface tension was measured using a Krüss K20 tensiometer (Krüss GmbH, Hamburg, Germany). The Wilhelmy plate method was applied for those measurements. A cleaned platinum plate was dipped into the examined solutions. The tensiometer detects the force that is needed to pull the plate out of the sample. This maximum value of the force is detected just before the plate leaves the solution, and this value is proportionate to the surface tension.

The density was determined using the same instrument equipped with the DE01 set.

The viscosity of the solutions was measured on an IKA ROTAVISC LO-VI viscometer (KA, Staufen, Germany) using spindle SP1.

All measurements were carried out in triplicate at room temperature (25 ± 1 °C).

2.5. Electrospinning Processing

Gelatin solutions (plain, with LPC, or with LPC and vitamin B12) were electrospun through a blunt stainless-steel needle (18 G) at a steady flow rate of 0.5 mL/h using a syringe pump (Razel Scientific Instruments, Stamford, CO, USA). An electric field (17 kV) was applied between the positively charged needle and the grounded metallic collector plate. The distance between the needle tip and the collector plate was 10 cm (Figure 1). The process occurred at ambient temperature (~25 °C).

2.6. Nanofibers Morphology and Size

The surface morphology of nanofibers was determined by scanning electron microscopy (SEM, model TESCAN MIRA3XMU, Brno, Czech Republic) using a voltage of 10 kV and a magnification of 50,000. The size distribution of the obtained nanofibers was determined from the SEM images using the Image J software V 1.8.0, which analyzed no less than 100 fibers.

2.7. Encapsulation Efficiency

The encapsulation efficiency of vitamin B12 into gelatin-based nanofibers was valued as described by Coelho et al. [26]. Namely, vitamin B12 released instantly at time zero corresponds to the vitamin outside of the fibers. Using this data, it is possible to calculate the encapsulation efficiency as the ratio between the amount of B12 in gelatin structures and the total amount of B12.

2.8. Fourier Transform Infrared Spectroscopy

Fourier transform infrared (FT-IR) spectroscopy was employed to evaluate the structural properties of the fibrous mats and the chemical interactions between their constituents. The as-prepared samples were subjected to analysis in attenuated total reflection mode

(ATR) in the wavenumber range of 4000–600 cm^{-1} with a resolution of 4 cm^{-1} and 100 accumulations per scan using an IRAffinitty-1S (Shimadzu, Kyoto, Japan).

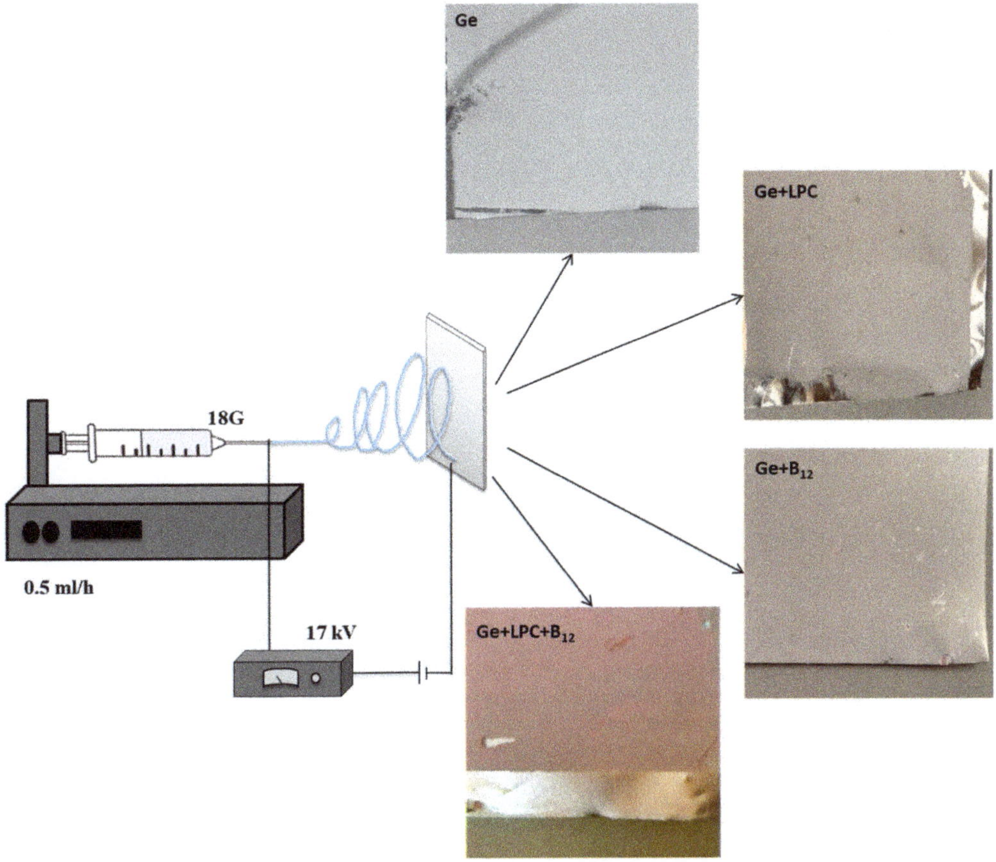

Figure 1. Schematic illustration of electrospinning set used in the study.

2.9. Thermogravimetric Analysis

Thermogravimetric analysis (TGA) was employed to show the mass losses of the samples upon heating from ambient temperature (30 ± 2 °C) to 250 °C. Keeping in mind that thermal treatments of foods in processing and preparation often involve temperatures up to 250 °C, this analysis projected the stability of the samples upon heating. The experiments were performed using the TG/DTA SetSys 2400 instrument (Setaram, Cailure, France). The heating rate was 10 °C/min and airflow was maintained at 20 mL/min.

2.10. Differential Scanning Calorimetry (DSC)

The thermal properties of the nanofibers were examined by the DSC technique. The samples were placed in DSC aluminum pans, which were hermetically sealed and analyzed using a DSC131 Evo (SETARAM Instrumentation, Caluire, France). The pens with samples were heated from 30 to 250 °C with a constant heating rate of 10 °C/min under a nitrogen atmosphere at a flow rate of 20 mL/min.

2.11. Water Activity

The water activity (Aw) values of the samples were measured using a water activity measurement device (LabSwift-aw, Novasina AG, Lachen, Switzerland).

2.12. Release Study

The release study was conducted by monitoring the vitamin B12 release directly into the water (pH 7), and the amount of released vitamin B12 was determined by measuring the absorbance (at 350 nm) up to the moment where the maximum occurred and stabilized. About 1 g of fiber was immersed in 20 mL of the medium at 25 °C under shaking at 300 rpm. Samples were taken from the release medium at predetermined time intervals. The release of vitamin B12 was analyzed using a microplate spectrophotometer (Multiskan™ GO, Thermo Scientific™, Waltham, MA, USA). The volume was kept constant by adding a fresh medium. The release measurements were carried out in triplicate.

The mechanisms governing the release of vitamin B12 from the as-prepared structures were studied using a few kinetic models, such as Higuchi (Equation (1)), Ritger–Peppas (Equation (2)), Kopcha (Equation (3)), and Peppas–Sahlin (Equation (4)):

$$m_t/m_\infty = k \cdot t^{1/2} \quad (1)$$

$$m_t/m_\infty = k \cdot t^n \quad (2)$$

$$m_t/m_\infty = A \cdot t^{0.5} + B \cdot t \quad (3)$$

$$m_t/m_\infty = k_1 \cdot t^m + k_2 \cdot t^{2m} \quad (4)$$

where m_t and m_∞ are the amounts of released vitamin B12 at time t and at infinite time, while k, and t correspond to the release constant and release time, respectively. Exponent n determines the mechanism of vitamin B12 release from the carrier: $0.45 < n < 0.89$ defines non–Fickian diffusion, $n \geq 0.89$ defines erosion mechanism, while $n \leq 0.45$ defines Fickian diffusion. Index m is the diffusion exponent. The parameters k_1 and A are Fickian constants, and the values k_2 and B are the erosion constants [19,30].

2.13. Statistical Analysis

The results were statistically analyzed using IBM SPSS Statistics 25 software (Armonk, New York, NY, USA). Depending on the homogeneity of variance test, significant differences between the samples were determined by one-way analysis of variance (ANOVA) with Tukey post hoc or Kruskal–Wallis H test with Mann–Whitney U test. A significance level of 0.05 was set.

3. Results and Discussion

In the electrospinning process, the competition among conductivity, tension force, and viscosity in a limited range affects fiber morphology and diameters. In this study, gelatin was spun from the aqueous acetic acid solution (30%, *v/v*) at a concentration of 20% (*w/v*). According to the literature [31,32], a 20–40% (*w/v*) concentration range of gelatin provides regular nanofiber formations under wide spectra of electrospinning process conditions (applied voltage and feed rate), while below a certain concentration, the particles will develop instead of fibers. In our experiments, the electrospinning of a sole LPC aqueous solution was not an effective process because the protein molecules in globular conformation do not display the necessary entanglements or interchain connections to electrospin.

Electrospinning under ambient conditions appeared to be a promising technique for encapsulating vitamin B12 with high encapsulation efficiency. The vitamin encapsulation efficiency (EE) of 97 and 99% was determined (for Ge+B12 and Ge+LPC+B12, respectively), which is in agreement with the result of Coelho et al., who used zein to encapsulate this vitamin by electrospinning and reported values between 61 and 100% depending on zein and vitamin concentrations and process parameters [26]. The electrospun formulations were evaluated by SEM to scrutinize their morphology and size. Ultra-thin nanofibers (81–109 nm) with homogeneous formats were developed, as shown in Figure 2. The use of green leaf protein corresponded to the formation of smoother and continuous fibers compared to somewhat choppy, branched fibers of solely gelatin (Figure 2B vs. Figure 2A). The formation of branched fibers in the case of sole gelatin indicates instability of the jet during electrospinning, which can split it into smaller jets or eject smaller jets, reducing local charge per surface area [33]. The absence of branching and more regular fibers' morphology induced by the LPC and vitamin addition could be due to the increased conductivity and electrical forces stabilizing the jet against splitting. This suggests improved molecular entanglements through the interactions between both constituents. When considering industrial applications, the size and polydispersity of the shapes are important factors to ensure the replicability and consistency of the products. LPC was more homogenous, filamentous, and thicker compared to pure gelatin. No obvious effect of vitamin B12 encapsulation on fiber morphology was observed.

The morphological results mainly corresponded to the properties of the blend solutions shown in Table 1. Pure gelatin solution (20%, *w/v*) had a high electrical conductivity of 4.10 mS/cm, and the value increased upon adding LPC since the LPC solution also expressed electrical conductivity. The addition of vitamin B12 further increased the electrical conductivity of the solution. This can be related to the decrease in viscosity, as it was reported that the decreased viscosity might affect the mobility of the charged species in the solutions, increasing conductivity [34]. This explains our morphological observations (Figure 2), proving that higher charge density generally generates smoother fibers because of the more intense whipping instability of the liquid jet. The surface tension of the polymer solution also plays a significant role in electrospinning. Leaf protein decreased the surface tension of the initial gelatin solution, which meant that viscoelastic forces easily overcame the surface tension, resulting in smoother fibers. The viscoelastic property of a feed solution is the most important parameter during the electrospinning process. However, measuring the viscoelastic properties is quite impractical, and instead, the viscosity of the solution could provide a solid indication of the spinnability and size of the electrospun fibers [31,32]. In general, the greater solution viscosity provides ticker fibers. A slight increase in viscosity was determined upon the addition of the leaf extract to the gelatin solution, which further resulted in thicker fibers.

Table 1. Viscosity, surface tension, density, and conductivity of the leaf protein concentrate (LPC) solution, gelatin solution (Ge), solution of vitamin B12 in 30% acetic acid (B12 solution), gelatin with the leaf protein concentrate solution (Ge+LPC), gelatin with vitamin B12 solution (Ge+B12) and gelatin with the leaf protein concentrate and vitamin B12 solution (Ge+LPC+B12).

Sample	Viscosity (mPa s)	Surface Tension (mN/m)	Density (g/mL)	Conductivity (mS/cm)
LPC	1.15 ± 0.08 ᵃ	33.4 ± 0.06 ᶜ	0.999 ± 0.000 ᵃ	2.76 ± 0.01 ᵇ
Ge	35.00 ± 0.04 ᶜ	33.5 ± 0.08 ᶜ	1.082 ± 0.000 ᵈ	4.10 ± 0.00 ᵈ
B12 solution *	1.20 ± 0.04 ᵃ	35.5 ± 0.02 ᵉ	1.037 ± 0.000 ᵇ	1.81 ± 0.07 ᵃ
Ge+LPC	40.00 ± 0.02 ᵉ	30.9 ± 0.01 ᵇ	1.085 ± 0.000 ᵉ	4.51 ± 0.02 ᵉ
Ge+B12	36.63 ± 0.05 ᵈ	29.1 ± 0.00 ᵃ	1.069 ± 0.001 ᶜ	3.83 ± 0.15 ᶜ
Ge+LPC+B12	30.17 ± 0.03 ᵇ	34.4 ± 0.47 ᵈ	1.070 ± 0.000 ᶜ	5.46 ± 0.07 ᶠ

* in 30% acetic acid, different letters within the same column indicate statistically significant differences ($p < 0.05$).

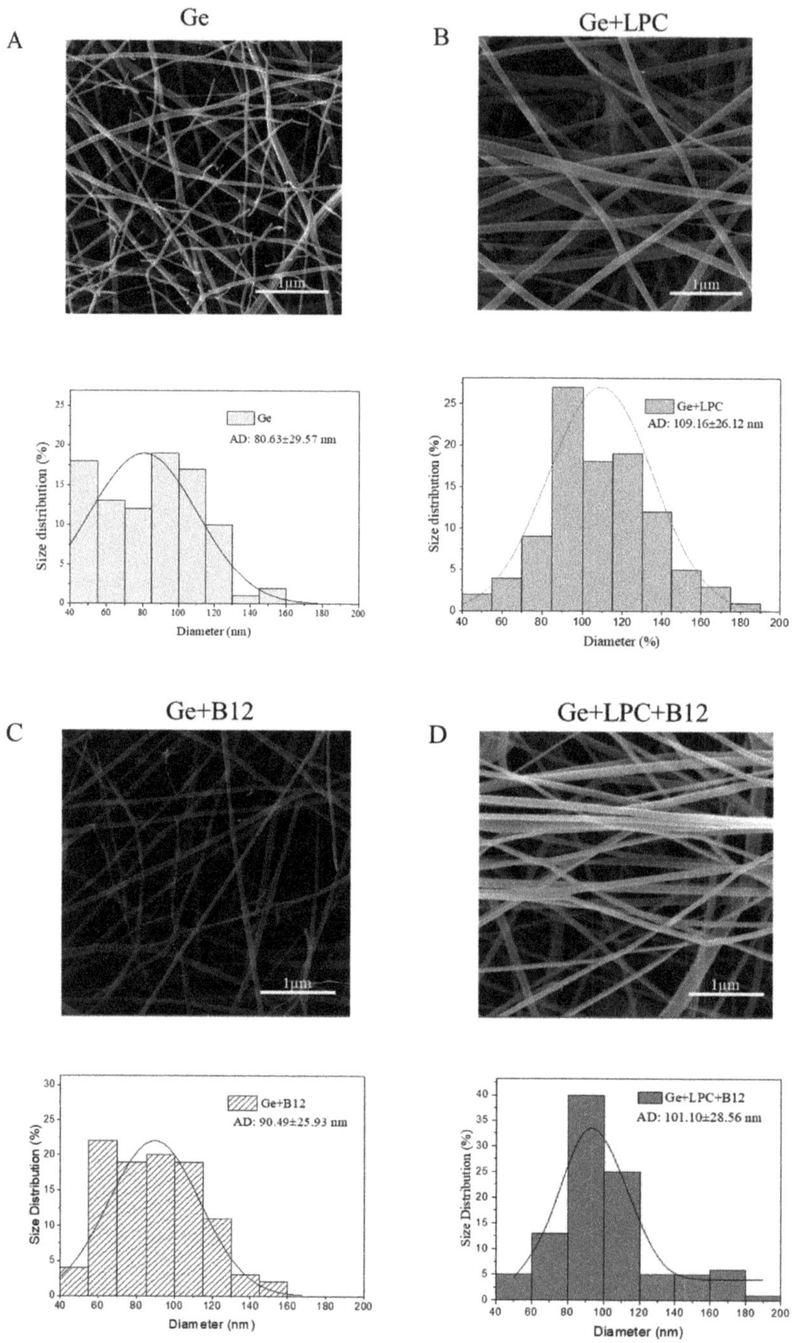

Figure 2. SEM images and fiber diameter distributions of (**A**) gelatin nanofibers (Ge), (**B**) gelatin nanofibers with LPC (Ge+LPC), (**C**) gelatin nanofibers with vitamin B12 (Ge+B12) and (**D**) gelatin nanofibers with LPC and vitamin B12 (Ge+LPC+B12). Data are expressed as mean ± SD (n = 100).

3.1. Fourier Transform Infrared Spectroscopy

Figure 3 shows the spectra of gelatin-based nanofibers. The neat gelatin nanofibers showed characteristic bands in agreement with those previously identified in the literature for gelatin-based materials [35,36]. These bands were as follows: 3294 (hydrogen bonding and N-H stretching, amide A), 2936 (C-H stretching vibrations of aliphatic groups), 1647 (C=O stretching vibrations, amide I), 1541 (C-N bending and stretching vibrations, amide II), 1456 (N–H bend and C-N stretching combination band, $NH3^+$ symmetric deformation), 1339 (C–H deformation of methyl group), 1238 (C-N stretching and N-H bending, amide III), and 1082 (C-O stretching) cm^{-1}.

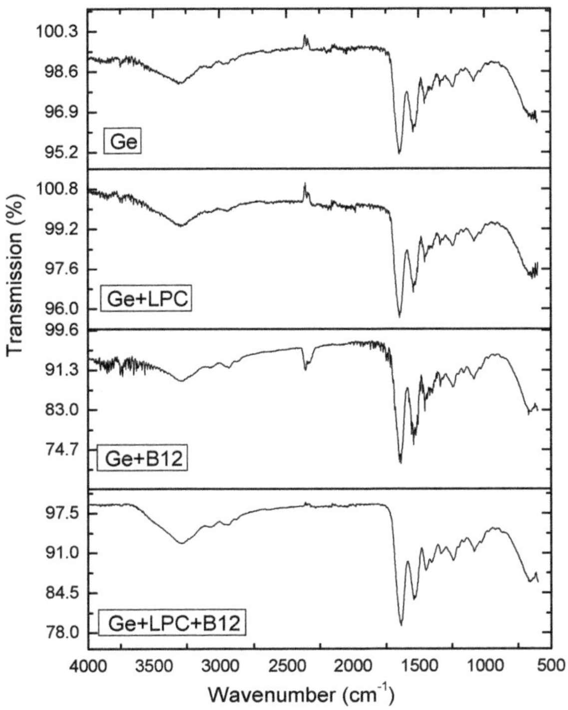

Figure 3. Infrared spectra of different electrospun nanofibers.

The nanofibers obtained by blending gelatin with pumpkin leaf extract showed a similar FT-IR spectrum as that previously discussed for neat gelatin nanofibers. The blending did not induce new bands nor shifts of gelatin-originating bands, which implies efficient, physical blending of gelatin and pumpkin leaf extract without phase separation or chemical interactions. The spectra obtained after vitamin B12 encapsulation, within the fibers based on the sole gelatin and blend of gelatin and pumpkin leaf extract were similar to those of the neat matrix. No new bands or band splits were observed after vitamin B12 encapsulation suggesting efficient vitamin B12 encapsulation within sole gelatin fibers and gelatin-pumpkin leaf extract blend-based fibers. Of note, vitamin B12 encapsulation shifted the band described to amide I. This band was shifted from 1647 cm^{-1} for the neat gelatin and gelatin-pumpkin leaf extract blend fibers to 1635 and 1636 cm^{-1} after vitamin B12 encapsulation within sole gelatin fibers and gelatin-pumpkin leaf extract blend-based fibers, respectively. This spectral change implies intermolecular interactions between the protein-based matrix and the active compound and changes in proteins' structure induced by the active compound incorporation [36]. According to the literature, cobalamin interacts with proteins with different binding affinity for different proteins by hydrogen bond, hydrophobic force, and π-π interactions [37].

3.2. Thermal Stability Analysis

Figure 4 presents the weight loss of LPC, gelatin powder, and the obtained fibers. The temperature range was chosen according to the intended final usage as a food ingredient. Keeping in mind that food is usually baked up to 250 °C, the weight loss was monitored up to this temperature. The first stage of weight loss for the fibers was starting from room temperature up to around 75 °C. This can be attributed to the bulk moisture of nanofibers and, thus, to the evaporation of free and weakly bonded water. The mass reduction was about 6%. The second weight loss started around 220 °C and went further with the temperature increase. This reduction was presumably related to protein degradation [38] and was also confirmed by DSC results. LPC exhibited degradation at lower temperatures compared to raw gelatine. The addition of vitamin B12 did not affect the thermal stability of the fibers; namely, cyanocobalamin is thermally stable in the investigated temperature range [39,40] Obviously, the prepared fibers have good thermal stability up to 220 °C, highlighting their potential for food applications, regardless of the thermal processing.

3.3. DSC

The thermal behavior was also studied by DSC analysis; thus, the effects of the addition of LPC and vitamin B12 encapsulation in nanofibers are presented in Figure 5 and summarized in Table 2. The thermogram of the unloaded LPC showed two endothermic peaks, the first around 96 °C with an enthalpy of 161.65 J/g, and the second around 220 °C. The first peak is presumably formed due to denaturation, while the second is related to the decomposition of LPC [7]. In the literature, it can be found that denaturation peaks for protein isolates vary between 69 °C and 109 °C for *Spirullina* protein isolates [41] and between 87 and 107 °C for legume protein isolates [42]. Namely, it is known that the thermal characteristics of globular proteins may be related to their thermal-induced aggregation and gelation [42–44]. As the denaturation temperature increases, the thermal stability of proteins increases, and the polypeptides have a more compact tertiary structure [42]. The DSC curve of gelatin powder shows the appearance of peaks at 118 °C and 228 °C. The first transition can be assigned to the devitrification of α-amino acid blocks, whereas the second one can be assigned to the devitrification of the blocks of amino acids (hydroxyproline and proline) [45]. Compared to gelatin powder, all the thermograms of fibers showed the first glass transition peak shifted to lower temperatures (90–95 °C) with higher values of enthalpy. The second peak is much weaker and also shifted to somewhat lower temperatures (211–223 °C). This occurrence can be explained by the electrospinning process, by which the segmental mobility of fibrous polymers is increased [8]. In addition, other authors also reported a decrease in the Tg of nanofibers due to interactions between the fiber constituents [46].

Table 2. The differential scanning calorimetry data for the leaf protein concentrate (LPC), gelatin powder (Ge powder), gelatin nanofibers (Ge), gelatin with the leaf protein nanofibers (Ge+LPC), Gelatin with vitamin B12 nanofibers (Ge+B12), and gelatin with the leaf protein concentrate and vitamin B12 nanofibers (Ge+LPC+B12).

Sample	Temperature			ΔH (J/g)
	Onset	Peak	Offset	
LPC	58.79	96.11	124.71	161.65
GE powder	76.25	118.24	168.86	251.32
Ge	67.75	90.24	161.72	490.02
Ge+LPC	66.38	95.53	130.93	310.53
Ge+B12	33.81	57.39	88.12	138.58
Ge+LPC+B12	63.33	89.99	124.32	232.39

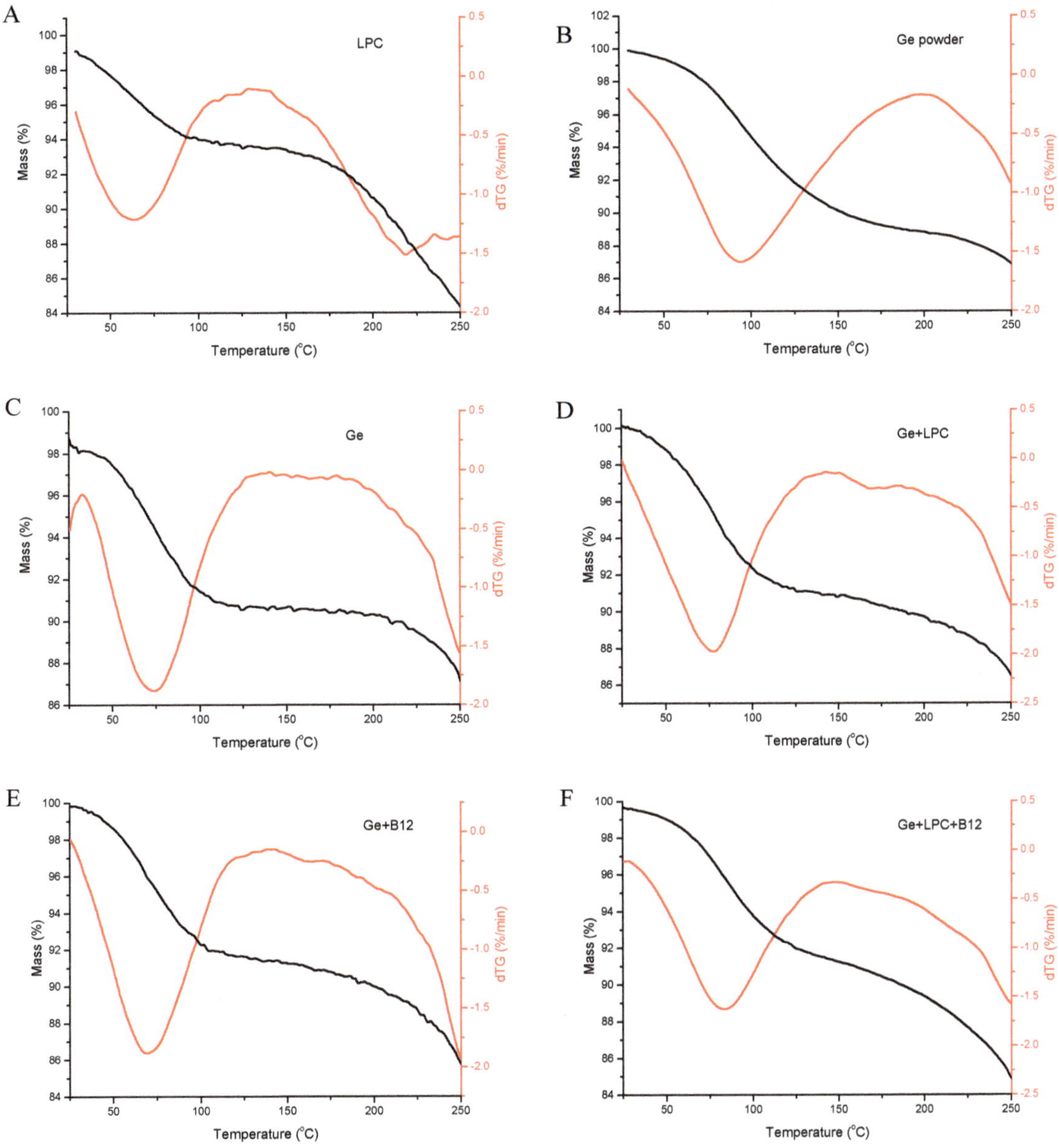

Figure 4. Thermogram and their first derivatives of (**A**) raw leaf protein concentrate (LPC), (**B**) gelatin powder (Ge powder), (**C**) gelatin nanofibers (Ge), (**D**) gelatin nanofibers with LPC (Ge+LPC), (**E**) gelatin nanofibers with vitamin B12 (Ge+B12), and (**F**) gelatin nanofibers with LPC and vitamin B12 (Ge+LPC+B12).

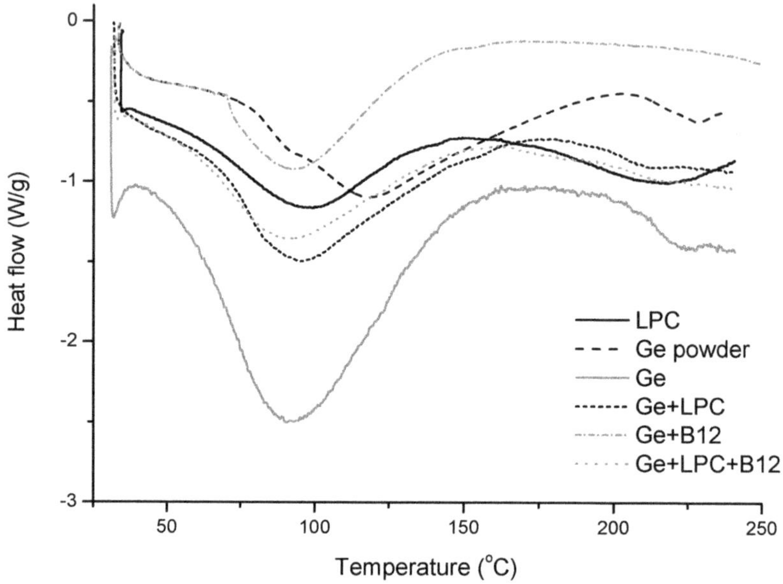

Figure 5. DSC curves of raw leaf protein concentrate (LPC), gelatin powder (Ge powder), gelatin nanofibers (Ge), gelatin nanofibers with LPC (Ge+LPC), gelatin nanofibers with vitamin B12 (Ge+B12) and gelatin nanofibers with LPC and vitamin B12 (Ge+LPC+B12).

3.4. Water Activity (Aw)

Water activity points to the available free water in the examined sample; thus, the parameter is important to define the microbial stability of the sample. For example, at optimum pH and temperature, the minimal Aw values required for the growth of pathogenic bacteria are 0.85–0.86. On the other hand, yeasts and molds tolerate lower Aw values, but usually not below 0.62 [47]. The results for the prepared fibers are shown in Table 3. The values obtained for native gelatin powder were similar to those reported by Rather et al. and also similar to the Aw values of other samples [48]. The Aw values of the prepared fibers ranged from 0.336 to 0.376, and there were no statistically significant differences among the samples, suggesting that the addition of protein or vitamin B12 did not affect the Aw values of gelatin fibers. The obtained Aw values of the developed fibers were lower than those of zein-based electrospun fibers incorporating gallic acid [49]. Accordingly, the newly developed fibers have good storage stability.

Table 3. Water activity values of the leaf protein concentrate (LPC), gelatin powder (Ge powder), gelatin nanofibers (Ge), gelatin with the leaf protein nanofibers (Ge+LPC), gelatin with vitamin B12 nanofibers (Ge+B12), and gelatin with the leaf protein concentrate and vitamin B12 nanofibers (Ge+LPC+B12).

Sample	**Aw**
LPC	0.393 ± 0.031^{a}
Ge powder	0.392 ± 0.042^{a}
Ge	0.336 ± 0.041^{a}
Ge+LPC	0.341 ± 0.001^{a}
Ge+B12	0.361 ± 0.001^{a}
Ge+LPC+B12	0.376 ± 0.033^{a}

Different letters within the same column indicate statistically significant differences ($p < 0.05$).

3.5. Release Study

The effect of the protein matrices on vitamin B12 release behavior was assessed. The release profile of Ge+B12 vs. Ge+LPC+B12 is depicted in Figure 6A. It is evident that the release of vitamin B12 from fibers fortified with LPC is significantly slower. For the purpose of comparison, after 15 min, vitamin B12 was almost completely released from solely gelatin fibers (>98%), while for the same period, only 60% of the vitamin was released from fibers with LPC. Due to high specific surface area, amorphous structure, and high porosity, neat gelatin nanofibers rapidly dissolve in water (unlike bulk gelatin) by forming a white gel [8,19,50]. However, the potential drug-polymer interactions may slow down matrix erosion and drug release by different mechanisms (shielding of matrix terminal residues, decreasing the porosity of the matrix, decreasing the drug partition to the water-filled micropores). Also, the addition of LPC increased the number of functional groups available for interactions.

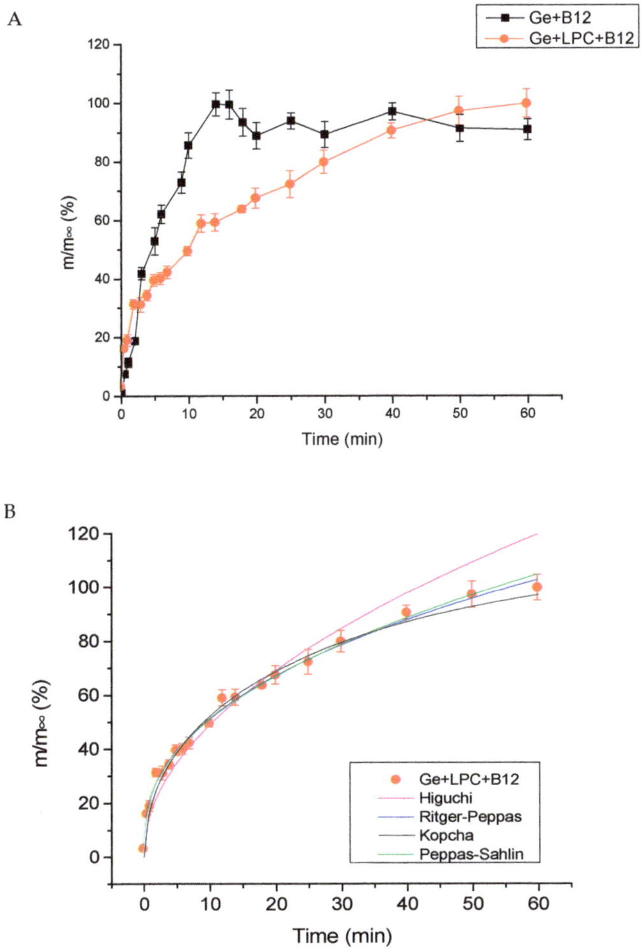

Figure 6. (**A**) Profiles of vitamin B12 release from gelatin and gelatin fortified with LPC nanofibers (each data represents the average value of three replicates). (**B**) Kinetics fitting of vitamin B12 release from the gelatin fortified with LPC nanofibers, fitting data are presented by lines.

The kinetic fitting of vitamin B12 released from the fibers obtained in this study was also performed. Four different models (Kopcha, Peppas–Sahlin, Ritger–Peppas, and Higuchi) were applied, and the results are presented in Figure 6B and Table 4. The Ritger–Peppas equation was found to be the most suitable one to describe the release mechanism of vitamin B12 from Ge+LPC+B12 nanofibers. The main criteria to calculate the goodness of fit are the values of the coefficient of determination (R2) and The Root Mean Square Error (RMSE). Generally, as the value of R2 is closer to 1, the model perfectly predicts values. On the other hand, as the RMSE value is lower than 10, the model fits the data superbly. For our sample, fitting parameter n is lower than 0.45 (Ritger–Peppas), indicating Fickian diffusion as a mechanism of vitamin B12 release. To confirm this, constant k_1 is higher than k_2, indicating the main mechanism governing the release is the same: Fickian diffusion (by Peppas–Sahlin).

Table 4. Model parameters of vitamin B12 release from gelatin-LPC-fortified fibers (Ge+LPC+B12) into aqueous medium.

Model	Parameter	Ge+LPC+B12
Higuchi	k	15.44
	R^2	0.935
	RMSE	5.476
Ritger-Peppas	k	20.93
	n	0.389
	R^2	0.947
	RMSE	4.943
Kopcha	A	19.32
	B	−0.877
	R^2	0.943
	RMSE	5.101
Pepas–Sahlin	k_1	20.156
	k_2	1.055
	m	0.354
	R^2	0.944
	RMSE	5.09

4. Conclusions

Food-grade nanofibers based on pumpkin leaf protein and gelatin for the delivery of vitamin B12 were fabricated using the electrospinning process, and it was an innovative approach in the field of food and nutrition. Extracting protein concentrate from pumpkin leaves as a by-product and incorporating it into nanofibers provided a homogeneous nanofiber structure and a sustainable and plant-derived carrier. This combination of gelatin and leaf protein concentrates in nanofibers delivered a biocompatible matrix with a promising potential for vitamin B12 delivery in the food industry. The obtained nanofibers ensured a slower release of vitamin B12 over time while sustaining thermal stability. This approach has several potential benefits, such as the prolonged shelf life of fortified foods and the development of functional food products with improved nutritional profiles. Additionally, using plant-based proteins from some crops aligns with the growing demand for sustainable and eco-friendly food technologies.

Author Contributions: Conceptualization, B.B. (Bojana Balanč), A.S.-J. and V.Đ.; Funding acquisition, Z.K.-J.; Methodology, B.B. (Bojana Balanč) and V.Đ.; Project administration, V.Đ. and Z.K.-J.; Software, A.S.-J.; Supervision, B.B. (Branko Bugarski), V.N. and Z.K.-J.; Validation, B.B. (Branko Bugarski), V.N.

and P.P.; Visualization, B.B. (Bojana Balanč); Writing—original draft, B.B. (Bojana Balanč), A.S.-J., V.Đ. and P.P.; Writing—review and editing, V.Đ. and Z.K.-J. All authors have read and agreed to the published version of the manuscript.

Funding: This work was supported by Science Fund of the Republic of Serbia, grant no. 7751519, Multifunctional leaf protein and assembled nanocarrier structures delivered by enzyme technology—MultiPromis and the Ministry of Science, Technological Development and Innovation of the Republic of Serbia (contract no. 451-03-65/2024-03/200135, contract no. 451-03-66/2024-03/200287 and contract no. 451-03-65/2024-03/200116).

Institutional Review Board Statement: Not applicable.

Informed Consent Statement: Not applicable.

Data Availability Statement: The original contributions presented in the study are included in the article, further inquiries can be directed to the corresponding author.

Conflicts of Interest: The authors declare that they have no known competing financial interests or personal relationships that could have appeared to influence the work reported in this paper.

References

1. Bhushani, J.A.; Anandharamakrishnan, C. Electrospinning and electrospraying techniques: Potential food based applications. *Trends Food Sci. Technol.* **2014**, *38*, 21–33. [CrossRef]
2. Alehosseini, A.; Gomez-Mascaraque, L.G.; Martínez-Sanz, M.; Lopez-Rubio, A. Electrospun curcumin-loaded protein nano fi ber mats as active/bioactive coatings for food packaging applications. *Food Hydrocoll.* **2019**, *87*, 758–771. [CrossRef]
3. Safwat, S.; Hathout, R.M.; Ishak, R.A.; Mortada, N.D. Elaborated survey in the scope of nanocarriers engineering for boosting chemotherapy cytotoxicity: A meta-analysis study. *Int. J. Pharm.* **2021**, *610*, 121268. [CrossRef] [PubMed]
4. SalehHudin, H.S.; Mohamad, E.N.; Mahadi, W.N.L.; Muhammad Afifi, A. Multiple-jet electrospinning methods for nanofiber processing: A review. *Mater. Manuf. Process* **2018**, *33*, 479–498. [CrossRef]
5. Xue, J.; Wu, T.; Dai, Y.; Xia, Y. Electrospinning and Electrospun Nanofibers: Methods, Materials, and Applications. *Chem. Rev.* **2019**, *119*, 5298–5415. [CrossRef] [PubMed]
6. Schiffman, J.D.; Schauer, C.L. A review: Electrospinning of biopolymer nanofibers and their applications. *Polymer Reviews* **2008**, *48*, 317–352. [CrossRef]
7. Tang, Y.; Zhou, Y.; Lan, X.; Huang, D.; Luo, T.; Ji, J.; Mafang, Z.; Miao, X.; Wang, H.; Wang, W. Electrospun gelatin nanofibers encapsulated with peppermint and chamomile essential oils as potential edible packaging. *J. Agric. Food. Chem.* **2019**, *67*, 2227–2234. [CrossRef] [PubMed]
8. Mosayebi, V.; Fathi, M.; Shahedi, M.; Soltanizadeh, N.; Emam-Djomeh, Z. Fast-dissolving antioxidant nanofibers based on Spirulina protein concentrate and gelatin developed using needleless electrospinning. *Food Biosci.* **2022**, *47*, 101759. [CrossRef]
9. Nieuwland, M.; Geerdink, P.; Brier, P.; van den Eijnden, P.; Henket, J.T.M.M.; Langelaan, M.L.P.; Stroeks, N.; van Deventer, H.C.; Martin, A.H. Food-grade electrospinning of proteins. *Innov. Food Sci. Emerg. Technol.* **2013**, *20*, 269–275. [CrossRef]
10. Tamayo Tenorio, A.; Gieteling, J.; de Jong, G.A.H.; Boom, R.M.; van der Goot, A.J. Recovery of protein from green leaves: Overview of crucial steps for utilisation. *Food Chem.* **2016**, *203*, 402–408. [CrossRef]
11. Santamaría-Fernández, M.; Lübeck, M. Production of leaf protein concentrates in green biorefineries as alternative feed for monogastric animals. *Anim. Feed Sci. Technol.* **2020**, *268*, 114605. [CrossRef]
12. Ducrocq, M.; Boire, A.; Anton, M.; Micard, V.; Morel, M.H. Rubisco: A promising plant protein to enrich wheat-based food without impairing dough viscoelasticity and protein polymerisation. *Food Hydrocoll.* **2020**, *109*, 106101. [CrossRef]
13. Martin, A.H.; Nieuwland, M.; De Jong, G.A.H. Characterization of heat-set gels from RuBisCO in comparison to those from other proteins. *J. Agric. Food. Chem.* **2014**, *62*, 10783–10791. [CrossRef] [PubMed]
14. Nieuwland, M.; Geerdink, P.; Engelen-Smit, N.P.; Van Der Meer, I.M.; America, A.H.P.; Mes, J.J.; Kootstra, A.M.J.; Henket, J.T.M.M.; Mulder, W.J. Isolation and gelling properties of duckweed protein concentrate. *ACS Food Sci. Technol.* **2021**, *1*, 908–916. [CrossRef]
15. Udousoro, I.; Ekanem, P. Assessment of proximate compositions of twelve edible vegetables in *Nigeria. Int. J. Modern Chem.* **2013**, *4*, 79–89.
16. Famuwagun, A.A.; Alashi, A.M.; Gbadamosi, S.O.; Taiwo, K.A.; Oyedele, J.D.; Adebooye, O.C.; Aluko, R.E. In Vitro Characterization of Fluted Pumpkin Leaf Protein Hydrolysates and Ultrafiltration of Peptide Fractions: Antioxidant and Enzyme-Inhibitory Properties. *Pol. J. Food Nutr. Sci.* **2020**, *70*, 429–443. [CrossRef]
17. Tanimowo Fadupin, G.; Ariyo, O. Effect of blanching on nutrient and anti-nutrient content of pumpkin (cucurbita pepo) leaves. *West Afr. J. Food Nutr.* **2014**, *12*.
18. Zhang, X.; Tan, L.; Taxipalati, M.; Deng, L. Fabrication and characterization of fast dissolving glycerol monolaurate microemulsion encapsulated gelatin nanofibers with antimicrobial activity. *J. Sci. Food Agric.* **2021**, *101*, 5660–5670. [CrossRef] [PubMed]

19. Rezaeinia, H.; Ghorani, B.; Emadzadeh, B.; Mohebbi, M. Prolonged-release of menthol through a superhydrophilic multilayered structure of balangu (Lallemantia royleana)-gelatin nanofibers. *Mater. Sci. Eng. C* **2020**, *115*, 111115. [CrossRef]
20. Tavassoli-Kafrani, E.; Hossein Goli, S.A.; Milad Fathi, F. Encapsulation of Orange Essential Oil Using Cross-linked Electrospun Gelatin Nanofibers. *Food Bioprocess Technol.* **2018**, *11*, 427–434. [CrossRef]
21. Maiorova, L.A.; Erokhina, S.I.; Pisani, M.; Barucca, G.; Marcaccio, M.; Koifman, O.I.; Salnikov, D.S.; Gromova, O.A.; Astolfi, P.; Ricci, V.; et al. Encapsulation of vitamin B12 into nanoengineered capsules and soft matter nanosystems for targeted delivery. *Colloids Surf. B.* **2019**, *182*, 110366. [CrossRef] [PubMed]
22. Estevinho, B.N.; Mota, R.; Leite, J.P.; Tamagnini, P.; Gales, L.; Rocha, F. Application of a cyanobacterial extracellular polymeric substance in the microencapsulation of vitamin B12. *Powder Technol.* **2019**, *343*, 644–651. [CrossRef]
23. Carlan, I.C.; Estevinho, B.N.; Rocha, F. Study of microencapsulation and controlled release of modified chitosan microparticles containing vitamin B12. *Powder Technol.* **2017**, *318*, 162–169. [CrossRef]
24. Baik, H.W.; Russell, R.M. Vitamin B_{12} deficiency in the elderly. *Annu. Rev. Nutr.* **1999**, *19*, 357–377. [CrossRef] [PubMed]
25. Allen, L.H. How common is vitamin B-12 deficiency? *Am. J. Clin. Nutr.* **2009**, *89*, 693S–696S. [CrossRef] [PubMed]
26. Coelho, S.C.; Laget, S.; Benaut, P.; Rocha, F.; Estevinho, B.N. A new approach to the production of zein microstructures with vitamin B12, by electrospinning and spray drying techniques. *Powder Technol.* **2021**, *392*, 47–57. [CrossRef]
27. Akbari, N.; Assadpour, E.; Kharazmi, M.S.; Jafari, S.M. Encapsulation of Vitamin B12 by Complex Coacervation of Whey Protein Concentrate–Pectin; Optimization and Characterization. *Molecules* **2022**, *27*, 6130. [CrossRef] [PubMed]
28. Marchianò, V.; Matos, M.; Serrano, E.; Álvarez, J.R.; Marcet, I.; Blanco-López, M.C.; Gutiérrez, G. Lyophilised nanovesicles loaded with vitamin B12. *J. Mol. Liq.* **2022**, *365*, 120129. [CrossRef]
29. Tan, Y.; Lee, P.W.; Martens, T.D.; McClements, D.J. Comparison of emulsifying properties of plant and animal proteins in oil-in-water emulsions: Whey, soy, and RuBisCo proteins. *Food Biophys.* **2022**, *17*, 409–421. [CrossRef]
30. Abbasnezhad, N.; Zirak, N.; Shirinbayan, M.; Kouidri, S.; Salahinejad, E.; Tcharkhtchi, A.; Bakir, F. Controlled release from polyurethane films: Drug release mechanisms. *J. Appl. Polym. Sci.* **2021**, *138*, 50083. [CrossRef]
31. Ratanavaraporn, J.; Rangkupan, R.; Jeeratawatchai, H.; Kanokpanont, S.; Damrongsakkul, S. Influences of physical and chemical crosslinking techniques on electrospun type A and B gelatin fiber mats. Int. *J. Biol. Macromol.* **2010**, *47*, 431–438. [CrossRef] [PubMed]
32. Okutan, N.; Terzi, P.; Altay, F. Affecting parameters on electrospinning process and characterization of electrospun gelatin nanofibers. *Food Hydrocoll.* **2014**, *39*, 19–26. [CrossRef]
33. Koombhongse, S.; Liu, W.; Reneker, D.H. Flat polymer ribbons and other shapes by electrospinning. *J. Polym. Sci.; Part B: Polym. Phys.* **2001**, *39*, 2598–2606. [CrossRef]
34. Salević, A.; Prieto, C.; Cabedo, L.; Nedović, V.; Lagaron, J.M. Physicochemical, antioxidant and antimicrobial properties of electrospun poly (ε-caprolactone) films containing a solid dispersion of sage (Salvia officinalis L.) extract. *Nanomaterials* **2019**, *9*, 270. [CrossRef] [PubMed]
35. Deng, L.; Zhang, X.; Li, Y.; Que, F.; Kang, X.; Liu, Y.; Feng, F.; Zhang, H. Characterization of gelatin/zein nanofibers by hybrid electrospinning. *Food Hydrocoll.* **2018**, *75*, 72–80. [CrossRef]
36. Salević-Jelić, A.; Lević, S.; Stojanović, D.; Jeremić, S.; Miletić, D.; Pantić, M.; Pavlović, V.; Ignjatović, I.S.; Uskoković, P.; Nedović, V. Biodegradable and active zein-gelatin-based electrospun mats and solvent-cast films incorporating sage extract: Formulation and comparative characterization. *Food Packag. Shelf Life* **2023**, *35*, 101027. [CrossRef]
37. Ghosh, R.; Thomas, D.S.; Arcot, J. Molecular Recognition Patterns between Vitamin B12 and Proteins Explored through STD-NMR and In Silico Studies. *Foods* **2023**, *12*, 575. [CrossRef]
38. Deng, L.; Li, Y.; Feng, F.; Zhang, H. Study on wettability, mechanical property and biocompatibility of electrospun gelatin/zein nanofibers cross-linked by glucose. *Food Hydrocolloids* **2019**, *87*, 1–10. [CrossRef]
39. Suryawanshi, D.; Wavhule, P.; Shinde, U.; Kamble, M.; Amin, P. Development, optimization and in-vivo evaluation of cyanocobalamin loaded orodispersible films using hot-melt extrusion technology: A quality by design (QbD) approach. *J. Drug Deliv. Sci. Technol.* **2021**, *63*, 102559. [CrossRef]
40. Guillot, A.J.; Merino-Gutierrez, P.; Bocchino, A.; O'Mahony, C.; Giner, R.M.; Recio, M.C.; Garrigues, T.M.; Melero, A. Exploration of Microneedle-assisted skin delivery of cyanocobalamin formulated in ultraflexible lipid vesicles. *Eur. J. Pharm. Biopharm.* **2022**, *177*, 184–198. [CrossRef]
41. Chronakis, I.S. Gelation of edible blue-green algae protein isolate (Spirulina platensis strain Pacifica): Thermal transitions, rheological properties, and molecular forces involved. *J. Agric. Food. Chem.* **2001**, *49*, 888–898. [CrossRef] [PubMed]
42. Ladjal-Ettoumi, Y.; Boudries, H.; Chibane, M.; Romero, A. Pea, chickpea and lentil protein isolates: Physicochemical characterization and emulsifying properties. *Food Biophys.* **2016**, *11*, 43–51. [CrossRef]
43. Delahaije, R.J.; Wierenga, P.A.; Giuseppin, M.L.; Gruppen, H. Comparison of heat-inducedaggregation of globular proteins. *J. Agric. Food. Chem.* **2015**, *63*, 5257–5265. [CrossRef] [PubMed]
44. Tang, C.H.; Sun, X. Physicochemical and structural properties of 8S and/or 11S globulins from mungbean [Vigna radiata (L.) Wilczek] with various polypeptide constituents. *J. Agric. Food. Chem.* **2010**, *58*, 6395–6402. [CrossRef] [PubMed]
45. Soliman, E.A.; Furuta, M. Influence of phase behavior and miscibility on mechanical, thermal and micro-structure of soluble starch-gelatin thermoplastic biodegradable blend films. *Food Nutr. Sci.* **2014**, *5*. [CrossRef]

46. Wang, P.; Li, Y.; Zhang, C.; Feng, F.; Zhang, H. Sequential electrospinning of multilayer ethylcellulose/gelatin/ethylcellulose nanofibrous film for sustained release of curcumin. *Food Chem.* **2020**, *308*, 125599. [CrossRef]
47. Chirife, J.; del Pilar Buera, M.; Labuza, T.P. Water activity, water glass dynamics, and the control of microbiological growth in foods. *Crit. Rev. Food Sci. Nutr.* **1996**, *36*, 465–513. [CrossRef]
48. Rather, J.A.; Majid, S.D.; Dar, A.H.; Amin, T.; Makroo, H.A.; Mir, S.A.; Barba, F.J.; Dar, B.N. Extraction of Gelatin From Poultry Byproduct: Influence of Drying Method on Structural, Thermal, Functional, and Rheological Characteristics of the Dried Gelatin Powder. *Front. Nutr.* **2022**, *9*, 895197. [CrossRef]
49. Neo, Y.P.; Ray, S.; Jin, J.; Gizdavic-Nikolaidis, M.; Nieuwoudt, M.K.; Liu, D.; Quek, S.Y. Encapsulation of food grade antioxidant in natural biopolymer by electrospinning technique: A physicochemical study based on zein–gallic acid system. *Food Chem.* **2013**, *136*, 1013–1021. [CrossRef]
50. Ghorani, B.; Emadzadeh, B.; Rezaeinia, H.; Russell, S.J. Improvements in gelatin cold water solubility after electrospinning and associated physicochemical, functional and rheological properties. *Food Hydrocoll.* **2020**, *104*, 105740. [CrossRef]

foods

MDPI

Article

Influence of Cuttlefish-Ink Extract on Canned Golden Seabream (*Sparus aurata*) Quality

Beatriz Martínez [1], Marcos Trigo [2], Alicia Rodríguez [3] and Santiago P. Aubourg [2,*]

1 Department of Food Technologies, CIFP Coroso, Avda. da Coruña, 174, 15960 Ribeira, Spain; bmartinezr@edu.xunta.gal
2 Department of Food Technology, Marine Research Institute (CSIC), c/Eduardo Cabello, 6, 36208 Vigo, Spain; mtrigo@iim.csic.es
3 Department of Food Science and Chemical Technology, Faculty of Chemical and Pharmaceutical Sciences, University of Chile, c/Carlos Lorca Tobar 964, Santiago 8380494, Chile; arodrigm@uchile.cl
* Correspondence: saubourg@iim.csic.es

Abstract: Four different concentrations of an aqueous extract of cuttlefish (*Sepia* spp.) ink (CI) were introduced, respectively, into the packing medium employed during golden seabream (*Sparus aurata*) canning. The quality parameters of the resulting canned fish were determined and compared to the initial fish and the control canned muscle. An important effect of the CI concentration introduced in the packing medium was proved. The presence in the packing medium of a relatively low CI concentration (CI-2 batch) led to a lower ($p < 0.05$) lipid oxidation development (fluorescent compound formation), lower ($p < 0.05$) changes of colour parameters (L^* and a^* values), and lower ($p < 0.05$) trimethylamine values in canned fish when compared to control canned samples. Additionally, the two lowest CI concentrations tested led to higher average values of C22:6ω3, ω3/ω6 ratios, and polyene index. On the contrary, the use of the most concentrated CI extract (CI-4 condition) led to a prooxidant effect (higher fluorescence ratio value). In agreement with environmental sustainability and circular economy requirements, the study can be considered the first approach to a novel and valuable use of the current marine byproduct for the quality enhancement of canned fish. On-coming research focused on the optimisation of the CI-extract concentration is envisaged.

Keywords: cuttlefish; ink; golden seabream; canning; lipid hydrolysis; phospholipids; ω3 fatty acids; lipid oxidation; colour; trimethylamine

1. Introduction

As a result of the processing of marine species, seafood industries generate a great quantity of discards [1,2]. Such substrates are known to include a wide variety of healthy and nutritional constituents, but they can also be considered as relevant environmental concerns of coastline areas [3,4]. Among cephalopod discards, ink sacs are generally discarded and thrown without appropriate management. Nevertheless, recent studies carried out on this byproduct have indicated several beneficial properties regarding human health [5–7], cosmetic purposes [8], and environmental sustainability [9]. Regarding seafood preservation, previous research has shown relevant antimicrobial [10,11] and antioxidant [12–15] properties of cephalopod-ink extracts when used in the preparation of seafood substrates.

The preservative behaviour of cephalopod ink has been related to melanin and melanin-free fractions [16]; thus, hydrogen atoms would be donated to any radical included in the reacting medium and would prevent the formation of free-radical formation [17]. The analysis of the melanin-free ink corresponding to splendid squid (*Loligo formosana*) indicated that the highest antioxidant behaviour was present in fractions presenting molecular weights lower than 3 kDa, whose activity was not modified after a heat treatment (30 min at 90 °C) [18]. Based on a chromatographic analysis, Senan [19] proved that the active principle in the methanolic extract obtained from cuttlefish (*Sepia pharaonis*) ink was a peptide molecule, which was shown to be stable during thermal treatment (30 min at 40–100 °C). Different molecular-weight fractions were prepared from water-soluble squid (*Ommastrephes bartrami*) melanin [20]; the authors proved that the highest in vitro antioxidant potency corresponded to fractions corresponding to molecular weights above 10 kDa. Recently, Xie et al. [21] indicated that water-extracted cuttlefish (*Sepia sculenta*) melanin contained a characteristic indole structure with irregular spherical structures and provided a protective effect on kojic and ascorbic acids from light irradiation; this preservative effect was maintained by prolonging the exposure time.

Canning is considered a traditional process that is widely employed and is able to inactivate enzymes and destroy pathogenic and spoilage microorganisms [22,23]. However, the high temperature involved may provoke the destruction of proteins, vitamins, and unsaturated fatty acids (FAs) and lead to a decrease in the nutritional and sensory values [24,25]. Therefore, searching for optimised conditions for the canning process [26] and the use of protective technologies [27] is nowadays considered mandatory. Additional concerns of fish canneries are the availability decrease of traditional species and the need for previous storage of the raw material to be processed [28,29]. In both cases, the use of farmed species can provide a more accurate and convenient raw material in any season of the year.

The current work focused on the employment of cuttlefish (*Sepia* spp.) ink (CI). The basic objective was to study the preservative properties of an aqueous CI extract during the canning process of farmed golden seabream (*Sparus aurata*). This fish species was chosen on the basis of its great production and the interest in introducing farmed species as raw material for canning. In this study, different concentrations of CI were added, respectively, to the packing medium. Quality changes regarding phospholipid content, lipid hydrolysis, FA profile, lipid oxidation, muscle colour and trimethylamine (TMA) value were measured in the initial and canned samples. The effect of the CI concentration employed was investigated.

2. Materials and Methods

2.1. Analysis of the Initial Ink and Preparation of the Ink Extract

Commercial ink from cuttlefish (*Sepia* spp.) was obtained from Sepink (Vilagarcía de Arousa, Pontevedra, Spain). Once in our laboratory, the ink was subjected to the following analytical determinations: proximate analysis, FA profile, and lipid quality determination.

For it, moisture content was determined as the weight difference before and after 4 h at 105 °C [30]; the results were calculated as g water·kg^{-1} ink. Crude protein content was measured using the Kjeldahl method [30], with a conversion factor of 6.25; the results were calculated as g protein·kg^{-1} ink. Ash content was measured according to the AOAC [30] method; results were calculated as g ash·kg^{-1} ink. Carbohydrates were estimated by difference and expressed as g carbohydrates·kg^{-1} ink.

The lipid content, the FA profile (individual FAs and FA groups and ratios) and the quality (presence of peroxides, thiobarbituric acid reactive substances, and free fatty acids) of the initial ink were determined according to the analytical methods described in Sections 2.3 and 2.4.

In order to obtain the ink extract, a mixture of CI (40 g) and distilled water (400 mL) was subjected to stirring (30 s), sonication (30 s), and centrifugation (3500× *g* at 4 °C for 30 min) according to previous research [14]. The resulting supernatant was taken. The extraction process was carried out three more times, with all supernatants being pooled together and made up to 1 L with distilled water.

The ink extract was subjected to the proximate analysis according to the methods previously mentioned. Additionally, the total volatile base-nitrogen (TVB-N) content was determined as described previously [31]; results were expressed as mg TVB-N·kg^{-1} ink extract).

2.2. Raw Fish and Fish Canning

Fresh farmed golden seabream (*S. aurata*) (44 specimens) (ca. 480–520 g each) were obtained in a local market and carried stored on ice to the laboratory. Then, four individuals were selected and considered as the initial fish. In order to obtain the fillets, the fish individuals were beheaded and eviscerated. After discarding the dark muscle from the fillets, the resulting white muscle was minced and analysed independently in each of the fish individuals (n = 4).

On the same day, the remaining fish were divided into four batches (ten individuals per batch) that were considered independently during the current study (n = 4). In all the batches, the fillets were obtained after the fish were subjected to beheading and evisceration. Then, 45 g portions of fillets were introduced into small flat cans (105 mm × 60 mm × 25 mm; 150 mL). Two cans were prepared from each individual fish. In each batch, 0, 10, 25, 40, and 80 mL of the above-mentioned CI extract were added, respectively, to the cans as packing media. Cans were filled by the addition of distilled water (95, 85, 70, 55, and 15 mL, respectively). As a result, CTR (canned control), and samples corresponding to CI-1, CI-2, CI-3, and CI-4 packing conditions were prepared, respectively.

The CI concentrations tested in the current work were based on preliminary tests. Thus, the CI-4 condition corresponds to the highest concentration that does not modify the sensory descriptors of canned golden seabream (namely odour and flavour or flesh colour). With the aim of analysing the effect of the CI-extract concentration, three conditions, including lower CI concentrations (i.e., CI-1, CI-2, and CI-3 packing conditions), were also checked.

The cans were vacuum-sealed (SOMME 222, Ezquerra, San Adrián, Navarra, Spain) and sterilised (115 °C, 45 min; F_o = 7 min) in a steam retort (Presoclave II 75L, JP Selecta, Barcelona, Spain). Once the sterilisation time was finished, the steam was cut off, the remaining steam was flushed away by the use of air, and the cans were cooled by using water at reduced pressure.

After a 3-month storage at room temperature (20 °C), the cans were opened. For it, the liquid part of the can was drained off gravimetrically and filtered by means of a filter paper. Once the dark muscle was discarded, the white muscle of the canned fish was wrapped in filter paper and employed for the different kinds of physico-chemical analyses.

According to the common practice carried out in canneries, a 3-month storage was accomplished. Canning manufacturers recommend that a 2–3-month storage ought to be carried out in order to optimise the sensory acceptability of the commercial canned product [28].

In all cases, chemical reagents and solvents corresponding to the reagent grade were used (Merck, Darmstadt, Germany).

2.3. Determination of Lipid Content and Composition in Fish Muscle

The lipid fraction was obtained by extracting the fish's white muscle according to the Bligh and Dyer [32] method. Quantification was carried out in agreement with Herbes and Allen [33]. Lipid content was calculated as $g \cdot kg^{-1}$ fish muscle.

The free fatty acid (FFA) value was analysed on the lipid extract of the fish muscle in agreement with the Lowry and Tinsley [34] method. Results were calculated as g $FFAs \cdot kg^{-1}$ lipids.

The phospholipid (PL) value was determined by measuring the organic phosphorus in the lipid extract according to the Raheja et al. [35] procedure. Results were calculated as g $PLs \cdot kg^{-1}$ lipids.

Fatty acid methyl esters (FAMEs) were prepared from the lipid extracts by employing acetyl chloride in methanol. The resulting FAMEs were then analysed by gas chromatography (Perkin-Elmer 8700 chromatograph, Madrid, Spain) in agreement with an established procedure [36]. Identification of peaks corresponding to FAMEs was achieved by a comparison of the retention times to those of standard mixtures (Qualmix Fish, Larodan, Malmo, Sweden; Supelco 37 Component FAME Mix, Supelco Inc., Bellefonte, PA, USA). For quantitative purposes, C19:0 was employed as an internal standard. The content of each FA was expressed as $g \cdot 100\ g^{-1}$ total FAs.

Values regarding FA groups (saturated FAs, STFAs; monounsaturated FAs, MUFAs; polyunsaturated FAs, PUFAs; $\omega 3$ and $\omega 6$ PUFAs) were calculated by means of values obtained in individual FAs. Additionally, the $\omega 3/\omega 6$ ratio and the polyene index (PI), considered as the C20:5ω3 + C22:6ω3/C16:0 concentration ratio, were calculated.

2.4. Assessment of Lipid Oxidation Development in Fish Muscle

The peroxide value (PV) was analysed spectrophotometrically (520 nm) (Beckman Coulter, DU 640; London, UK) on the lipid extract according to the previous research [37]. Results were calculated as meq. active oxygen·kg^{-1} lipids.

The thiobarbituric acid index (TBA-i) was analysed in agreement with Vyncke [38]. Thus, the thiobarbituric acid reactive substances (TBARSs) value was spectrophotometrically measured at 532 nm. Results were calculated as mg malondialdehyde·kg^{-1} muscle.

The formation of interaction compounds produced by the reaction of oxidised lipids and protein-type molecules was determined by fluorescence spectroscopy (Fluorimeter LS 45; Perkin Elmer España; Tres Cantos, Madrid, Spain) in agreement with previous research [39]. The fluorescence ratio (FR) was measured in the aqueous phase obtained from the lipid extraction of the fish muscle.

2.5. Determination of Muscle Colour and TMA Value in Fish Muscle

The assessment of colour parameters (L^*, a^* and b^*) was carried out on the fish-muscle surfaces of the raw and canned samples. For it, an instrumental determination (CIE 1976 Lab), performed with a tristimulus Hunter Labscan 2.0/45 colourimeter (Reston, VA, USA), was undertaken. For the analysis of each sample, colour scores were averaged over four determinations, which were taken by rotating the measuring head 90° between triplicate measurements per position.

The TMA-nitrogen (TMA-N) content was analysed using the picrate colourimetric method [40]. Results were calculated as mg TMA-N·kg^{-1} muscle.

2.6. Statistical Analysis

As previously indicated, four replicates (n = 4) were considered in this work. Data obtained were evaluated by analysis of variance (ANOVA) to explore differences resulting from the effect of the sterilisation and canned storage steps, and the effect of the CI-extract concentration added to the packing medium. For carrying out the comparison of average values, the Tukey HSD procedure was applied. A confidence interval at the 95% level ($p < 0.05$) was taken into account to establish significant differences among batches. The PASW Statistics 18 software for Windows (SPSS Inc., Chicago, IL, USA) was employed.

3. Results and Discussion

3.1. Composition and Quality of the Initial Ink

The initial ink exhibited the following proximate composition (g·100 kg^{-1}): 77.28 ± 0.89 (moisture), 6.52 ± 0.19 (protein), 0.13 ± 0.01 (lipids), 8.93 ± 0.34 (ash), and 7.14 ± 0.45 (carbohydrates). A different composition was reported by Xie et al. [21] for *S. esculenta* ink: 13.9% (proteins), 0.3% (carbohydrates), 0.02% (lipids), and 6.1% (ash).

The FA profile of the initial ink indicated the following composition of individual FAs (g·100 g^{-1} total FAs): 2.45 ± 0.13 (C14:0), 0.69 ± 0.06 (C15:0), 26.68 ± 1.77 (C16:0), 1.38 ± 0.12 (C16:1ω7), 2.63 ± 0.09 (C17:0), 18.52 ± 1.29 (C18:0), 8.34 ± 0.33 (C18:1ω9), linebreak 1.91 ± 0.11 (C18:1ω7), 4.01 ± 0.16 (C18:2ω6), 2.99 ± 0.19 (C20:1ω9), 0.38 ± 0.01 (C20:2ω6), 6.36 ± 0.34 (C20:4ω6), 0.42 ± 0.01 (C22:1ω9), 6.17 ± 0.53 (C20:5ω3), 1.55 ± 0.12 (C22:4ω6), 0.56 ± 0.03 (C24:1ω9), 1.54 ± 0.13 (C22:5ω3), and 13.27 ± 1.35 (C22:6ω3). Therefore, the most abundant FA was C16:0, followed by C18:0 and C22:6ω3; other relatively abundant FAs were C18:1ω9, C20:4ω6, and C20:5ω3.

The content of the FA groups was also determined (g·100 g^{-1} total FAs). According to the individual FA profile, STFAs were shown to be the most abundant group (51.03 ± 3.31), while MUFAs depicted the lowest levels (15.62 ± 0.75). The values obtained for PUFAs and ω3 PUFAs were 33.35 ± 2.63 and 21.05 ± 2.01, respectively. Additionally, the following FA ratio values were obtained: 1.71 ± 0.08 (ω3/ω6 ratio) and 0.74 ± 0.12 (PI).

Regarding the lipid quality of the initial ink, the following lipid oxidation values were observed: 0.81 ± 0.11 (PV) and 0.07 ± 0.01 (TBA-i). According to general research on seafood quality, these values correspond to a low-rancidity substrate susceptible to being employed in further activities [39,41]. A relatively high FFA value was detected (i.e., 327.13 ± 6.21 g·kg^{-1} lipids); this high value can be explained on the basis of the low total lipid content and the inverse ratio reported between total lipid and FFA values [42].

The ink extract showed moisture and ash contents of 996.04 ± 0.12 and 0.01 ± 0.00 g·kg^{-1}, respectively; no presence of lipids and proteins was detected. The determination of the TVB-N content indicated a 13.1 ± 0.4 mg·kg^{-1} score.

3.2. Determination of Total Lipid, FFA, and PL Values in Fish Muscle

Values obtained for total lipid content in fish muscle are depicted in Table 1. Canning and canned storage led to a general lipid content increase that can be explained on the basis of moisture loss from the muscle during the heating treatment [29,36]. No differences ($p > 0.05$) among the canned samples were obtained, all values being included in the 20.79–24.03 g·kg^{-1} muscle.

A comparison between the initial fish and canned fish corresponding to the control condition indicated a marked increase ($p < 0.05$) in the FFA value as a result of the sterilisation and canned storage steps (Table 1). All CI-treated fish showed higher mean values of FFAs than the control canned fish. However, the differences were not found significant ($p > 0.05$).

In the present work, the FFA value can be influenced by different factors. First, the sterilisation process would provoke the hydrolysis of higher-molecular-weight lipid classes such as triacylglycerols (TAGs) and PLs [24,43]. Then, FFAs can be oxidised or broken-down during the sterilisation process, providing a greater accessibility to oxygen and other pro-oxidant molecules when compared to the TAG and PL classes [44,45]; consequently, this effect would lead to a decrease in the FFA content. The current results have shown that the first effect (i.e., thermal hydrolysis of TAGs and PLs) was more important than the second one (i.e., thermal breakdown of FFAs). All CI-treated fish showed higher average FFA levels than the canned control condition; however, the differences were not found significant ($p > 0.05$).

According to the present study, a preservative effect on FFAs was inferred by Barbosa et al. [36] during Atlantic Chub mackerel (*Scomber colias*) canning by the presence of an aqueous extract of *F. spiralis* in the packing medium. The same authors [31] proved a higher FFA retention in canned Atlantic mackerel (*Scomber scombrus*) by the presence of *Bifurcaria bifurcata* extracts in the packing medium. On the contrary, a lower FFA content was detected in canned Chub mackerel (*S. colias*) by including an aqueous extract of octopus (*Octopus vulgaris*) cooking liquor [46]. In the meantime, no effect was proven in canned Chub mackerel (*S. colias*) by the addition of *Ulva lactuca* extracts [36].

Regarding the PL fraction in the current study, the content revealed a marked decrease ($p < 0.05$) as a result of the sterilisation treatment and canned storage (comparison between the initial fish and control canned fish) (Table 1). However, all CI-treated fish showed higher average values than the control canned condition. Remarkably, canned muscle corresponding to the CI-1 condition did not show a significant loss ($p > 0.05$) of the PL content when compared to the initial fish.

Great attention has been accorded to PLs according to their high bioavailability and preserving effect as delivery systems for different kinds of diseases [47,48]. Based on the needs and requirements expressed by the food and pharmaceutical industries, great efforts are being made toward the retention of the PL fraction in seafood as it presents high PUFA levels [49,50].

Previous studies regarding the effect of antioxidant compounds included in the packing system on the PL content in canned fish are scarce. The addition of an aqueous extract of *Fucus spiralis* to the packing medium of brine-canned Chub mackerel (*S. colias*) led to a preserving effect on the PL composition [36]. Similarly, Méndez et al. [51] proved that there was a retention of the PL content in brine-canned horse mackerel (*Trachurus trachurus*) by an addition to the packing medium of an aqueous extract of octopus (*O. vulgaris*) cooking liquor.

Table 1. Total lipid (TL), free fatty acid (FFA), and phospholipid (PL) values * in initial and canned fish, including different concentrations of cuttlefish ink (CI) in the packing medium **.

Initial or Canned Sample	Lipid Determination		
	TL ($g\cdot kg^{-1}$ Muscle)	FFA ($g\cdot kg^{-1}$ Lipids)	PL ($g\cdot kg^{-1}$ Lipids)
Initial fish	13.62 a (1.64)	0.49 a (0.05)	39.69 b (3.84)
CTR	21.15 b (2.71)	1.58 b (0.31)	29.30 a (3.74)
CI-1	20.79 b (1.76)	1.90 b (0.26)	35.01 ab (3.39)
CI-2	24.03 b (3.09)	1.87 b (0.19)	27.01 a (4.51)
CI-3	22.19 b (3.60)	1.96 b (0.60)	28.75 a (3.46)
CI-4	21.47 b (2.82)	1.97 b (0.27)	29.42 a (3.14)

* Mean values of four replicates ($n = 4$); standard deviations are indicated in brackets. In each column, different letters (a, b) denote significant differences ($p < 0.05$). ** Abbreviations: CTR (canned control), CI-1, CI-2, CI-3, and CI-4 (canned samples including increasing concentrations of the CI extract in the packing medium, according to Section 2).

3.3. *Analysis of the FA Profile of Fish Muscle*

The individual FA composition of the initial and canned fish is shown in Table 2. In all cases, C18:2ω6 (linoleic acid) (26.88–28.27% range) and C18:1ω9 (oleic acid) (23.67–26.77% range) were the most abundant FAs, followed by C16:0 (palmitic acid) (15.02–16.53% range) and C22:6ω3 (docosahexaenoic acid, DHA) (8.62–11.94% range).

Table 2. Fatty acid (FA) composition * of initial and canned fish, including different concentrations of cuttlefish ink (CI) in the packing medium **.

FA	Initial Fish	CTR	CI-1	CI-2	CI-3	CI-4
14:0	1.33 a (0.07)	1.84 c (0.02)	1.67 b (0.04)	1.85 c (0.11)	1.74 bc (0.11)	1.73 c (0.20)
15:0	0.25 a (0.02)	0.25 a (0.01)	0.28 a (0.02)	0.28 a (0.03)	0.26 a (0.02)	0.26 a (0.02)
16:0	16.53 c (0.45)	15.80 ab (0.31)	15.76 b (0.13)	15.58 ab (0.46)	15.02 ab (0.52)	15.02 a (0.46)
16:1ω7	2.30 a (0.06)	2.67 b (0.18)	2.63 b (0.19)	3.00 bc (0.35)	3.20 c (0.31)	2.96 bc (0.23)
17:0	0.35 a (0.02)	0.37 a (0.02)	0.37 a (0.02)	0.38 a (0.03)	0.36 a (0.02)	0.37 a (0.02)
18:0	4.72 a (0.25)	4.63 a (0.24)	4.71 a (0.15)	4.91 a (0.19)	4.76 a (0.17)	4.89 a (0.22)
18:1ω9	23.67 a (1.18)	25.37 ab (1.04)	25.58 ab (0.94)	25.69 ab (1.14)	26.77 b (1.03)	26.00 b (1.01)
18:1ω7	3.10 a (0.26)	3.44 a (0.35)	3.20 a (0.12)	3.21 a (0.14)	3.35 a (0.13)	3.25 a (0.13)
18:2ω6	26.88 a (0.31)	28.17 c (0.44)	27.57 bc (0.39)	26.67 ab (0.94)	27.73 bc (0.71)	28.27 bc (0.79)
20:1ω9	0.95 a (0.07)	1.34 a (0.12)	1.20 b (0.04)	1.38 b (0.14)	1.28 b (0.11)	1.24 b (0.10)
20:2ω6	1.00 a (0.07)	1.04 c (0.08)	0.97 a (0.07)	1.11 a (0.08)	0.94 a (0.10)	0.96 a (0.12)

Table 2. *Cont.*

FA	Initial Fish	CTR	CI-1	CI-2	CI-3	CI-4
20:4ω6	1.20 a (0.09)	1.10 b (0.10)	1.13 a (0.08)	1.14 a (0.07)	1.01 a (0.09)	1.08 a (0.10)
22:1ω9	0.28 a (0.03)	0.31 a (0.04)	0.28 a (0.01)	0.34 a (0.05)	0.29 a (0.02)	0.27 a (0.03)
20:5ω3	2.26 b (0.12)	1.99 a (0.10)	2.06 ab (0.20)	2.10 ab (0.16)	2.00 ab (0.12)	2.06 ab (0.24)
22:4ω6	0.27 a (0.02)	0.25 a (0.01)	0.24 a (0.01)	0.28 a (0.01)	0.26 a (0.02)	0.26 a (0.04)
24:1ω9	0.50 a (0.04)	0.48 a (0.03)	0.49 a (0.02)	0.51 a (0.04)	0.47 a (0.05)	0.48 a (0.05)
22:5ω3	2.39 b (0.08)	2.15 a (0.10)	2.12 a (0.12)	2.36 ab (0.12)	2.09 a (0.07)	2.16 ab (0.20)
22:6ω3	11.94 b (0.74)	8.68 a (1.14)	9.66 a (0.95)	9.11 a (0.34)	8.35 a (1.13)	8.62 a (0.95)

* Mean values of four replicates ($n = 4$); standard deviations are indicated in brackets. In each row, different letters (a, b, c) denote significant differences ($p < 0.05$). ** Abbreviations of sample names as expressed in Table 1.

According to the nutritional and healthy properties of seafood in general, quality changes related to the FA values in the present study will be addressed to the ω3 PUFA fraction (eicosapentaenoic acid, EPA; DHA; docosapentaenoic acid, DPA; total ω3 PUFAs) and the FA ratios (ω3/ω6 and PI).

A comparison of the samples corresponding to the initial fish and the canned control condition revealed a loss ($p < 0.05$) of EPA and DHA values (Table 2) as a result of the sterilisation and the canned storage steps. Regarding the CI treatment, all CI-treated fish showed higher average EPA values than the canned control; a significant loss could not be inferred ($p > 0.05$) in treated samples when compared to the initial fish. For the DHA value, the samples corresponding to the CI-1 and CI-2 conditions showed higher average values than the canned control fish. On the contrary, samples corresponding to CI-3 and CI-4 conditions revealed lower average values. Regarding the DPA values, a remarkable loss ($p < 0.05$) was detected after the sterilisation and canned storage steps (Table 2). The highest DPA average value was detected in the canned fish corresponding to the CI-2 condition and did not provide differences with the initial fish.

The total ω3 PUFA content showed a similar profile to the DHA value (Table 3). Thus, all canned samples (control and CI treated) underwent a content decrease ($p < 0.05$) in comparison to the initial fish, and no significant effect ($p > 0.05$) was concluded as a result of the CI addition to the packing medium. Notably, the samples corresponding to CI-1 and CI-2 conditions showed higher average values of total ω3 FAs than the counterpart canned samples.

The analysis of the ω3/ω6 ratio indicated a marked decrease ($p < 0.05$) as a result of the sterilisation and canned storage steps (Table 3). Therefore, it could be inferred that ω3 FAs have been more susceptible to heat treatment than those belonging to the ω6 series. A comparison among the canned samples (CI treated and control) did not show a significant effect ($p > 0.05$) of the presence of the CI extract in the packing medium. However, and in agreement with current results on the DHA content (Table 3), the CI-1 and CI-2 conditions led to the highest average values of the ω3/ω6 ratio.

Table 3 also includes values of the STFA, MUFA, and PUFA groups in the initial and canned fish samples. In all cases, the PUFA group was shown to be the most abundant ($p < 0.05$). On the contrary, the STFA group provided the lowest ($p < 0.05$) levels in all the kinds of fish samples. As a result of the sterilisation step and canned storage, a decrease ($p < 0.05$) of the PUFA content was detected, but no change ($p > 0.05$) could be inferred for the STFA and MUFA groups. Regarding the CI effect, no differences ($p > 0.05$) could be observed between the canned control and the CI-treated samples for any of the FA groups.

Table 3. Fatty acid (FA) group and ratio values * of initial and canned fish, including different concentrations of cuttlefish ink (CI) in the packing medium **.

	Initial Fish	CTR	CI-1	CI-2	CI-3	CI-4
ω3 PUFAs	16.62 b (0.94)	12.88 a (1.26)	13.89 a (1.23)	13.58 a (0.19)	12.50 a (1.15)	12.90 a (1.30)
ω3/ω6	0.57 b (0.03)	0.42 a (0.05)	0.46 ab (0.04)	0.47 a (0.01)	0.42 a (0.05)	0.42 a (0.05)
STFAs	23.20 a (0.64)	22.91 a (0.56)	22.78 a (0.16)	23.02 a (0.61)	22.16 a (0.61)	22.29 a (0.52)
MUFAs	30.82 a (1.56)	33.63 ab (1.48)	33.41 ab (1.16)	34.18 ab (1.69)	35.40 b (1.43)	34.23 b (1.10)
PUFAs	45.98 b (0.93)	43.34 a (1.01)	43.81 ab (1.20)	42.79 a (1.09)	42.45 a (0.85)	43.48 a (0.71)
PI	0.86 b (0.03)	0.68 a (0.06)	0.75 ab (0.07)	0.72 a (0.01)	0.69 a (0.05)	0.71 a (0.06)

* Mean values of four replicates ($n = 4$); standard deviations are indicated in brackets. In each row, different letters (a, b) denote significant differences ($p < 0.05$). ** Abbreviations of sample names as expressed in Table 1. Other abbreviations: PUFAs (polyunsaturated FAs), STFAs (saturated FAs), MUFAs (monounsaturated FAs), and PI (polyene index).

The PI is considered a valuable and practical tool for measuring the possible loss of the PUFA value during marine-species canning or processing in general [36,41,51]. In the present study, this quality parameter showed an average decrease as a result of the thermal treatment and the canned storage (Table 3). Notably, canned fish corresponding to the CI-1 condition did not reflect a significant loss ($p > 0.05$) when compared to the initial fish. Compared to the control canned condition, all CI-treated samples provided higher than average values for this index; however, the differences were not found to be significant ($p > 0.05$). No previous research accounts for the effect of cuttlefish ink on PUFA compounds in canned fish. However, a preservative effect of the cuttlefish (*Sepia officinalis*) ink on PUFA compounds (i.e., higher PI) was already proven by Trigo et al. [14] during a heating treatment of the fish in a model system. The addition to the packing medium of preservative compounds obtained from marine sources has shown a preservative effect on the PUFA content. Thus, Ortiz et al. [52] showed a PI retention of canned Atlantic salmon (*Salmo salar*) muscle when packed in a medium including an ulte (basal part of alga *Durvillaea antarctica*) extract; however, no differences were obtained in such a study when other algae (cochayuyo, frond of *D. antarctica*; *U. lactuca*; and *Pyropia columbina*) extracts were included in the packing system. Higher PI scores were observed in canned Atlantic mackerel (*S. scombrus*) by the addition of *B. bifurcata* extracts [31] and in canned Chub mackerel (*S. colias*) by the addition of *F. spiralis* or *U. lactuca* extracts [36]. Recently, the addition of an aqueous extract of octopus (*O. vulgaris*) cooking liquor to the packing medium led to a higher PI in canned Chub mackerel (*S. colias*) [46].

3.4. Determination of Lipid Oxidation in Fish Muscle

The presence of peroxides (primary oxidation compounds) was shown to be low (i.e., < 2.07) in all the samples included in the current study (Table 4). In most cases, differences among the samples were not significant ($p > 0.05$). However, a higher ($p < 0.05$) peroxide content in canned samples corresponding to the CI-1 condition than in the canned control was observed. No effect ($p > 0.05$) of the sterilisation and canned storage steps could be inferred.

As in the case of the peroxide determination, the levels detected for the TBARSs were low, with all values being below the 0.14 score (Table 4). No effect ($p > 0.05$) of the presence of the CI extract in the packing medium could be outlined on the TBARS formation. No effect ($p > 0.05$) was also inferred as a result of the sterilisation and canned storage steps.

Interaction compound formation is depicted in Figure 1. A comparison between the initial fish and fish corresponding to the canned control condition showed a remarkable ($p < 0.05$)

increase as a result of the canning process and canned storage. Among canned samples, the lowest average values were obtained in fish corresponding to the CI-1 and CI-2 packing conditions. An inhibitory effect ($p < 0.05$) on the fluorescent compound formation was proven by employing the CI-2 packing condition when compared to the canned control samples. On the contrary, the CI-4 condition led to an increased formation ($p < 0.05$) of such kinds of compounds. This dependency on the CI concentration employed is in agreement with the results previously shown regarding the DHA (Table 2) and $\omega3/\omega6$ ratio (Table 3) values.

During seafood thermal processing in general, the levels detected for the different lipid oxidation compounds can be considered the result of heating formation and also of heating breakdown [24,25]. According to the low presence of primary and secondary lipid oxidation compounds in the present canned samples, no effect of the CI extract could be inferred on the content of such deteriorative compounds. However, an inhibitory effect on the formation of fluorescent compounds was observed if the CI-2 condition is considered. On the contrary, a prooxidant effect was obtained if the CI-4 condition is taken into account. Therefore, a selective effect of the CI concentration added to the packing medium could be concluded.

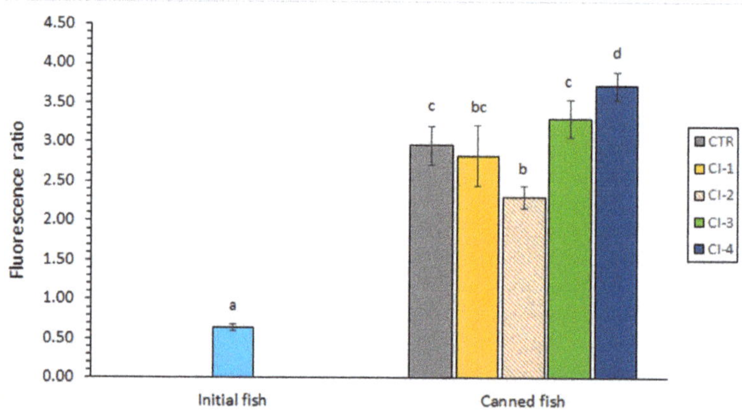

Figure 1. Fluorescence ratio in initial and canned fish, including different concentrations of cuttlefish ink (CI) in the packing medium. Mean values of four replicates ($n = 4$); standard deviations are indicated by bars. Different letters (a, b, c, d) denote significant differences ($p < 0.05$). Abbreviations of sample names as expressed in Table 1.

Table 4. Primary and secondary lipid oxidation * in initial and canned fish, including different concentrations of cuttlefish ink (CI) in the packing medium **.

Initial or Canned Sample	Lipid Oxidation Index	
	Peroxide Value (Meq. Active Oxygen·kg^{-1} Lipids)	Thiobarbituric Acid Index (mg Malondialdehyde·kg^{-1} Muscle)
Initial fish	1.65 ab (0.34)	0.08 a (0.02)
CTR	1.10 a (0.18)	0.07 a (0.03)
CI-1	2.06 b (0.53)	0.08 a (0.03)
CI-2	1.37 ab (0.63)	0.09 a (0.03)
CI-3	1.53 ab (0.19)	0.10 a (0.02)
CI-4	1.98 ab (0.85)	0.13 a (0.03)

* Mean values of four replicates ($n = 4$); standard deviations are indicated in brackets. In each column, different letters (a, b) denote significant differences ($p < 0.05$). ** Abbreviations of sample names as expressed in Table 1.

An explanation for this contradictory effect found in the CI-2 and CI-4 batches is difficult to give, since an analysis of the molecules responsible and included in the CI extract was not carried out in the current study. However, previous research has proven that the antioxidant activity of an antioxidant compound may depend on different kinds of factors, such as the lipid composition, antioxidant concentration, temperature, oxygen pressure, and the presence of other antioxidants and food constituents [53,54]. Closely related to the present research, Agustini et al. [13] subjected salted-boiled milkfish to two different concentrations (0.75 and 1.50%) of melanin-free ink extract; a prolonged shelf-life time of up to 9 days was obtained with the less-concentrated treatment. Regarding concrete antioxidant compounds, α-tocopherol may behave as a prooxidant at higher concentrations, when prooxidants such as transition metals are present [55]. In the presence of added iron, the effectiveness of different natural antioxidants was reduced and, in some cases, resulted in a prooxidant effect [56]. During a linoleic acid-model system, α-tocopherol showed both antioxidant and prooxidant effects depending on the concentration and the simultaneous presence of Cu (II) and ascorbate [57]. A remarkable antioxidant effect was proven for ascorbate in steam- and microwave-cooked fish [58]; however, above critical values and in agreement with the present research, ascorbate showed pro-oxidant properties. Additionally, Medina et al. [59] proved a remarkable effect of concentration on the antioxidant capacity of polyphenol compounds during the thermal treatment of tuna (*Thunnus alalunga*).

No previous results are available, to the best of our knowledge, regarding the effect on the canned seafood quality of ink extracts obtained from cephalopod species. However, an aqueous extract of cuttlefish (*S. officinalis*) ink led to an inhibitory effect on the formation of conjugated dienes and trienes and fluorescent compounds during a heating treatment of fish in a model system [14]. Additionally, a preservative effect of cephalopod ink has already been proven on non-thermally treated seafood. Thus, the addition of melanin-free ink from splendid squid (*L. formosana*) led to a remarkable decrease in primary and secondary lipid oxidation development in chilled (15 days in ice) mackerel (*Rastrelliger kanagurta*) muscle [18]. Vate et al. [12] indicated a reduction of the peroxide and TBARS formation in refrigerated (20 days at 4 °C) surimi gel sardine (*Sardinella albella*) by the addition of melanin-free ink from splendid squid (*L. formosana*); additionally, a content reduction of nonanal and 2-decenal was reported. Tilapia (*Sparus latus*) fillets were treated with different dosages of melanin-free extract obtained from *S. esculenta* ink [60]; the authors detected an inhibitory effect on TBARS formation in fish fillets during the refrigerated storage (4 °C). Agustini et al. [13] observed an increased shelf-life time of soft-boned milkfish by previous spraying with squid-ink solutions; an antioxidant effect was inferred during storage at 10 °C, 20 °C and 30 °C of fish. An inhibitory effect on peroxide content was detected by Essid et al. [61] in smoked sardines (*Sardinella aurita*) during cold storage (35 days at 4 °C) by previous soaking in a cuttlefish (*S. officinalis*)-ink solution; furthermore, an extension of the shelf-life time was observed.

The presence of macroalgae extracts in the packing medium of canned fish has shown preservative effects. Thus, a decrease in the TBARS level was observed in canned Atlantic salmon (*S. salar*) by the addition of algae (*D. antarctica*, *U. lactuca*, or *P. columbina*) extracts [52] and a decrease in the fluorescent compound formation was proven in canned Atlantic mackerel (*S. scombrus*) by the presence of *B. bifurcata* extracts [31] and in canned Chub mackerel (*S. colias*) by including algae (*F. spiralis* or *U. lactuca*) aqueous extracts [36]. Regarding the employment of waste substrates produced during seafood processing, a lower fluorescent compound formation was detected in canned horse mackerel (*T. trachurus*) by including an aqueous extract of octopus (*O. vulgaris*) cooking liquor in the packing medium [51].

3.5. Assessment of Colour Changes in Fish Muscle

A remarkable increase ($p < 0.05$) of the L^* value was detected as a result of the thermal treatment and canned storage (Table 5). This increase was partially avoided in all cases by the presence of the CI extract in the packing medium. Thus, all CI-treated fish revealed lower ($p <$

0.05) L^* values than the control canned fish. Additionally, canned fish corresponding to the CI-1 condition did not provide differences ($p > 0.05$) with the initial fish.

Table 5. Colour changes * in initial and canned fish, including different concentrations of cuttlefish ink (CI) in the packing medium **.

Initial or Canned Sample	Colour Parameter		
	L^*	a^*	b^*
Initial fish	46.35 a (0.96)	2.21 b (0.76)	−3.76 a (0.24)
CTR	81.40 c (2.03)	−1.07 a (0.51)	8.98 b (0.68)
CI-1	55.54 ab (8.39)	2.18 b (0.67)	6.81 b (1.79)
CI-2	63.48 b (6.09)	1.38 b (0.83)	7.31 b (1.06)
CI-3	62.46 b (4.58)	1.53 b (0.36)	8.66 b (1.95)
CI-4	60.39 b (5.50)	1.91 b (0.36)	7.39 b (0.99)

* Mean values of four replicates ($n = 4$); standard deviations are indicated in brackets. In each column, different letters (a, b, c) denote significant differences ($p < 0.05$). ** Abbreviations of sample names as expressed in Table 1.

An average decrease in the a^* value was observed by comparing the initial fish and all kinds of the canned samples (Table 5). This decrease was significant ($p < 0.05$) in the case of the canned control samples but not ($p > 0.05$) in the CI-treated canned fish. No effect ($p > 0.05$) of the CI concentration added to the packing medium could be inferred from this colour parameter.

A comparison between the initial fish and all the kinds of canned samples indicated a relevant increase of the average b^* value by means of the sterilisation process and the canned storage (Table 5). The addition of the CI extract to the packing medium led to lower mean b^* values than the canned control fish; however, the differences were not found to be significant ($p > 0.05$). The lowest average value was observed for canned fish corresponding to the CI-1 condition.

In agreement with the great influence on the appearance and acceptability of all kinds of seafood, the determination of changes in muscle colour during processing has attracted great attention [62]. The current evolution of colour parameters during the canning process is in agreement with previous studies regarding the heating treatment of seafood. In general, the canning process has led to increased L^* and b^* values and to a^* value decreases [63–65].

No previous research has focused on the influence of cephalopod ink on colour changes in canned seafood. However, the presence in the packing medium of preservative compounds has proven valuable effects. Thus, the addition of an aqueous extract of *B. bifurcata* to the packing medium of canned Atlantic mackerel (*S. scombrus*) led to decreased L^* and b^* values [31]. Recently, Méndez et al. [51] obtained lower L^* and b^* values in canned horse mackerel (*T. trachurus*) by the addition of an aqueous extract of octopus (*O. vulgaris*) cooking liquor to the packing medium.

3.6. Determination of TMA Content in Fish Muscle

A relevant ($p < 0.05$) TMA content increase was observed in all canned samples when compared to the initial fish (Figure 2). Lower average values were found in CI-treated fish when compared to the counterpart canned control. The lowest mean values were observed in the fish samples corresponding to the CI-2 condition, which were significantly lower ($p < 0.05$) than those corresponding to the canned control samples. As for the lipid quality retention (FR, Figure 1), a selective effect of the CI concentration could be inferred.

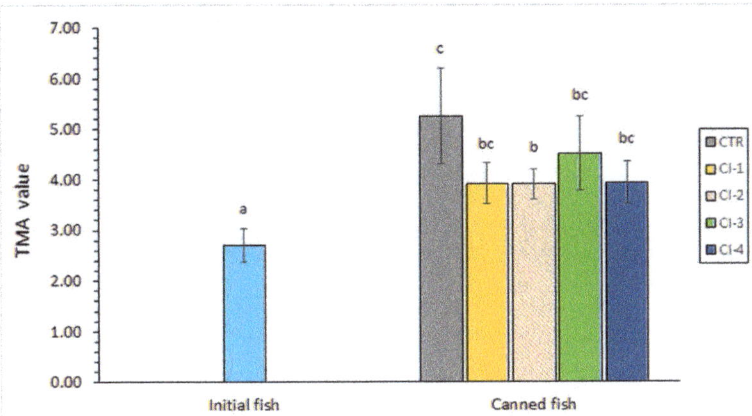

Figure 2. Trimethylamine (TMA) value in initial and canned fish, including different concentrations of cuttlefish ink (CI) in the packing medium. Average values of four replicates ($n = 4$); standard deviations are indicated by bars. Different letters (a, b, c) denote significant differences ($p < 0.05$). Abbreviations of sample names as expressed in Table 1.

TMA is recognised as a remarkable deteriorative compound whose value can reflect the quality degradation of seafood subjected to different kinds of technological procedures [66]. Unfortunately, no previous studies are available, to the best of our knowledge, regarding the effect of cephalopod-ink extracts on the TMA value of canned seafood. However, an inhibitory effect has already been proven in non-thermally treated seafood. Thus, Karim et al. [11] demonstrated that a previous squid melanin-free ink soaking from the squid (*Loligo duvaucelli*) led to an increased shelf-life time and a decrease of the TMA and TVB values during refrigerated (4 °C for 15 days) storage. Sadok et al. [67] proved that a previous cuttlefish (*S. officinalis*)-ink soaking led to an inhibitory effect on the formation of off-odour compounds (TMA and TVB values) in cold-stored (0 °C for 12 days and −2 °C for 23 days) peeled shrimp (*Penaeus kerathurus*). Shi et al. [10] subjected yellowfin seabream (*S. latus*) to a marinated treatment with a melanin-free extract from squid ink; as a result, an inhibitory effect on TVB formation was detected in the fish during refrigerated storage at 4 °C. Tilapia (*S. latus*) fillets were subjected to melanin-free extracts obtained from *S. esculenta* ink [60]; the authors observed an inhibitory effect on TVB formation in fish fillets during refrigerated storage (4 °C). Recently, an inhibitory effect on the formation of TVB and TMA was detected by Essid et al. [61] in smoked sardines (*S. aurita*) during cold storage (35 days at 4 °C) by previous soaking in a cuttlefish (*S. officinalis*)-ink solution.

Previous research has also addressed the effect resulting from the addition to the packing medium of preservative compounds obtained from marine sources on the TMA value of canned seafood. As in the present results, an inhibitory effect on TMA formation was detected in canned Chub mackerel (*S. colias*) by the addition of an aqueous *F. spiralis* extract to the packing medium [68]. On the contrary, no effect on the TVB and TMA values was detected in Atlantic mackerel (*S. scombrus*) [31] and Chub mackerel (*S. colias*) [36] by the presence of aqueous extracts of several macroalgae.

4. Conclusions

The present research provides a first approach for the employment of a CI extract for the quality enhancement of canned fish. However, the degree of this effect was shown to be notably influenced by the concentration employed. Thus, the employment of the CI concentration corresponding to the CI-2 condition led to a lower lipid oxidation development (fluorescent compound formation), lower colour changes (L^* and a^* parameters) and a lower TMA value in canned fish when compared to the control canned samples. Although differences were not significant, higher average values of DHA, ω3/ω6 ratio, and PI were detected in canned fish corresponding to the CI-1 and CI-2 conditions. On the contrary, the use of the most concentrated CI extract (CI-4 condition) led to a prooxidant effect (higher FR value) when compared to the control canned fish.

On the basis of the notable effect of the CI concentration employed, further research that takes into account an optimisation design (i.e., response surface methodology) is considered necessary in order to employ the most convenient CI concentration for the canning processing of the current farmed species. Additionally, further research is envisaged to carry out the analysis of the CI extract and, therefore, the molecules involved in the preservative properties detected in the present study. This research is in agreement with the current requirements for the search for new and natural sources of preservative compounds and also agrees with general commitments for environmental sustainability and circular economy.

Author Contributions: Conceptualization, B.M. and S.P.A.; methodology, B.M. and M.T.; data curation, M.T. and A.R.; writing—original draft preparation, S.P.A.; writing—review and editing, A.R. and S.P.A. All authors have read and agreed to the published version of the manuscript.

Funding: This research received no external funding.

Institutional Review Board Statement: Not applicable.

Informed Consent Statement: Not applicable.

Data Availability Statement: The original contributions presented in the study are included in the article, further inquiries can be directed to the corresponding author.

Acknowledgments: The cuttlefish ink was kindly provided by Sepink (Vilagarcía de Arousa, Pontevedra, Spain).

Conflicts of Interest: The authors declare no conflicts of interest.

References

1. Rustad, T.; Storro, I.; Slizyte, R. Possibilities for the utilisation of marine by-products. *Int. J. Food Sci. Technol.* **2011**, *46*, 2001–2014. [CrossRef]
2. Olsen, R.L.; Toppe, J.; Karunasagar, I. Challenges and realistic opportunities in the use of by-products from processing of fish and shellfish. *Trends Food Sci. Technol.* **2014**, *36*, 144–152. [CrossRef]
3. Atef, M.; Ojagh, M. Health benefits and food applications of bioactive compounds from fish byproducts: A review. *J. Funct. Foods* **2017**, *35*, 673–681. [CrossRef]
4. Vázquez, J.A.; Meduiña, A.; Durán, A.I.; Nogueira, M.; Fernández-Compás, A.; Pérez-Martín, R.I.; Rodríguez-Amado, I. Production of valuable compounds and bioactive metabolites from by-products of fish discards using chemical processing, enzymatic hydrolysis, and bacterial fermentation. *Mar. Drugs* **2019**, *17*, 139. [CrossRef] [PubMed]
5. Fahmy, S.R.; Ali, E.M.; Ahmed, N.S. Therapeutic effect of Sepia ink extract against invasive pulmonary aspergillosis in mice. *J. Basic App. Zool.* **2014**, *67*, 196–204. [CrossRef]
6. Solano, F. Melanin and melanin-related polymers as materials with biomedical and biotechnological applications—Cuttlefish ink and mussel foot proteins as inspired biomolecules. *Int. J. Mol. Sci.* **2017**, *18*, 1561. [CrossRef] [PubMed]
7. Momenzadeh, N.; Hajian, S.; Shabankare, A.; Ghavimi, R.; Kabiri-Samani, S.; Kabiri, H.; Hesami-Zadeh, K.; Shabankareh, A.N.T.; Nazaraghay, R.; Nabipour, I.; et al. Photothermic therapy with cuttlefish ink-based nanoparticles in combination with anti-OX40 mAb achieve remission of triple-negative breast cancer. *Int. Immunopharm.* **2023**, *115*, 109622. [CrossRef] [PubMed]
8. Neifar, A.; Abdelmalek, I.B.; Bouajila, G.; Kolsi, R.; Bradai, M.N.; Abdelmouleh, A.; Gargouri, A.; Ayed, N. Purification and incorporation of the black ink of cuttlefish *Sepia officinalis* in eye cosmetic products. *Color. Technol.* **2013**, *129*, 150–154. [CrossRef]
9. Wang, H.; Xu, T.; Zheng, B.; Cao, M.; Gao, F.; Zhou, G.; Ma, C.; Dang, J.; Yao, W.; Wu, K.; et al. Cuttlefish ink loaded polyamidoxime adsorbent with excellent photothermal conversion and antibacterial activity for highly efficient uranium capture from natural seawater. *J. Hazard. Mat.* **2022**, *433*, 128789. [CrossRef] [PubMed]
10. Shi, L.; Liu, H.; Zhong, J.; Pan, J. Fresh-keeping effects of melanin-free extract from squid ink on yellowfin sea bream (*Sparus latus*) during cold storage. *J. Aquat. Food Prod. Technol.* **2015**, *24*, 199–212. [CrossRef]
11. Karim, N.U.; Sadzali, N.L.; Hassan, M. Effects of squid ink as edible coating on squid sp. (*Loligo duvauceli*) spoilage during chilled storage. *Int. Food Res. J.* **2016**, *23*, 1895–1901.
12. Vate, N.K.; Benjakul, S.; Agustini, T.W. Application of melanin-free ink as a new antioxidative gel enhancer in sardine surimi gel. *J. Sci. Food Agric.* **2015**, *95*, 2201–2207. [CrossRef] [PubMed]

13. Agustini, T.W.; Hadiyanto, H.; Amalia, U.; Wijayantia, I. Potential of melanin free ink as antioxidant and its application for preserving and predicting shelf life of salted-boiled milkfish. *Int. J. Postharv. Technol. Innov.* **2019**, *6*, 57–69. [CrossRef]
14. Trigo, M.; Paz, D.; Bote, A.; Aubourg, S.P. Antioxidant activity of an aqueous extract of cuttlefish (*Sepia officinalis*) ink during fish muscle heating. *Antioxidants* **2023**, *12*, 1996. [CrossRef] [PubMed]
15. Liang, Y.; Han, Y.; Dan, J.; Li, R.; Sun, H.; Wang, J.; Zhang, W. A high-efficient and stable artificial superoxide dismutase based on functionalized melanin nanoparticles from cuttlefish ink for food preservation. *Food Res. Int.* **2023**, *163*, 112211. [CrossRef] [PubMed]
16. Derby, C.D. Cephalopod Ink: Production, chemistry, functions and applications. *Mar. Drugs* **2014**, *12*, 2700–2730. [CrossRef] [PubMed]
17. Chen, S.; Xue, C.; Li, Y.; Gao, Z.; Ma, Q. Studies on the free radical scavenging activities of melanin from squid ink. *China J. Mar. Drugs* **2007**, *26*, 24–27.
18. Vate, N.K.; Benjakul, S. Antioxidative activity of melanin-free ink from splendid squid (*Loligo formosana*). *Int. Aquat. Res.* **2013**, *5*, 9. [CrossRef]
19. Senan, V.P. Antibacterial activity of methanolic extract of the ink of cuttlefish, *Sepia pharaonis*, against pathogenic bacterial strains. *Int. J. Pharm. Sci. Res.* **2015**, *6*, 1705–1710.
20. Guo, X.; Chen, S.; Hu, Y.; Li, G.; Liao, N.; Ye, X.; Liu, D.; Xue, C. Preparation of water-soluble melanin from squid ink using ultrasound-assisted degradation and its antioxidant activity. *J. Food Sci. Technol.* **2014**, *51*, 3680–3690. [CrossRef] [PubMed]
21. Xie, J.; Li, H.; Che, H.; Dong, X.; Yang, X.; Xie, W. Extraction, physicochemical characterisation, and bioactive properties of ink melanin from cuttlefish (*Sepia esculenta*). *Int. J. Food Sci. Technol.* **2021**, *56*, 3627–3640. [CrossRef]
22. Horner, W. Canning fish and fish products. In *Fish Processing Technology*, 2nd ed.; Hall, G., Ed.; Blackie Academic and Professional, Chapman and Hall: London, UK, 1997; pp. 119–159.
23. Veiga, A.; Martínez, E.; Ojea, G.; Caride, A. Principles of thermal processing in canned seafood. In *Quality Parameters in Canned Seafoods*; Cabado, A.G., Vieites, J.M., Eds.; Nova Science Publishers, Inc.: New York, NY, USA, 2008; pp. 83–103.
24. Lukoshkina, M.; Odoeva, G. Kinetics of chemical reactions for prediction of quality of canned fish during storage. *App. Biochem. Microb.* **2003**, *39*, 321–327. [CrossRef]
25. Ling, B.; Tang, J.; Kong, F.; Mitcham, E.J.; Wang, S. Kinetics of food quality changes during thermal processing: A review. *Food Bioprocess Technol.* **2015**, *8*, 343–358. [CrossRef]
26. Pitarch, J.L.; Vilas, C.; de Prada, C.; Palacín, C.G.; Alonso, A.A. Optimal operation of thermal processing of canned tuna under product variability. *J. Food Eng.* **2021**, *304*, 110594. [CrossRef]
27. Tokur, B.; Korkmaz, K. Novel thermal sterilization technologies in seafood processing. In *Innovative Technologies in Seafood Processing*; Özoğul, Y., Ed.; CRC Press: Boca Raton, FL, USA; Taylor and Francis Group: Abingdon, UK, 2020; pp. 303–322.
28. Ruiz-Roso, B.; Cuesta, I.; Pérez, M.; Borrego, E.; Pérez-Olleros, L.; Varela, G. Lipid composition and palatability of canned sardines. Influence of the canning process and storage in olive oil for five years. *J. Sci. Food Agric.* **1998**, *77*, 244–250. [CrossRef]
29. Prego, R.; Trigo, M.; Martínez, B.; Aubourg, S.P. Effect of previous frozen storage, canning process and packing medium on the fatty acid composition of canned mackerel. *Mar. Drugs* **2022**, *20*, 666. [CrossRef] [PubMed]
30. AOAC. *Official Methods for Analysis of the Association of Analytical Chemistry*, 15th ed.; Association of Official Chemists, Inc.: Arlington, VA, USA, 1990; pp. 931–937.
31. Barbosa, R.G.; Trigo, M.; Fett, R.; Aubourg, S.P. Impact of a packing medium with alga *Bifurcaria bifurcata* extract on canned Atlantic mackerel (*Scomber scombrus*) quality. *J. Sci. Food Agric.* **2018**, *98*, 3462–3467. [CrossRef] [PubMed]
32. Bligh, E.; Dyer, W. A rapid method of total extraction and purification. *Can. J. Biochem. Physiol.* **1959**, *37*, 911–917. [CrossRef] [PubMed]
33. Herbes, S.E.; Allen, C.P. Lipid quantification of freshwater invertebrates: Method modification for microquantitation. *Can. J. Fish. Aquat. Sci.* **1983**, *40*, 1315–1317. [CrossRef]
34. Lowry, R.; Tinsley, I. Rapid colorimetric determination of free fatty acids. *J. Am. Oil Chem. Soc.* **1976**, *53*, 470–472. [CrossRef]
35. Raheja, R.; Kaur, C.; Singh, A.; Bhatia, A. New colorimetric method for the quantitative determination of phospholipids without acid digestion. *J. Lipid Res.* **1973**, *14*, 695–697. [CrossRef] [PubMed]
36. Barbosa, R.G.; Trigo, M.; Campos, C.A.; Aubourg, S.P. Preservative effect of algae extracts on lipid composition and rancidity development in brine-canned Atlantic Chub mackerel (*Scomber colias*). *Eur. J. Lipid Sci. Technol.* **2019**, *121*, 1900129. [CrossRef]
37. Chapman, R.; McKay, J. The estimation of peroxides in fats and oils by the ferric thiocyanate method. *J. Am. Oil Chem. Soc.* **1949**, *26*, 360–363. [CrossRef]
38. Vyncke, W. Direct determination of the thiobarbituric acid value in trichloracetic acid extracts of fish as a measure of oxidative rancidity. *Fette Seifen Anstrichm.* **1970**, *72*, 1084–1087. [CrossRef]
39. Aubourg, S.P.; Medina, I.; Pérez-Martín, R.I. A comparison between conventional and fluorescence detection methods of cooking-induced damage to tuna fish lipids. *Z. Lebensm. Unters. Forsch.* **1995**, *200*, 252–255. [CrossRef] [PubMed]
40. Tozawa, H.; Erokibara, K.; Amano, K. Proposed modification of Dyer's method for trimethylamine determination in codfish. In *Fish Inspection and Quality Control*; Kreuzer, R., Ed.; Fishing News Books Ltd.: London, UK, 1971; pp. 187–190.
41. Rustad, T. Lipid oxidation. In *Handbook of Seafood and Seafood Products Analysis*; Nollet, L.M., Toldrá, F., Eds.; CRC Press: Boca Raton, FL, USA; Taylor and Francis Group: Abingdon, UK, 2010; pp. 87–95.
42. Piclet, G. Le poisson aliment. Composition-Intérêt nutritionnel. *Cah. Nutr. Diététique* **1987**, *XXII*, 317–335.
43. Medina, I.; Sacchi, R.; Aubourg, S.P. A ^{13}C-NMR study of lipid alterations during fish canning: Effect of filling medium. *J. Sci. Food Agric.* **1995**, *69*, 445–450. [CrossRef]

44. Labuza, T. Kinetics of lipid oxidation in foods. *CRC Crit. Rev. Food Technol.* **1971**, *2*, 355–405. [CrossRef]
45. Miyashita, K.; Tagagi, T. Study on the oxidative rate and prooxidant activity of free fatty acids. *J. Am. Oil Chem. Soc.* **1986**, *63*, 1380–1384. [CrossRef]
46. Malga, J.M.; Trigo, M.; Martínez, B.; Aubourg, S.P. Preservative effect on canned mackerel (*Scomber colias*) lipids by addition of octopus (*Octopus vulgaris*) cooking liquor in the packaging medium. *Molecules* **2022**, *27*, 739. [CrossRef]
47. Küllenberg, D.; Taylor, L.A.; Schneider, M.; Massing, U. Health effects of dietary phospholipids. *Lipids Health Dis.* **2012**, *11*, 3. [CrossRef] [PubMed]
48. Li, J.; Wang, X.; Zhang, T.; Huang, Z.; Luo, X.; Deng, Y. A review on phospholipids and their main applications in drug delivery systems. *Asian J. Pharm. Sci.* **2015**, *10*, 81–98. [CrossRef]
49. Aubourg, S.P.; Medina, I.; Pérez-Martín, R. Polyunsaturated fatty acids in tuna phospholipids: Distribution in the *sn-2* location and changes during cooking. *J. Agric. Food Chem.* **1996**, *44*, 585–589. [CrossRef]
50. Takahashi, K.; Inoue, Y. Marine by-product phospholipids as booster of medicinal compounds. *Adv. Food Nutr. Res.* **2012**, *65*, 31–46. [PubMed]
51. Méndez, L.; Trigo, M.; Zhang, B.; Aubourg, S.P. Antioxidant effect of octopus by-products in canned horse mackerel (*Trachurus trachurus*) previously subjected to different frozen storage times. *Antioxidants* **2022**, *11*, 2091. [CrossRef] [PubMed]
52. Ortiz, J.A.; Vivanco, J.P.; Aubourg, S.P. Lipid and sensory quality of canned Atlantic salmon (*Salmo salar*): Effect of the use of different seaweed extracts as covering liquids. *Eur. J. Lipid Sci. Technol.* **2014**, *116*, 596–605. [CrossRef]
53. Nawar, W.W. Lipids. In *Food Chemistry*, 3rd ed.; Fennema, O., Ed.; Marcel Dekker: New York, NY, USA, 1996; pp. 225–314.
54. Pokorný, J. Introduction. In *Antioxidants in Foods. Practical Applications*; Pokorný, J., Yanishlieva, N., Gordon, M., Eds.; CRC Press: Boca Raton, FL, USA, 2001; pp. 1–3.
55. Afonso, C.; Bandarra, N.M.; Nunes, L.; Cardoso, C. Tocopherols in seafood and aquaculture products. *Crit. Rev. Food Sci. Nutr.* **2016**, *56*, 128–140. [CrossRef] [PubMed]
56. Noon, J.; Mills, T.B.; Norton, I.T. The use of natural antioxidants to combat lipid oxidation in O/W emulsions. *J. Food Eng.* **2020**, *281*, 110006. [CrossRef]
57. Bakir, T.; Beker, B.Y.; Sönmezoğlu, I.; Imer, F.; Apak, R. Antioxidant and prooxidant effects of alpha-tocopherol in a linoleic acid-copper(II) ascorbate system. *Eur. J. Lipid Sci. Technol.* **2013**, *115*, 372–376. [CrossRef]
58. Ramanathan, L.; Das, N. Studies on the control of lipid oxidation in ground fish by some polyphenolic natural products. *J. Agric. Food Chem.* **1992**, *40*, 17–21. [CrossRef]
59. Medina, I.; Satué, M.T.; German, J.B.; Frankel, E. Comparison of natural polyphenols antioxidant from extra virgin olive oil with synthetic antioxidants in tuna lipids during thermal oxidation. *J. Agric. Food Chem.* **1999**, *47*, 4873–4879. [CrossRef] [PubMed]
60. Duan, Z.H.; Liu, H.Z.; Luo, P.; Gu1, Y.P.; Li, Y.Q. The effect of melanin-free extract from *Sepia esculenta* ink on lipid peroxidation, protein oxidation and water-holding capacity of tilapia fillet during cold storage. *Chem. Central J.* **2018**, *12*, 30. [CrossRef] [PubMed]
61. Essid, I.; Aroussia, H.; Soufi, E.; Bouriga, N.; Gharbi, S.; Bellagha, S. Improving quality of smoked sardine fillets by soaking in cuttlefish ink. *Food Sci. Technol.* **2022**, *42*, e65020. [CrossRef]
62. Schubring, R. Quality Assessment of Fish and Fishery Products by Color Measurement. In *Handbook of Seafood and Seafood Products Analysis*; Nollet, L.M., Toldrá, F., Eds.; CRC Press, Francis and Taylor Group: Boca Raton, FL, USA, 2010; pp. 395–424.
63. Mohan, C.O.; Remya, S.; Murthy, L.N.; Ravishankar, C.N.; Kumar, K.A. Effect of filling medium on cooking time and quality of canned yellowfin tuna (*Thunnus albacares*). *Food Cont.* **2015**, *50*, 320–327. [CrossRef]
64. Mohan, C.O.; Remya, S.; Ravishankar, C.N.; Vijayan, P.K.; Srinivasa Gopal, T.K. Effect of filling ingredient on the quality of canned yellowfin tuna (*Thunnus albacares*). *Int. J. Food Sci. Technol.* **2014**, *49*, 1557–1564. [CrossRef]
65. Gómez-Limia, L.; Carballo, J.; Rodríguez-González, M.; Martínez, S. Impact of the filling medium on the colour and sensory characteristics of canned European eels (*Anguilla anguilla* L.). *Foods* **2022**, *11*, 1115. [CrossRef] [PubMed]
66. Özoğul, Y. Methods for freshness quality and deterioration. In *Handbook of Seafood and Seafood Products Analysis*; Nollet, L.M., Toldrá, F., Eds.; CRC Press: Boca Raton, FL, USA; Taylor and Francis Group: Abingdon, UK, 2010; pp. 189–214.
67. Sadok, S.; Abdelmoulah, A.; El Abed, A. Combined effect of sepia soaking and temperature on the shelf life of peeled shrimp *Penaeus kerathurus*. *Food Chem.* **2004**, *88*, 115–122. [CrossRef]
68. Aubourg, S.P.; Trigo, M.; Martínez, B.; Rodríguez, A. Effect of prior chilling period and alga-extract packaging on the quality of a canned underutilised fish species. *Foods* **2020**, *9*, 1333. [CrossRef] [PubMed]

foods

MDPI

Article

A Preliminary Study on the Effect of Adding Sugarcane Syrup on the Flavor of Barley Lager Fermentation

Hechao Lv [1], Yusheng Jia [1], Chaoyi Liu [1], Jia Xu [1], Caifeng Xie [1,2,3], Kai Li [1,2,3], Kai Huang [4] and Fangxue Hang [1,2,3,*]

1 College of Light Industry and Food Engineering, Guangxi University, Nanning 530004, China; lhcgzyx@126.com (H.L.); jys1562022@163.com (Y.J.); liuchaoyi32@163.com (C.L.); 15191008212@163.com (J.X.); fcx11@163.com (C.X.); gxlikai@gxu.edu.cn (K.L.)
2 Provincial and Ministerial Collaborative Innovation Center for Sugar Industry, Nanning 530004, China
3 Engineering Research Center for Sugar Industry and Comprehensive Utilization, Ministry of Education, Nanning 530004, China
4 Guangxi Institute of Industrial Technology, Nanning 530001, China; kaige@gxgy.edu.cn
* Correspondence: hangfx@163.com

Abstract: This study focuses on the diversified utilization of the sugarcane industry, and sugarcane syrup, as a by-product of the sugarcane industry, is a good raw material for fermentation. Bringing sugarcane syrup into beer is conducive to the enrichment of the sugar industry, and it can improve the flavor of beer and make it more aromatic. This study determined the optimal fermentation process for beer. By analyzing the consumption rate of the carbon and nitrogen sources of raw materials, the nutrient utilization of yeast, and the causes of differences in flavor substances, the flavor composition and flavor stability of beer were determined by SPME-HS-GC-MS technology. The results showed that beer brewed with sugarcane syrup as an auxiliary raw material met the basic specifications of beer. The addition of sugarcane syrup to the wort base increased the utilization of amino acids by the yeast, and LS (lager with added cane syrup) increased the nine flavor compounds of the beer, which constituted the basic flavor of the beer, bringing new flavor compounds compared with the normal all-barley beer. Forced aging experiments showed that LS produced fewer aging compounds than OWBL. Various experiments have shown that it is feasible to ferment beer with sugarcane syrup instead of partial wort.

Keywords: sugarcane syrup; beer; process optimization; volatile substance; aroma stabilization

1. Introduction

Lager beer is one of the world's oldest and most commonly consumed alcoholic beverages. It is made from barley wort fermented by yeast at 10–15 °C [1]. The final aroma composition of the beer depends on the materials required for brewing and the parameters of the brewing process [2]. Volatile components in beer have a much greater impact on flavor profiles than in other alcohols [3]. Also, the detection of volatile substances is an important part of the process of the comprehensive evaluation of lager quality [4] and is the most intuitive way to analyze lager quality. Gas chromatography–mass spectrometry (GC-MS) coupled with headspace solid-phase microextraction (HS-SPME) has been widely used for the study of beer aroma [5], and electronic noses (e-noses) equipped with a series of electrochemical sensors have been shown to provide a comprehensive odor assessment [6]. Also, gas chromatography (GC) can be used to determine beer aging components. Studies have shown that carbonyl compounds are the main component of the "aging" taste of beer [7]. This means that measuring the amount of change in carbonyl compounds during the aging of beer is an important measure of flavor stability.

Unfavorable growing conditions for barley in some parts of the world (including those attributable to climate change) have led lager producers to use more versatile local raw materials to replace barley in conventional recipes. These materials are called adjuncts [8].

Syrup is one of the most commonly used liquid adjuncts. The use of syrup in lagers results in a lower production cost and a shorter fermentation cycle [9]. This fermentation has also been associated with altered (abnormal) patterns of sugar uptake and altered production of some flavor compounds [10].

China is one of the main sugarcane-producing countries, and Guangxi is the main sugarcane-growing area in China, accounting for more than 60% of China's planting area. In the sugar production process, sugar cane is pressed into juice and then heated and concentrated to produce sugar cane syrup (sugar content 65–68°BX). Sugar cane syrup is rich in nutrients and is a quality brewing ingredient [11]. Sugarcane syrup is rich in amino acids, and its amino acid composition is different from that of wort. Changes in nutrient content can trigger different yeast reactions to produce beers with different flavors [2]. Sugar cane also contains many types of higher alcohols, which give beer a greater variety of flavor substances [6]. Moreover, studies have shown that sugarcane syrup contains a large number of polyphenolic components, including phenolic acids, flavonoids, and quinones [12], which have antioxidant properties. It is assumed that the polyphenolic components in sugarcane syrup can inhibit the oxidation of unsaturated fatty acids, isorhythmic ketones, higher alcohols, and other substances to form aldehydes and produce an "aging odor" [13]. Carbonyl compounds cause flavor changes in many foods, including milk, butter, vegetables, and oils. In 1966, Japanese researchers first found a significant increase in volatile carbonyl compounds in beer during storage and found that the increase in the aging flavor of stored beer coincided with an increase in the concentration of carbonyl compounds [14].

2. Materials and Methods

2.1. Experimental Material

Saccharomyces cerevisiae var. diastaticus with the trade name "BF 16" from Angie's Yeast Company (China) was used for wort fermentation. The yeast was purchased as a solid fermenter, and the initial concentration of the medium after activation was 9×10^7 cells/mL. The yeast liquid was added at 1% of the total wort volume according to the manufacturer's recommendations. The yeast strain used was low flocculating and highly fermentable (76–80%) with an optimal fermentation temperature of 10–14 °C.

The materials for this study were three variants of Chinese lager: ordinary whole barley lager (OWBL), lager with added cane syrup (LS), commercially available lager 1 (CAL_1), and commercially available lager 2 (CAL_2). Ingredients include barley malt, wheat malt, light caramel malt, Munich malt, and red malt. All malts were sourced from Jinan Shuangmai Company (Jinan, Shandong, China). Citra (Qingdao, Shandong, China) and Cascade hops (Portland, OR, USA) were purchased from Guangzhou Dejussi Trading Company Limited (Guangzhou, China). Sugarcane syrup (Guangxi Baiguitang Co., Ltd., Chongzuo, Guangxi, China) was also added during the production process. The sugarcane syrup added during the production process was different from the traditional syrup made by chemical methods; after being pressed from sugarcane into sugarcane juice, it was pressed through a 50 nm membrane, which was evaporated and the filtrate was concentrated to 60–65°BX.

2.2. Physicochemical Parameters

The pH of finished beer was measured using a precision pH meter pH-3C. The extract content was measured at 20 °C using an Abbe refractometer, residual sugar was measured by DNS colorimetry [15], acidity was measured by sodium hydroxide titration, and the color of the beer was measured by a spectrophotometer at a wavelength of 420 nm. These measurements were repeated three times.

2.3. Determination of Bittering Value of Beer

After the beer was degassed, 10 mL of degassed beer was acidified by adding 0.5 mL of 6 mol/L hydrochloric acid and then 20 mL of iso-octane, extracted by oscillation, centrifuged, and the absorbance was measured at 275 nm using a UV-visible spectrophotometer (Agilent 8453, Agilent, Santa Clara, CA, USA) [16].

The content of bitter substances in beer was calculated by the following formula.

$$X = A_{275} \times 50$$

where X is the content of bitter substances in the sample, expressed in units of "BU"; A_{275} is the absorbance of the sample measured at 275 nm; and 50 is the conversion factor.

2.4. Beer Sensory Appetite Experiment

Twenty students (male:female = 1:1) who were physically fit and had a good sense of smell and taste were randomly selected to participate in the sensory evaluation according to the scoring rules table. The specific evaluation criteria can be found in the Supplementary Information.

2.5. Determination of Fermentable Sugars

A high-performance liquid chromatographic method was used for the quantification of fermentable sugars in wort and beer fermentation broth by an external standard [17]. Detection methods: chromatographic column: NH_2 Analytical HPLC Column (4.6 × 250 mm, 5 μm); column temperature: 35 °C; differential refractive index detector; detector temperature: 35 °C; mobile phase: 75% acetonitrile; flow rate: 1.0 mL/min; injection volume: 20 μL. Standard preparation: quantitative glucose, fructose, maltose, sucrose, and maltotriose standards were dissolved in ultrapure water for gradient dilution and passed through a 0.22 μm membrane. Sample pretreatment: The wort and beer fermentation broth were centrifuged at 7000 rpm and then passed through a 0.22 μm membrane.

2.6. Determination of Free Amino Acids

The beer fermentation broth was analyzed using an automated amino acid analyzer and quantified using the external standard method [18]. Sample treatment: the samples were diluted 5 times with 4% sulfosalicylic acid solution, left at 20 °C for 30 min, centrifuged, and the supernatant was filtered with 0.22 μm nylon 66 membrane.

2.7. Determination of Volatile Aroma Substances in Lager Beer

Electric nose: 1 mL of beer was diluted 10-fold and transferred into a sealed 100 mL conical flask and held at 25.0 °C ambient temperature for 60 min to allow the evaporation of volatile components. The volatile gas in the beaker was drawn into the sensor at a rate of 400 mL/min, and the computer recorded a time of 150 s. The sample was washed for 120 s before each measurement to remove the residual volatile gas and return the response value to the baseline state. The response value was selected from 145 to 150 s, and the average value was calculated. The measurement operation was repeated three times [19].

HS-SPME-GC-MS: HS-SPME conditions: a 5 mL glass vial containing 30 mL of beer sample was taken, 10 μL of 2-octanol at a concentration of 82.2 μg/mL was added as an internal standard, and the vial was closed with a cap with a PTFE cushion until analysis. 2-octanol was used as an internal standard, 2 g NaCl was added and rotated, and the vial was closed with a lid with a PTFE gasket until analysis. Equilibrium was achieved by pre-incubation for 10 min at 40 °C with magnetic stirring at 300 rpm. Experiments were conducted using a composite 50/30 m DVB/CAR/PDMS extraction head, and the head was aged for 20 min before each sample's extraction to remove residues attached to the extraction head. The SPME fiber was extracted in the headspace of the sample for 40 min at 40 °C, then transferred to the injection port (230 °C) and left for 5 min for desorption before detection. The mixed standards were four alcohols: n-propanol, isoamyl alcohol, phenylethanol, and isobutanol. The standards were prepared as mixed standards and added to the internal standard injection, and their retention times and ion fragments were compared with those of the compounds in the samples. GC conditions: column: Agilent DB-Wax capillary column (60 m × 0.25 mm); carrier gas: helium, 99.999%; constant flow mode: 1.0 mL/min; etc. The temperature was initially 35 °C for 4 min, raised to 60 °C at 3 °C/min, held for 5 min, and finally raised to 230 °C for 15 min at 4 °C/min. The inlet temperature was 230 °C, and the sample was injected manually without splitting. The temperature was raised to 230 °C and finally held at 230 °C for 15 min. The inlet temperature was 230 °C, and the sample was injected manually without a split. MS conditions: ionization mode: EI; ionization energy: 70 eV; ion source temperature: 230 °C; quadrupole temperature: 150 °C; transfer line temperature: 270 °C; scan mode: SCAN; mass scan range: 32 to 300 amu; no solvent delay [13,20].

2.8. Determination of Antioxidant Power of Beer

Determination of antioxidant power of beer: Take 0.5 mL of beer sample diluted twice, add 2.5 mL of 6.5×10^{-5} mol/L DPPH ethanol solution, and react for 1 h at room temperature, then dilute the reaction solution 4-fold and move to a spectrophotometer for colorimetric comparison at a wavelength of 517 nm, and then compare its absorbance [21].

2.9. Determination of Volatile Aging Substances

In this experiment, OWBL and LS were used as research subjects, while LS and OWBL without sugarcane juice were used as control samples for the forced aging experiments. The aging aroma substances were stored at 45 °C and protected from light for a period of 9 d, and were characterized by GC-MS [22]. Standards: 2-furaldehyde, glutaraldehyde, phenylacetaldehyde, benzaldehyde, acetaldehyde, isobutyraldehyde, 2-methylbutyraldehyde, 3-methylbutyraldehyde, and trans-dinonenal. According to the chromatographic peaks, transfer 0.5 mL of each of the anti-dinonenal according to the different preparation methods and neutralize with ethanol to 200 mL to produce a mixed standard first-class reserve solution. Take 10 mL of 5% ethanol solution (pH 4.6–4.7 with 0.1% phosphoric acid) as the base, and add 1 μL, 2 μL, 4 μL, 8 μL, 15 μL, and 25 μL of the primary reserve solution to produce the standard sample solution. A 65 μm (PDMS-DVB) SPME fiber extraction head was used. HS-SPME conditions: a dilute solution of PFBOA was prepared by mixing 100 μL (6 g/L) of PBFOA solution with 10 mL of water in 30 mL glass vials and sealed with PTFE gasketed lids, and each vial containing diluted PFBOA was first equilibrated for 5 min at 50 °C and 250 rpm. The PDMS-DVB fibers were exposed to the headspace of the diluted PFBOA for 10 min. The SPME fibers with PFBOA were exposed to the headspace of the beer samples and extracted at 50 °C and 250 rpm for 45 min. The fibers were resolved in the GC inlet for 5 min [23]. GC conditions: gas chromatography was performed on an Agilent DB-Wax capillary column (60 m × 0.25 mm × 0.25 μm); carrier gas: helium, 99.999%; constant-flow mode: 1.0 mL/min; column warming program: initial oven temperature was maintained at 40 °C for 10 min, 10 °C/min up to 140 °C for 10 min, and 5 °C/min up to 250 °C for 10 min. The temperature of the inlet port was 250 °C, and the sample was injected manually without splitting [14]. MS conditions: ionization mode: EI; ionization energy: 70 eV, ion source temperature: 230 °C, quadrupole temperature: 150 °C; conversion line set at 270 °C; mass range: m/z 50–350; scanning mode: 4.65 scans/s. Carbonyl PFBOA derivatives were identified and the m/z 181 fragment was identified as the major fragment, and all aldehyde analyses were performed in the single-ion monitoring (SIM) mode, monitoring the ion m/z 181. Beer was also analyzed by gas chromatography/mass spectrometry (GC/MS) without PFBOA derivatization [24].

2.10. Methods of Data Analysis

Data were processed and plotted using Excel, SPSS 24.0, Xcalibur 4.0, and Origin 9.0.

Experimental GC-MS data of the samples were processed using Xcalibur 4.0 software and compared by computerized search and spectral library. The mixed C8–C40 standards were analyzed under the same GC-MS conditions and the retention indices (RIs) of the components to be measured were calculated according to the following equation.

Retention index

$$RI = 100z + 10\ [TR(x) - TR(z)]\ TR(z + 1) - TR]\ RI = 100z + 10\ [TR(x) - TR(z)]\ TR(z + 1) - TR$$

TR(x): retention time of the component to be measured;
TR(z): Retention time of n-alkanes with carbon atom number z;
TR(z + 1): Retention time of n-alkanes with z + 1 carbon atoms.

3. Results and Discussion

3.1. *Influence of Process Parameters on the Finished Beer*

In order to discuss the effects of wort concentration and sucrose syrup addition ratio on the indexes during beer fermentation, a brewing one-factor optimization experiment was carried out to ensure that other conditions were the same.

The above Table 1 compares all LSs brewed at different concentrations. The sugar cane juice and the wort itself were close in color, both being clear amber–yellow, and there was little difference in the appearance of gloss and color when comparing the different percentages of sugar juice added during the pre-fermentation period. The raw wort concentration had a significant effect on the alcohol, total acid, residual sugar, color, and bitterness values of the beers, which derived their bitterness mainly from hop resins and, to a much lesser extent, from non-hop components, such as bittering peptides, amino acids, and polyphenols produced during the brewing process [25]. Saccharomyces cerevisiae takes up the amino acids present in the wort, which they bring with them from the wort with an amino group so that they can be integrated into its own structure. The remaining amino acids (α-keto acids) enter into an irreversible chain reaction that culminates in the formation of the byproducts, higher alcohols [3]. A moderate amount of higher alcohol in beer will give it a rich, full-bodied feel, but if the level is too high, it will give the beer a serious "after-bitter" taste and cause the drinker to go "on the head". Esters are the most important aroma substances produced by yeast. They have a very low odor threshold in beer and greatly determine the final aroma [26]. Beer produces organic acids under aerobic conditions, a pathway thought to be a byproduct of the tricarboxylic acid cycle [26]. As the amount of organic acids produced by this process increases, the taste of the beer develops an unpleasant "sourness" that degrades the quality of the beer [27].

Table 1. Effect of different sugarcane juice additions on fermentation of sugarcane beer (original wort of the beer is 12°P, fermentation at 12 °C).

Addition of Sugar Cane Syrup	Physical and Chemical Indicators Related to Sugar Cane Beer						
	Alcoholic Strength (%)	Total Acidity (mL/100 mL)	pH	Residual Sugar (g/L)	Color (EBC)	Bitterness Value (BU)	Senior Alcohol (mg/L)
OWBL (10°P)	$5.6^{d} \pm 0.3$	$1.96^{c} \pm 0.10$	$4.77^{c} \pm 0.02$	$9.41^{b} \pm 0.02$	$13.35^{bc} \pm 0.32$	$19.8^{a} \pm 1.0$	$116.67^{a} \pm 22.53$
5%	$6.0^{c} \pm 0.1$	$1.53^{f} \pm 0.11$	$4.78^{bc} \pm 0.01$	$9.55^{b} \pm 0.04$	$12.88^{c} \pm 0.36$	$18.0^{b} \pm 0.5$	$94.39^{b} \pm 18.64$
10%	$6.5^{b} \pm 0.1$	$1.53^{f} \pm 0.11$	$4.77^{bc} \pm 0.00$	$9.19^{b} \pm 0.13$	$13.61^{b} \pm 0.16$	$17.4^{b} \pm 1.4$	$97.00^{b} \pm 20.01$
15%	$6.1^{c} \pm 0.2$	$1.57^{ef} \pm 0.03$	$4.78^{bc} \pm 0.01$	$9.40^{b} \pm 0.21$	$12.78^{c} \pm 0.20$	$16.6^{bc} \pm 0.9$	$100.94^{bc} \pm 19.05$
20%	$6.2^{c} \pm 0.1$	$1.63^{de} \pm 0.01$	$4.80^{b} \pm 0.01$	$7.65^{c} \pm 1.23$	$13.05^{bc} \pm 0.67$	$16.3^{bc} \pm 0.5$	$100.83^{bc} \pm 24.56$
25%	$7.2^{a} \pm 0.1$	$1.67^{d} \pm 0.03$	$4.61^{d} \pm 0.00$	$8.72^{b} \pm 0.04$	$13.24^{bc} \pm 0.24$	$16.6^{bc} \pm 1.2$	$119.72^{a} \pm 16.88$
30%	$6.6^{b} \pm 0.1$	$2.30^{a} \pm 0.01$	$4.63^{d} \pm 0.01$	$9.25^{b} \pm 0.51$	$13.04^{bc} \pm 0.21$	$15.4^{c} \pm 0.6$	$108.45^{c} \pm 24.23$
50%	$6.7^{b} \pm 0.1$	$2.20^{b} \pm 0.00$	$4.83^{a} \pm 0.01$	$12.50^{a} \pm 0.51$	$17.6^{a} \pm 0.20$	$13.6^{d} \pm 0.9$	$105.88^{c} \pm 19.65$

(Note: The data in this table are expressed as the mean (n = 3) ± standard deviation, using Duncan's test, and the difference is significant if the superscript letters are different ($p < 0.05$)).

As can be seen from the Table 2, the beer with 25% addition had the highest alcohol content of 7.2%, and the sugar cane beer with 30% and 50% addition had alcohol contents of 6.6% to 6.7%, with higher residual sugar contents of 9.25 g/L and 12.50 g/L, respectively. The yeast had limited metabolic capacity in the post-fermentation period (uncoordinated carbon-to-nitrogen ratio) and could not fully utilize the supplemented carbon source, resulting in a high residual sugar content in the beer. Experiments have shown that the addition of juice in the range of less than 25% has good fermentability and low residual sugar. The total acidity does not correspond quantitatively with the pH value because the beer contains a lot of H^+ in a non-free state. With a sugar cane juice pH of 5.13 and the addition of 5% to 25% juice, the total acidity of the beer ranged from 1.53 to 1.67 mL/100 mL, which was lower than the 1.96 mL/100 mL of OWBL, and exceeding 25% juice addition increase the acidity of the beer. From the table, it seems that the amount of juice addition was not linearly related to the bitterness value and color of the beer, which were related to the fermentation metabolic state of the brewer's yeast. After the sensory panel tasting, the fruit aroma of the added sugar cane juice increased with the addition ratio; as the fruit aroma increased, the sweetness of the beer became stronger, and the corresponding hop flavor became lighter. Sugarcane syrup addition and advanced alcohol content were nonlinearly correlated. We were unable to determine the relationship between the amount of syrup added and the content of higher alcohols. When 20% sugar cane juice was added, the fruit aroma and hop aroma were more harmonious, and the taste was suitable. Comparing the different concentrations of LSs in the table, the raw wort concentration had a significant effect on the alcoholic strength, total acid, residual sugar, color value, and bitterness value of the beers. The bitterness value decreased with increasing concentration because the wort concentration affected the leaching of bitter substances from hops during boiling. As the wort concentration increased, the ethanol content of the brew increased, and the higher alcohol content also increased. Measuring the sensory data and the advanced alcohol content, the preference was higher for 10°P pale lager.

Table 2. Effect of different wort concentrations on fermentation of sugarcane beer (beer with 20% syrup addition, fermentation at 12 °C).

Wort Consistency	Physical and Chemical Indicators Related to Sugar Cane Beer						
	Alcoholic Strength (%)	Total Acidity (mL/100 mL)	pH	Residual Sugar (g/L)	Color (EBC)	Bitterness Value (BU)	Senior Alcohol (mg/L)
6°P	$3.6^{e} \pm 0.1$	$1.21^{c} \pm 0.10$	$4.48^{b} \pm 0.13$	$5.60^{e} \pm 0.41$	$7.54^{c} \pm 0.59$	$17.1^{a} \pm 0.1$	$92.32^{a} \pm 17.75$
8°P	$4.4^{d} \pm 0.0$	$1.23^{c} \pm 0.03$	$4.54^{b} \pm 0.04$	$7.52^{d} \pm 0.23$	$10.40^{bc} \pm 1.13$	$15.6^{b} \pm 0.0$	$97.67^{a} \pm 20.54$
10°P	$5.2^{c} \pm 0.0$	$1.49^{bc} \pm 0.41$	$4.62^{ab} \pm 0.10$	$10.13^{c} \pm 0.43$	$12.42^{b} \pm 0.66$	$14.2^{c} \pm 0.1$	$103.48^{ab} \pm 25.12$
12°P	$6.2^{b} \pm 0.0$	$1.88^{ab} \pm 0.12$	$4.73^{a} \pm 0.01$	$12.54^{b} \pm 0.30$	$13.65^{b} \pm 0.42$	$11.9^{d} \pm 1.1$	$106.35^{ab} \pm 18.04$
14°P	$7.2^{a} \pm 0.0$	$2.19^{a} \pm 0.24$	$4.63^{ab} \pm 0.10$	$15.41^{a} \pm 0.69$	$18.58^{a} \pm 4.48$	$9.2^{e} \pm 0.5$	$119.2^{b} \pm 18.91$
OWBL (10°P)	$5.6^{d} \pm 0.3$	$1.96^{c} \pm 0.10$	$4.77^{c} \pm 0.02$	$9.41^{b} \pm 0.02$	$13.35^{bc} \pm 0.32$	$19.8^{a} \pm 1.0$	$114.25^{b} \pm 22.06$

(Note: The data in this table are expressed as the mean (n = 3) ± standard deviation, using Duncan's test, and the difference is significant if the superscript letters are different ($p < 0.05$)).

The fermentation cycle of pale lagers fermented at different temperatures varied, As expressed in Table 3. With the shortest fermentation time at 14 °C requiring only 6 d for primary fermentation (high-foam phase), while 8 °C and 12 °C required 8–12 d for primary fermentation. The higher the temperature, the better the utilization of the carbon source, and the lowest residual sugar was obtained, with a residual sugar content of 8.06 g/L for the beer at 14 °C and 11.28 g/L for the beer at 8 °C. The temperature had an effect on the production of higher alcohols in the beer. The fermentation at 8 °C was slow, and the highest residual sugar content was 11.28 g/L. Temperature had an effect on the production of higher alcohols in beer, and from the table, it can be seen that higher temperatures produced higher levels of higher alcohols than OWBLs. At the same time, the concentration of higher alcohols also increases, and too many higher alcohols reduce the quality of beer. Similarly, high fermentation temperatures can lead to the production of organic acids, which can reduce the quality of beer.

Table 3. Effect of different fermentation temperatures on fermentation of sugarcane beer (beer had a wort strength of 10°P and syrup addition of 20%).

Temp	Physical and Chemical Indicators Related to Sugar Cane Beer						
	Alcoholic Strength (%)	Total Acidity (mL/100 mL)	pH	Residual Sugar (g/L)	Color (EBC)	Bitterness Value (BU)	Higher Alcohol (mg/L)
10 °C	$5.3^{b} \pm 0.0$	$1.60^{b} \pm 0.01$	$4.82^{a} \pm 0.01$	$11.28^{a} \pm 0.51$	$14.3^{b} \pm 0.30$	$20.5^{a} \pm 1.0$	$90.24^{a} \pm 19.23$
12 °C	$5.3^{b} \pm 0.1$	$1.60^{b} \pm 0.01$	$4.68^{a} \pm 0.01$	$10.13^{b} \pm 0.25$	$14.7^{ab} \pm 0.30$	$17.3^{b} \pm 1.5$	$109.15^{b} \pm 18.69$
14 °C	$5.7^{a} \pm 0.0$	$1.80^{a} \pm 0.10$	$4.86^{a} \pm 0.01$	$8.06^{c} \pm 0.31$	$15.20^{a} \pm 0.10$	$19.05^{ab} \pm 1.1$	$124.49^{c} \pm 22.72$

(Note: The data in this table are expressed as the mean (n = 3) ± standard deviation, using Duncan's test, and the difference is significant if the superscript letters are different ($p < 0.05$)).

3.2. Determination of the Optimal Fermentation Process

The addition of hops can bring bitterness to the beer and enrich its flavor hierarchy. The bitterness of beer mainly comes from hop resins, and a small portion of it comes from non-hop components, such as bittering peptides, amino acids, polyphenols, and so on, during the brewing process, among which iso-alpha-acids contribute the most to the bitterness of beer [28]. And the addition of hops can clarify the wort by complexing and precipitating the proteins in the wort during the boiling process. Various concentrations of hops were added to each brew, and the sensory evaluation results showed that the bittering value of the beer was most acceptable in the range of 13–15, which corresponded to hops addition of 0.04%.

The eligible samples were selected for response orthogonal experiments, which required the basic indexes of brewing beer: the advanced alcohol value was between 50 and 120 mg/L, the normal value of diacetyl was between 0 mg/L and 0.1 mg/L, the acidity was $\leq$2.6 mg/100 mL, the mass fraction of carbon dioxide was above 3.5, and the alcohol content was $\geq$3.7%. In the sensory evaluation experiments, we found higher sensory evaluation scores for beers that better met the beer rating criteria, so we chose the total sensory tasting score as the only criterion for evaluating the quality of beer for the response surface experiments, and we conducted the response surface experiments with the wort concentration, the amount of syrup added, amount of hops added, and the fermentation temperature as the factors for the experimental experiments. (The sensory evaluation criteria can be found in the Supplementary Information). The results of the response orthogonal experiment are as Table 4:

Table 4. Orthogonal experimental factors and levels.

Level	Factors			
	Amount of Syrup Added (%)	Wort Consistency (°P)	Fermentation Temperature (°C)	Hops Added (%)
1	10	8	8	0.02
2	20	10	10	0.04
3	30	12	12	0.06

According to the R values in Table 5, the main and secondary influences of individual factors on sensory evaluation are: D > A > B > C, i.e., hop addition > wine addition > wine concentration. That is to say, the amount of hop addition > the amount of juice addition > the concentration of original wort > the fermentation temperature. According to the K of each factor, the optimized condition is $A_2B_2C_2D_2$. According to the experimental verification, it was unanimously evaluated that 10°P LS had better color and taste, which meant that $A_2B_2C_2D_2$ was the optimal condition, which is in line with the results of the one-way experiment. $A_2B_2C_2D_2$ was concluded to be the optimal fermentation condition. The final results were: 10°P wort concentration as the base, addition of 20% sugarcane syrup, fermentation temperature of 12 °C, and an ale hops addition of 0.04%.

Table 5. Orthogonal experiment results.

Experiment Number	A	B	C	D	Sensory Evaluation
1	1	1	1	1	76
2	1	2	2	2	91
3	1	3	3	3	69
4	2	1	2	3	72
5	2	2	3	1	87
6	2	3	1	2	86
7	3	1	3	2	78

Table 5. *Cont.*

Experiment Number	A	B	C	D	Sensory Evaluation
8	3	2	1	3	62
9	3	3	2	1	73
K_1	236	226	224	236	
K_2	245	240	236	255	
K_3	213	228	234	203	
k_1	78.667	75.333	74.667	78.667	
k_2	81.667	80.000	78.667	85.000	
k_3	71.000	76.000	78.000	67.667	
R	10.667	4.667	4.000	17.333	

The basic physicochemical indexes of sugarcane lager beer and OWBL under the same conditions are listed in the following Table 6.

Table 6. Final physical and chemical indicators for Ls and OWBL.

Project	Standards	LS	OWBL (10°P)
Alcoholic strength (%)	10.1°P–11.0°P $\geq$ 3.7	4.9–5.3	4.7–4.9
Total acidity (mg/100 mL)	10.1°P–14.0°P $\leq$ 2.6	1.0–2.0	1.6–2.4
Diacetyl (mg/L)	$\leq$0.10	0.07	0.09
CO_2 (mass fraction)	0.35–0.65	0.40	0.45
Higher alcohol (mg/L)	50–120	109.15	105.42
Amount of hops added (%)	Bitterness value $13 \leq x \leq 15$	14.01	14.9

3.3. Changes in Substances during Fermentation

As the same fermentation conditions were used (different fermentation substrates (carbon and nitrogen sources)), the resulting beers had different basic indicators. The carbon and nitrogen sources directly influence the rate of yeast fermentation and the expression of the entire flavor profile of beer (alcohols, esters, aldehydes, and ketones), which in turn affect the overall flavor profile of beer [1]. The analysis of the fermentation substrate depletion in lagers is an essential step in judging the quality of lager beers. Two main changes occur during the fermentation of lager: the conversion of nitrogenous compounds and the fermentation of sugars. Other by-products are also produced along with these two changes.

3.3.1. Fermentable Sugars

Comparing the sugar consumption during the fermentation of normal beer and LS, the results of the sugar spectrum analysis for 10°P wort were 2.43 g/L for fructose, 8.39 g/L for glucose, 1.42 g/L for sucrose, 80.60 g/L for maltose, and 5.28 g/L for maltotriose. High-performance liquid-phase sugar analysis for 10°P sugarcane syrup juice raw material showed that the fructose content was 0.79 g/L, the glucose content was 0.63 g/L, and the sucrose content was 11.58 g/L. The main sugar in the sugarcane syrup was sucrose.

As shown in Table 7, yeast itself will preferentially utilize sucrose [29], which breaks down the non-reducing sugar (sucrose) in wort into one molecule of fructose and one molecule of glucose under the action of sucrose converting enzyme. The addition of sucrose from sucrose juice increases the amount of glucose, and glucose deterrence prevents the expression of the MAL gene (which affects the uptake of maltose and maltotriose) when the glucose concentration exceeds 1% and allows the yeast to fully utilize the glucose, with a small amount of fructose remaining at the end of fermentation; fructose is twice as sweet as sucrose per mole, and the higher the fructose residue, the sweeter the final beer product; 10°P LS has a lower maltose content than OWBL, and the addition of cane syrup results in a change in the ratio of sugars that causes the residual fructose and sucrose to rise.

Table 7. Measurement of the amount of fermentable sugars before and after fermentation.

	Glucose (g/L)	Fructose (g/L)	Maltose (g/L)	Sucrose (g/L)	Maltotriose (g/L)
10°P Wort	8.39 ± 0.26	2.43 ± 0.11	80.6 ± 6.8	1.42 ± 0.02	5.28 ± 0.42
10°P Cane Juice	0.63 ± 0.03	0.79 ± 0.05	0	11.58 ± 1.35	0
10°P OWBL (mg/100 mL)	0.68 ± 0.02	0.12 ± 0.01	3.07 ± 0.22	0.87 ± 0.03	0.29 ± 0.01
LS (10°P substrate) (mg/100 mL)	1.22 ± 0.03	0.36 ± 0.01	2.62 ± 0.12	0.92 ± 0.12	0.38 ± 0.01

3.3.2. Free Amino Nitrogen

Wort contains a large amount of nitrogenous substances, but not all of these substances are utilized by yeast in cellular metabolic pathways. Nitrogen-containing compounds in wort are mainly composed of free amino nitrogen, ammonia ions, and small molecular peptides (dipeptides and tripeptides), and yeast produces different volatile aromas due to the utilization of different ratios of amino acids, i.e., amino nitrogen is the main source of variations in aroma substances in beer.

The Table 8 shows the amino acid contents in LS, OWBL, wort, and sugarcane syrup. It can be inferred that LS has a higher and more comprehensive utilization of amino nitrogen as compared with OWBL. The amino nitrogen content of LS was significantly lower than that of normal whole-barley beer. It is hypothesized that this is due to the addition of sugarcane syrup at the fermentation stage altering the yeast metabolic pathway, allowing the yeast to newly utilize amino acids (e.g., histidine) that are not readily available in OWBL. LS utilizes amino acids at a higher rate than OWBL. Fewer amino acids remain, which makes LS less susceptible to contamination by spoilage bacteria that utilize amino acids for growth, such as lactobacilli and lactococci. Fruit juices contain 19 of the 20 essential amino acids (without cysteine), and just as wort utilizes sugar, the amino acids are taken up by yeast cells in the order in which they are taken up [30]. The amino acids of yeast are classified into two groups (a, b) based on the percentage of amino acids utilized by the yeast; the a group of amino acids is more readily utilized by the yeast and utilized in a higher percentage, and the b group of amino acids is utilized in a smaller percentage. It can be said that the addition of sugarcane syrup can promote the utilization of b-amino acids by yeast and change the fermentation pattern of yeast, thus producing different flavor substances.

Table 8. Amount of change in free amino acids and available nitrogen sources before and after fermentation.

Amino Acid Type	LS (10°P Substrate) (mg/100 mL)	10°P OWBL (mg/100 mL)	Control Wort (mg/100 mL)	Sugar Cane Juice (mg/100 mL)
Asp	5.85 ± 0.47	7.22 ± 0.64	9.17 ± 0.78	13.45 ± 1.04
Thr	0.93 ± 0.04	2.13 ± 0.18	6.79 ± 0.47	2.92 ± 0.34
Ser	1.91 ± 0.16	2.35 ± 0.28	6.88 ± 5.35	8.41 ± 0.96
Glu	14.29 ± 1.54	8.03 ± 0.78	13.24 ± 1.64	81.56 ± 8.13
Gly	2.96 ± 0.13	4.16 ± 0.31	3.88 ± 0.28	0.79 ± 0.04
Ala	8.99 ± 0.68	12.2 ± 1.51	11.2 ± 1.67	7.35 ± 0.76
Cys	0.24 ± 0.03	0.37 ± 0.02	0.54 ± 0.06	0.11 ± 0.01
Val	2.54 ± 0.03	8.72 ± 0.79	12.32 ± 1.36	5.25 ± 0.04
Met	0.47 ± 0.04	2.12 ± 0.31	4.33 ± 0.42	0.64 ± 0.05
Ile	0.76 ± 0.06	4.02 ± 0.36	7.54 ± 0.69	2.83 ± 0.29
Leu	0.78 ± 0.08	7.12 ± 0.82	16.60 ± 1.72	2.16 ± 0.23
Tyr	6.06 ± 0.53	10.78 ± 1.32	12.96 ± 1.32	2.26 ± 0.19
Phe	2.77 ± 0.31	11.26 ± 1.41	16.33 ± 1.75	2.12 ± 0.18
Lys	0 ± 0	5.56 ± 0.53	10.07 ± 0.99	0.79 ± 0.09
NH3	1.46 ± 0.14	2.56 ± 0.28	3.12 ± 0.35	0.68 ± 0.07
His	1.49 ± 0.11	4.15 ± 0.40	5.61 ± 0.61	2.67 ± 0.19
Arg	1.19 ± 0.09	8.49 ± 0.72	11.7 ± 1.25	1.34 ± 0.13
Hyp	1.53 ± 0.08	2.81 ± 0.19	7.94 ± 0.81	3.19 ± 0.27
Pro	35.81 ± 2.67	41.71 ± 5.96	41.98 ± 3.78	3.16 ± 0.21

3.4. Detection of Aroma Substances in Beer

3.4.1. The Electronic Nose

Principal component analysis (PCA) and latent Dirichlet allocation (LDA) were chosen for the analysis of e-nose data, where beer has a wide variety of volatiles, and PCA was used to analyze different broad categories of flavor substances in lagers, while LDA was used to determine the variability of the broad categories of substances among beers.

From the figure, it can be seen that the contribution rates of the first and second principal components were 73.51% and 20.49%, respectively, and the total contribution rate was 94%, which can represent the characteristic information of the broad categories of volatiles of the samples. The flavor differences between the laboratory lager and the commercially available lagers were evident in PC_2, while PC_1 was close to that of the commercially available lager. The difference between PC_1 and PC_2 of the sugarcane lager and those of the ordinary and commercially available lagers was significant. This proves that the aroma of sugarcane lager is different from that of commercially available lager.

The aroma of sugarcane lager is different from that of commercially available lager and laboratory lager, which is mainly due to the different raw materials and fermentation processes, as represented in Figure 1. The direction chosen by PCA maintains maximum structure between data in lower dimensions, while the direction chosen by LDA achieves maximum separation between the given classes [30]. The LDA classification results are more representative of the variability than the PCA results. The data collection points of the same type of lager under the same conditions in the ellipse in Figure 1b represent the fingerprint profile of that product type, and the more dense data points represent the higher repetition rate of that sample. As can be seen from the figure, the contribution of LDA discriminant LD_1 and LD_2 for lager samples were 98.64% and 0.93%, respectively, and the total sum was 99.57%, which can represent all the characteristic information of the samples. The data collected from each type of lager samples were distributed in different areas without overlapping and far away from each other; this indicates that there were significant differences in odor. This indicates that there were differences in the flavor quality of lagers brewed from different raw materials.

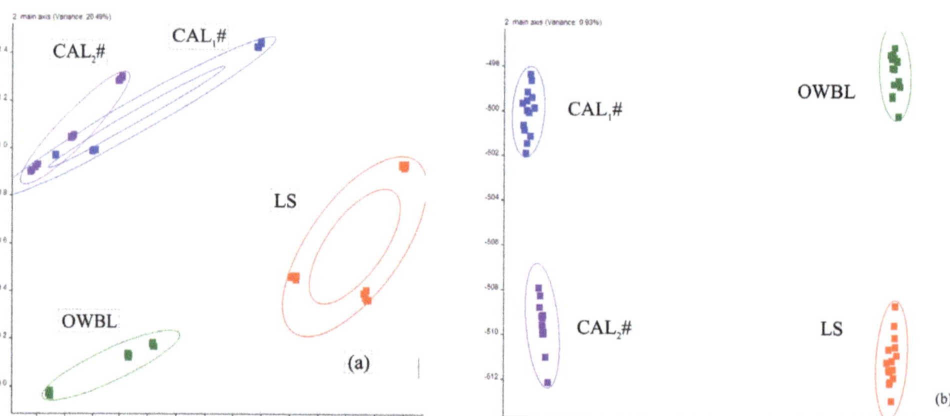

Figure 1. Electronic nose fingerprints of laboratory lager and market lager. (**a**) Principal component analysis results; (**b**) LDA analysis results.

As can be seen from the Figure 2, all the lighter samples had the highest response values for the three sensor types W5S, W1S, W1W, and W2W. The high response value for ammonia–oxygen compounds is due to the high sensitivity of the detector itself. The substances corresponding to inorganic sulfides in lager are sulfides, such as SO_2, which is a metabolic product of yeast in the fermentation process. The amount of SO_2 produced is related to the yeast strain, wort concentration, and fermentation process [31]. The high response value of the corresponding methyl group sensor in the figure is mainly due to the high content of 2-methyl sulfide substances in lager. In nature, 2-methyl sulfide is often produced by the decomposition of proteins and has a "green" and "fruity" aroma, i.e., sugarcane syrup can provide lager a high "green" and "fruity" aroma, and "fruity" aroma.

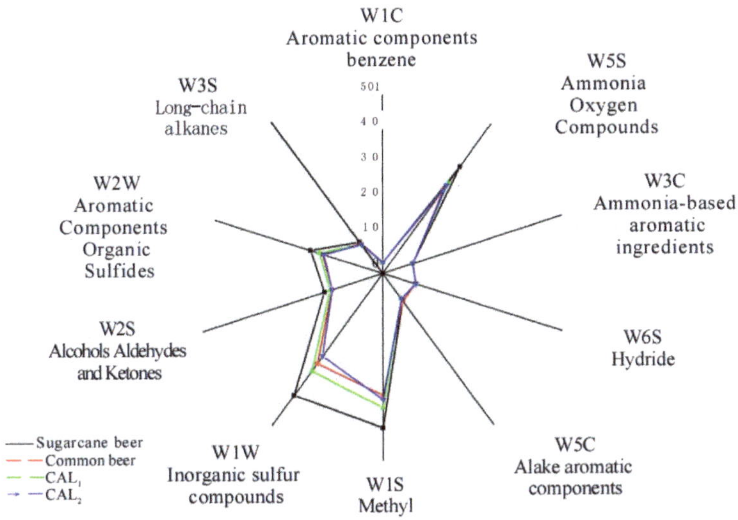

Figure 2. Electronic nose radar map of home-brewed lager in the laboratory and lager on the market.

3.4.2. GC-MS

There were a total of 77 compounds in the GC-MS assay, and the beer compounds detected by GC-MS were divided into three categories: compounds with no significant difference, compounds with more LS than OWBL, and compounds with less LS than OWBL, which were added to the commercially available beer assay to test for volatiles of roughly the same composition as the commercially available beer. The reason for adding the commercially available beer was to test the volatiles to ensure that the brewing process was approximately the same as the commercially available pale lager. In this experiment, substances specific to a single commercially available lager were ignored, thus minimizing the effects of the use of different hop types and different fermentation processes, as well as different flavor substances produced by different brewing yeasts. The results are shown in Table 9.

Table 9. Volatile components of homemade LS, OWBL, and market beer.

NO	Volatile Compounds	RI	CAS	Relative Content (%)				Aroma Characteristic
				OWBL	LS	CAL_1	CAL_2	
1	Acetaldehyde	710	75-07-0	0.082 [a] ± 0.014	NQ	0.031 [b] ± 0.006	NQ	Pungent, ethereal, aldehydic, fruity
2	Isobutyl acetate	997	110-19-0	0.038 [a] ± 0.007	NQ	0.068 [b] ± 0.014	0.141 [c] ± 0.020	Sweet, fruity, ethereal, banana, tropical
3	Ethyl valerate	1118	539-82-2	0.028 [a] ± 0.004	NQ	NQ	NQ	Sweet, fruity, apple, pineapple, green, tropical
4	2-Heptanone	1126	110-43-0	0.011 [a] ± 0.001	NQ	NQ	NQ	Fruity, spicy, sweet, herbal, coconut, woody
5	Dipentene	1140	7705-14-8	0.046 [a] ± 0.007	NQ	0.033 [b] ± 0.009	NQ	Citrus, herbal, terpene, camphor
6	Methylheptenone	1295	110-93-0	0.024 [a] ± 0.004	NQ	NQ	0.024 [a] ± 0.005	Citrus, green, musty Lemongrass, apple
7	2-Nonanone	1385	821-55-6	0.045 [a] ± 0.007	0.021 [b] ± 0.004	0.026 [b] ± 0.003	0.026 [b] ± 0.003	Fresh, sweet, green, weedy, earthy, herbal
8	Decanal	1545	112-31-2	1.152 [a] ± 0.326	0.141 [b] ± 0.003	0.187 [b] ± 0.002	0.511 [c] ± 0.011	Sweet, aldehydic, waxy, orange peel, citrus, floral
9	Octanol	1597	111-87-5	0.895 [a] ± 0.178	NQ	0.791 [a] ± 0.136	0.867 [a] ± 0.147	Waxy, green, orange, aldehydic, rose, mushroom
10	Undecanal	1639	112-44-7	0.062 [a] ± 0.014	NQ	NQ	NQ	Waxy, soapy, floral, aldehydic, citrus, green, fatty, fresh laundry
11	Ethyl 4-trans-decenoate	1729	76649-16-6	0.372 [a] ± 0.046	NQ	NQ	NQ	Green, fruity Waxy, cognac
12	Acetic acid, decyl ester	1744	112-17-4	0.036 [a] ± 0.005	NQ	NQ	0.057 [b] ± 0.009	Waxy, clean, fresh laundered cloths, citrus, soapy
13	(E)-methyl geranate	1759	1189-09-9	0.617 [a] ± 0.092	0.386 [b] ± 0.063	0.127 [c] ± 0.021	0.325 [b] ± 0.057	Waxy, green Fruity, flower
14	Alpha-terpineol	1762	10482-56-1	0.043 [a] ± 0.006	0.02 [b] ± 0.003	0.038 [a] ± 0.004	0.095 [c] ± 0.013	Lilac, floral, terpenic
15	Cyclooctane	1799	292-64-8	0.987 [a] ± 0.167	NQ	0.377 [b] ± 0.073	NQ	Camphor odor
16	Phenethyl isobutyrate	1956	103-48-0	0.057 [a] ± 0.011	NQ	NQ	NQ	Floral, fruity, rose, tea, peach, pastry
17	2-Acetylpyrrole	1978	1072-83-9	0.046 [a] ± 0.008	NQ	NQ	NQ	Musty, nut, skin, maraschino cherry, coumarinic, licorice, walnut, bready
18	Gamma-nonanolactone	2025	104-61-0	0.053 [a] ± 0.012	NQ	NQ	NQ	Coconut, creamy, waxy, sweet, buttery, oily
19	Ethyl acetate	891	141-78-6	1.638 [a] ± 0.347	1.761 [a] ± 0.331	2.472 [b] ± 0.421	2.375 [b] ± 0.397	Ethereal, fruity, sweet, weedy, green
20	Ethanol	895	64-17-5	14.774 [b] ± 2.358	12.073 [a] ± 1.872	18.836 [d] ± 2.468	13.821 [ab] ± 2.181	Strong alcoholic, ethereal, medical
21	Ethyl butyrate	1045	105-54-4	0.248 [a] ± 0.026	0.214 [b] ± 0.031	0.268 [c] ± 0.038	0.217 [b] ± 0.0025	Fruity, juicy, fruit, pineapple, cognac
22	Propan-1-ol	1059	71-23-8	0.080 [a] ± 0.012	0.065 [b] ± 0.011	0.083 [a] ± 0.012	0.090 [c] ± 0.021	Alcoholic, fermented fusel, musty
23	Ethyl isovalerate	1062	108-64-5	NQ	NQ	NQ	0.011 [a] ± 0.001	Fruity, sweet, apple, pineapple, tutti frutti
24	Isobutyl isobutyrate	1068	97-85-8	NQ	NQ	NQ	0.02 [a] ± 0.002	Ethereal, fruity, tropical fruit, pineapple, grape skin, banana
25	Isobutyl alcohol	1078	78-83-1	0.506 [a] ± 0.083	0.620 [b] ± 0.096	0.504 [a] ± 0.092	0.632 [b] ± 0.101	Ethereal, winey, cortex
26	Isoamyl acetate	1110	123-92-2	4.213 [a] ± 0.7235	6.493 [b] ± 0.925	10.603 [c] ± 1.532	12.447 [cd] ± 1.826	Sweet, fruity, banana, solvent
27	Myrcene	1122	123-35-3	0.025 [a] ± 0.003	0.019 [b] ± 0.002	0.029 [c] ± 0.003	0.048 [d] ± 0.005	Peppery, terpene, spicy, balsam, plastic
28	Isopentyl isobutyrate	1176	2050-01-3	0.018 [a] ± 0.002	NQ	0.123 [b] ± 0.025	0.166 [c] ± 0.037	Fruity, ethereal, tropical, green grape, cherry, unripe banana, apple, cocoa
29	3-Methyl-1-butanol	1200	123-51-3	13.623 [a] ± 1.793	11.607 [b] ± 1.467	13.146 [a] ± 1.983	11.17 [b] ± 1.249	Fusel, oil, alcoholic, whiskey, fruity, banana
30	2-Pentylfuran	1213	3777-69-3	NQ	NQ	NQ	0.01 [a] ± 0.001	Fruity, green, earthy, beany, vegetable, metallic
31	Ethyl caproate	1219	123-66-0	4.851 [a] ± 0.865	3.433 [b] ± 0.547	6.19 [c] ± 1.223	4.319 [ab] ± 0.792	Sweet, fruity, pineapple, waxy, green, banana
32	3,7-Dimethyl-1	1221	13877-91-3	NQ	NQ	NQ	0.018 [a] ± 0.003	Citrus, tropical, green, terpene, woody, green
33	Hexyl acetate	1228	142-92-7	0.079 [a] ± 0.013	0.135 [b] ± 0.028	0.149 [bc] ± 0.033	0.02 [d] ± 0.003	Fruity, green, apple, banana, sweet
34	2-Methylbutyl 2-methylbutyrate	1230	2445-78-5	NQ	NQ	NQ	0.024 [a] ± 0.003	Sweet, fruity, ester, berry, green, waxy, apple
35	Hexyl methyl ketone	1241	111-13-7	0.075 [a] ± 0.013	0.018 [b] ± 0.003	0.066 [c] ± 0.012	0.066 [c] ± 0.013	Earthy, weedy, natural, woody, herbal
36	Isopentyl isopentanoate	1273	659-70-1	NQ	NQ	NQ	0.035 [a] ± 0.005	Sweet, fruity, green, ripe, apple, jammy, tropical
	Ethyl heptanoate	1281	106-30-9	0.151 [a] ± 0.024	0.100 [b] ± 0.013	0.116 [ba] ± 0.015	0.274 [c] ± 0.044	Fruity, pineapple, cognac, rum wine
38	Hexanol	1299	111-27-3	0.059 [a] ± 0.009	0.039 [b] ± 0.006	0.037 [b] ± 0.005	NQ	Ethereal, fusel, oil, fruity, alcoholic, sweet, green
39	(Z)-4-Heptenal	1331	6728-31-0	NQ	NQ	0.027 [a] ± 0.004	0.035 [b] ± 0.006	Oily, fatty, green, dairy, milky, creamy
40	2-Hthylhexyl acetate	1369	103-09-3	0.021 [a] ± 0.003	0.020 [a] ± 0.003	NQ	NQ	Earthy, herbal, humus, undergrowth

Table 9. *Cont.*

NO	Volatile Compounds	RI	CAS	Relative Content (%)				Aroma Characteristic
				OWBL	LS	CAL_1	CAL_2	
41	Nonyl aldehyde	1391	124-19-6	NQ	NQ	$0.073^{a} \pm 0.012$	$0.197^{b} \pm 0.037$	Waxy, aldehydic, rose, fresh, orris, orange peel, fatty, peely
42	Ethyl caprylate	1430	106-32-1	$19.674^{a} \pm 3.85$	$19.791^{a} \pm 4.03$	$17.016^{ab} \pm 3.42$	$15.357^{b} \pm 2.86$	Fruity, wine, waxy, sweet, apricot, banana, brandy, pear
43	Furfural	1482	98-01-1	NQ	NQ	$0.321^{a} \pm 0.064$	$0.144^{b} \pm 0.022$	Sweet, woody, almond, fragrant, baked bread
44	Octyl acetate	1503	112-14-1	$0.197^{a} \pm 0.033$	$0.217^{a} \pm 0.029$	$0.414^{b} \pm 0.061$	$0.687^{c} \pm 0.106$	Green, earthy, mushroom, herbal, waxy
45	Camphor	1558	76-22-2	NQ	$0.019^{a} \pm 0.003$	NQ	NQ	Camphoreous
46	2-Nonanol	1567	628-99-9	$0.140^{a} \pm 0.028$	$0.056^{b} \pm 0.009$	$0.034^{c} \pm 0.005$	NQ	Waxy, green, creamy, citrus, orange, cheese, fruity
47	Ethyl nonanoate	1578	123-29-5	$0.269^{a} \pm 0.051$	$0.321^{a} \pm 0.063$	$0.098^{b} \pm 0.0156$	$0.054^{c} \pm 0.008$	Fruity, rose, waxy, rum, wine, natural, tropical
48	Linalool	1588	78-70-6	$0.539^{a} \pm 0.088$	$0.453^{b} \pm 0.072$	$0.474^{b} \pm 0.078$	$0.680^{c} \pm 0.117$	Citrus, floral, sweet, bois de rose, woody, green, blueberry
49	Octanoicacid	1592	5461-6-3	$0.017^{a} \pm 0.003$	$0.035^{b} \pm 0.006$	NQ	NQ	Fruity, green, oily, floral
50	Carbonochl	1633	7452-59-7	NQ	$0.586^{a} \pm 0.089$	NQ	NQ	Sugar cane aroma, fruity
51	2-Decanol	1656	1120-06-5	$0.030^{a} \pm 0.005$	$0.017^{b} \pm 0.002$	NQ	$0.051^{c} \pm 0.009$	NF
52	Ethyl caprate	1690	110-38-3	$6.257^{a} \pm 1.331$	$4.795^{b} \pm 0.874$	$1.401^{c} \pm 0.258$	$2.275^{c} \pm 0.367$	Sweet, waxy, fruity, apple, grape, oily, brandy
53	Acetophenone	1705	98-86-2	NQ	$0.032^{a} \pm 0.005$	NQ	NQ	Sweet, pungent, hawthorn, mimosa, almond, acacia, chemical
54	Myrcene	1721	123-35-3	NQ	NQ	$0.357^{a} \pm 0.053$	$0.191^{b} \pm 0.031$	Peppery, terpene, spicy, balsam, plastic,
55	Ethyl benzoate	1739	93-89-0	$0.082^{a} \pm 0.012$	$0.091^{b} \pm 0.014$	$0.056^{c} \pm 0.008$	$0.064^{d} \pm 0.009$	Fruity, dry, musty, sweet, wintergreen
56	ethyl 9-decenoate	1748	67233-91-4	$4.062^{a} \pm 0.879$	$3.96^{a} \pm 0.697$	$0.369^{b} \pm 0.652$	$0.725^{b} \pm 0.117$	Fruity, fatty
57	2-Dodecanol	1771	10203-28-8	$0.205^{a} \pm 0.031$	$0.139^{b} \pm 0.024$	NQ	NQ	NF
58	4-Ethylbenzaldehyde	1782	4748-78-1	NQ	$0.038^{a} \pm 0.005$	NQ	NQ	Bitter, almond, sweet, anise
59	Geranyl acetate	1790	105-87-3	$0.031^{a} \pm 0.005$	$0.029^{a} \pm 0.004$	$0.034^{a} \pm 0.005$	$0.030^{a} \pm 0.006$	Floral, rose, lavender, green, waxy
60	Decyl alcohol	1800	112-30-1	NQ	$0.678^{a} \pm 0.098$	NQ	$0.488^{b} \pm 0.078$	Fatty, waxy, floral, orange, sweet, clean, watery
61	Citronellol	1804	106-22-9	$0.508^{a} \pm 0.089$	$0.346^{b} \pm 0.064$	$0.352^{b} \pm 0.059$	$0.174^{c} \pm 0.031$	Floral, leather, waxy, rose bud, citrus
62	Ethyl phenylacetate	1807	101-97-3	$0.027^{a} \pm 0.004$	$0.049^{b} \pm 0.001$	$0.029^{a} \pm 0.005$	$0.036^{c} \pm 0.004$	Sweet, floral, honey, rose, balsam, cocoa
63	Nerol	1810	106-25-2	$0.028^{a} \pm 0.004$	$0.032^{ab} \pm 0.006$	NQ	$0.009^{c} \pm 0.001$	Sweet, natural, neroli, citrus, magnolia
64	Beta-damascenone	1816	23726-93-4	$0.087^{a} \pm 0.012$	NQ	$0.275^{b} \pm 0.040$	$0.333^{c} \pm 0.058$	Apple, rose, honey, tobacco, sweet
65	Ethyl laurate	1819	106-33-2	$0.159^{a} \pm 0.027$	$0.219^{b} \pm 0.031$	$0.019^{c} \pm 0.003$	NQ	Sweet, waxy, floral, soapy, clean
66	Geraniol	1821	106-24-1	$0.103^{a} \pm 0.027$	$0.054^{b} \pm 0.009$	$0.066^{c} \pm 0.010$	$0.095^{d} \pm 0.012$	Sweet, floral, fruity, rose, waxy, citrus
67	Geranylacetone	1825	689-67-8	$0.170^{a} \pm 0.030$	$0.068^{b} \pm 0.011$	$0.075^{b} \pm 0.013$	NQ	Fresh, rose, leaf, floral, green, magnolia, aldehydic, fruity
68	Trimethyl pentanyl diisobutyrate	1828	6846-50-0	NQ	$0.077^{a} \pm 0.013$	$0.039^{b} \pm 0.006$	NQ	NF
69	Ethyl hydrocinnamate	1930	2021-28-5	$0.070^{a} \pm 0.010$	$0.108^{b} \pm 0.017$	$0.072^{a} \pm 0.013$	$0.068^{a} \pm 0.013$	Hyacinth, rose, honey, fruity, rum
70	Butylated hydroxytoluene	1936	128-37-0	NQ	$0.013^{a} \pm 0.002$	$0.486^{b} \pm 0.071$	NQ	Mild, phenolic, camphor
71	DMS	767	75-18-3	$0.019^{ba} \pm 0.003$	$0.085^{c} \pm 0.011$	$0.010^{a} \pm 0.001$	$0.025^{b} \pm 0.004$	Sulfury, onion, sweet, corn, vegetable, cabbage, tomato, green, radish
72	Heptyl acetate	1356	112-06-1	$0.066^{a} \pm 0.012$	$0.113^{b} \pm 0.017$	$0.183^{c} \pm 0.030$	$0.134^{bc} \pm 0.023$	Fresh, green, rum, ripe, fruit, pear, apricot, woody
73	Benzyl alcohol	1571	100-51-6	$0.033^{a} \pm 0.005$	$0.077^{b} \pm 0.012$	$0.117^{c} \pm 0.020$	NQ	Floral, rose, phenolic, balsamic
74	Isoamyl octanoate	1725	2035-99-6	$0.101^{a} \pm 0.017$	$0.302^{b} \pm 0.048$	NQ	$0.015^{c} \pm 0.002$	Sweet, oily, fruity, green, soapy, pineapple, coconut
75	Phenethyl acetate	1813	103-45-7	$2.611^{a} \pm 0.415$	$8.201^{cb} \pm 1.311$	$6.086^{b} \pm 0.985$	$9.984^{c} \pm 1.837$	Floral, rose, sweet, honey, fruity, tropical,
76	Phenethyl alcohol	1941	60-12-8	$9.061^{a} \pm 1.364$	$12.49^{b} \pm 2.074$	$9.913^{a} \pm 1.693$	$10.503^{ab} \pm 1.795$	Floral rose, dried rose, flower, rose water
77	2,4-Di-t-butylphenol	2341	96-76-4	$0.127^{a} \pm 0.016$	$0.357^{b} \pm 0.063$	$0.125^{a} \pm 0.020$	$0.105^{c} \pm 0.018$	Phenolic

Note: Relative content data in this table are expressed as mean ($n = 5$) ± standard deviation, using Duncan's test, and differences in superscript letters are significant ($p < 0.05$), NQ means that the substance is not detectable.

It is the aroma profile of the LS that is illustrated in Table 9. The combination of GC-MS and an electronic nose can only characterize the composition of the flavor substances in beer and determine its flavor components. The characteristic aroma varies from lager to lager, and this aroma comes partly from the aroma of the raw materials and adjuncts themselves and partly from the yeast or metabolites produced during their fermentation. There are a total of 77 compounds in Table 9, and 51 substances were detected in sugarcane lager. Fifty-two compounds were found to be the same in the two commercially available lagers compared, of which esters and alcohols accounted for 80%, indicating that in sugarcane lager, alcohols and esters constitute the basic flavor; thirty-two compounds were common to the four lagers, and eleven substances were unique to sugarcane lager compared with the other three lagers, namely camphor, carbonochl, ethyl(E) cinnamate, acetophenone, 4-ethylbenzaldehyde, cyclododecane, ethyl myristate, m-phthalaldehyde, phenethyl camphor, carbonochl, and ethyl(E) cinnamate, which were found to be contained in the sugarcane syrup itself after comparison with the sugarcane syrup and wort components. These are the flavor substances of the sugarcane syrup itself and the flavor substances obtained by yeast fermentation of the sugarcane syrup. At the same time, compared with commercially available whole-barley lagers, laboratory whole-barley lagers produce unique substances, such as 2-acetylpyrrole, gamma-nonanolactone, etc. These substances can be considered to be produced by the overfermentation of wort by yeast under laboratory conditions, but they are not present in laboratory sugarcane lagers, and the possible reason for this analysis is because brewer's yeast can produce substances such as acetophenone by the joint action of wort and sugarcane syrup, but it is not possible to conclude that acetophenone, 4-ethylbenzaldehyde, cyclododecane, ethyl myristate, m-phthalaldehyde, phenethyl hexanoate, methanone, (4-ethylphenyl)phenyl-, and ethyl palmitate are produced by the yeast metabolism of sugarcane syrup or by yeast synergism of sugarcane syrup and wort. It is presumed that it is mainly produced by the yeast using amino acids (such as histidine) that are not readily available in the post-fermentation stage, combined with the sugar in the sugar cane syrup.

3.5. Determination of Lager Aging Characteristic Aroma

The comparison of aldehydes in lagers before and after aging is shown in Table 10: 2-furaldehyde, acetaldehyde, isobutyraldehyde, glutaraldehyde, phenylacetaldehyde, and 3-methylbutyraldehyde. 2-Furaldehyde was the main heat load indicator and was not detected in commercially available lager, sugar cane lager, or OWBL before aging, while 2-furaldehyde increased to varying degrees in all lagers after aging, which can be attributed to residual carbohydrate reactions, with sugar cane lager having significantly lower residual sugar than OWBL, which had the lowest production of furfuraldehyde. From the data in the table, it can be seen that the degree of accumulation of 2-furaldehyde in sugarcane lager was lower than those of OWBL and commercially available lager, i.e., sugarcane lager had a higher level of resistance to aging than ordinary lager, which contributed to its flavor stability. However, in an experiment to measure the antioxidant power of beer, we found that the antioxidant power of LS was about 10% lower than that of OWBL. It is hypothesized that this was due to the fact that the free amino acid residue at the completion of fermentation of LS was smaller than that of OWBL, resulting in a weaker oxidation of amino acids. The increase in phenylacetaldehyde in sugarcane lager from 1.861 μg/L to 10.744 μg/L was higher due to Strecker degradation of amino acids to form aldehydes, such as 2-methylpropionaldehyde (isobutyraldehyde) and phenylacetaldehyde [32]. But, whether it is the effect of sugarcane syrup remains to be studied. The increase in glutaraldehyde during lager aging is then due to the autoxidation of linoleic acid. During forced aging, the fruity aroma of the lager gradually fades at first and the hops flavor gradually disappears, followed by a strong "currant flavor" and "soy sauce flavor", and the longer the aging time, the heavier the "currant flavor". Saison [33] described the aging flavors after sensory panel evaluation as "cardboard taste", "metallic taste", solvent taste", "old hops taste", "cool chestnut taste", "merlot taste", "thioether taste ", "acetaldehyde" ("green apple"), and "white wine". The main substance of "cardboard flavor" is trans-2-nonenal (T2N) [23], but it was not detected in the present study. Some researchers have suggested that aging flavors differ by lager type [34], and the "cardboard taste" in lagers is not the only characteristic aging taste of lager.

Table 10. Comparison of major aged aldehydes in five kinds of lager after 9 days of aging storage at 45 °C.

	Aging-Related Substance Content (μg/L)					
	2-Furaldehyde	**Acetaldehyde**	**Isobutyraldehyde**	**Valeraldehyde**	**Phenylacetaldehyde**	**Total Aldehyde**
Unaged commercially available lager (CAL1)	NQ	3.04 ± 0.413	38.385 ± 3.125	5.386 ± 0.281	6.133 ± 0.818	52.944 [b] ± 4.696
Aged commercially available lagers (CAL_1)	66.887 ± 8.423	75.414 ± 9.052	49.914 ± 5.747	17.257 ± 1.425	8.996 ± 0.755	218.468 [a] ± 25.402
Unaged commercially available lager (CAL_2)	NQ	7.08 ± 0.629	15.394 ± 1.724	16.841 ± 2.218	4.823 ± 0.293	44.138 [b] ± 4.864
Unaged commercially available lager (CAL_2)	48.792 ± 5.466	92.348 ± 10.133	38.266 ± 2.635	25.978 ± 3.935	12.759 ± 0.828	218.143 [a] ± 22.997
Unaged LS	NQ	28.592 ± 3.981	2.7126 ± 0.416	14.197 ± 1.325	1.861 ± 0.252	47.366 [b] ± 5.78
Aged LS	11.412 ± 2.793	77.862 ± 9.051	8.758 ± 1.520	13.886 ± 2.047	10.744 ± 0.973	122.662 [c] ± 16.384
Unaged OWBL	NQ	69.306 ± 7.248	7.872 ± 0.923	9.162 ± 1.325	3.743 ± 0.287	90.083 [d] ± 11.125
Aged OWBL	22.503 ± 2.83	139.78 ± 18.723	14.657 ± 1.937	16.035 ± 1.692	9.622 ± 0.983	202.597 [a] ± 26.165

Note: Total Aldehyde data in this table are expressed as mean (n = 5) ± standard deviation, using Duncan's test, and differences in superscript letters are significant ($p < 0.05$), NQ means that the substance is not detectable.

4. Conclusions

The present study was a preliminary investigation of the effect of brewing lager beer with sugarcane syrup instead of some malt on the quality of the resulting beer. The optimal fermentation process for sugarcane beer was confirmed by one-way and orthogonal experiments, and it was found that the effect of syrup addition on the organoleptic flavor of the beer was smaller than that of hops addition, and larger than that of fermentation temperature and wort concentration. Analysis of the fermentation process showed that the beer with partial replacement of wort with sugarcane syrup had a higher utilization of free amino nitrogen and more complete fermentation. The experiments showed that sugarcane lager flavor substances constituted the basic flavor of lager beer, while because of the different ratio of amino acids, some flavor substances in sugarcane lager beer had differentiation from OWBL.

Volatile aldehydes are the most intuitive indicator for evaluating the quality and degree of aging of lagers, but judging the quality and aging rate of lagers by volatile matter is incomplete. Although lagers with sucrose syrup as an adjunct in volatile content detection have unique flavor substances and higher flavor stability compared with whole-barley lagers, the specific mechanism of sucrose syrup's effect on the fermentation of lagers is not yet clear, and the research direction brought by sucrose syrup as an adjunct to lager beer is still extensive. In addition to the detection of volatile substances, it is necessary to determine the mechanism by which sugarcane syrup promotes yeast utilization of amino acids that are not readily available and to comprehensively evaluate the quality of lagers with sugarcane syrup as an adjunct so that sugarcane, a saccharide, can be used in a more diverse range of products.

Supplementary Materials: The following supporting information can be downloaded at: https://www.mdpi.com/article/10.3390/foods13152339/s1. 1. Purpose of this study. 2. Beer fermentation process. 3. Sugarcane syrup production method. Table S1: Sources of production of commercially available beer. Table S2: Sources of raw materials for beer brewing. Table S3: Sensory evaluation table of sugarcane beer.

Author Contributions: H.L.: Data curation, Writing—original draft, Writing—review and editing, Investigation, Formal analysis. Y.J.: Methodology, Investigation, Formal analysis. C.L.: Data curation, Validation, Formal analysis, Investigation. C.X.: Conceptualization, Resources. Project administration, K.L.: Supervision, Project administration, Funding acquisition. F.H.: Conceptualization, Methodology, Resources. J.X.: Conceptualization, Data curation, Validation. K.H.: revised, Validation. All authors have read and agreed to the published version of the manuscript.

Funding: Guangxi Science and Technology Program (Item Number: AA22117015), Guangxi Science and Technology Agency.

Institutional Review Board Statement: Ethical review and approval were waived for this study due to the experiment does not require ethical approval, all the beer samples are not dangerous to the subjects, its in compliance with all Food Regulatory Agency regulations, and food safety testing has shown those samples to be safe.

Informed Consent Statement: Informed con-sent was obtained from all subjects involved in the study.

Data Availability Statement: The original contributions presented in the study are included in the article/Supplementary Materials, further inquiries can be directed to the corresponding author.

Conflicts of Interest: The authors declare that they have no conflicts of interest.

References

1. Lin, C.L.; García-Caro, R.D.L.C.; Zhang, P.; Carlin, S.; Gottlieb, A.; Petersen, M.A.; Vrhovsek, U.; Bond, U. Packing a Punch: Understanding How Flavours Are Produced in Lager Fermentations. *FEMS Yeast Res.* **2021**, *21*, foab040. [CrossRef]
2. Pires, E.J.; Teixeira, J.A.; Brányik, T.; Vicente, A.A. Yeast: The Soul of Beer's Aroma—A Review of Flavour-Active Esters and higher alcohols Produced by the Brewing Yeast. *Appl. Microbiol. Biotechnol.* **2014**, *98*, 1937–1949. [CrossRef] [PubMed]
3. Gasiński, A.; Kawa-Rygielska, J.; Szumny, A.; Czubaszek, A.; Gąsior, J.; Pietrzak, W. Volatile Compounds Content, Physicochemical Parameters, and Antioxidant Activity of Beers with Addition of Mango Fruit (*Mangifera indica*). *Molecules* **2020**, *25*, 3033. [CrossRef] [PubMed]
4. Inui, T.; Tsuchiya, F.; Ishimaru, M.; Oka, K.; Komura, H. Different Beers with Different Hops. Relevant Compounds for Their Aroma Characteristics. *J. Agric. Food Chem.* **2013**, *61*, 4758–4764. [CrossRef]
5. Lehnhardt, F.; Becker, T.; Gastl, M. Flavor Stability Assessment of Lager Beer: What We Can Learn by Comparing Established Methods. *Eur. Food Res. Technol.* **2020**, *246*, 1105–1118. [CrossRef]
6. Wang, L.; Wang, P.; Deng, W.; Cai, J.; Chen, J. Evaluation of Aroma Characteristics of Sugarcane (*Saccharum officinarum* L.) Juice Using Gas Chromatography-Mass Spectrometry and Electronic Nose. *LWT* **2019**, *108*, 400–406. [CrossRef]
7. Wang, H.M.; Ding, C.H. Application of Syrup in Beer Brewing. *Liquor. Mak. Sci. Technol.* **2004**, *03*, 58–60. [CrossRef]
8. Dabija, A.; Ciocan, M.E.; Chetrariu, A.; Codină, G.G. Maize and Sorghum as Raw Materials for Brewing: A Review. *Appl. Sci.* **2021**, *11*, 3139. [CrossRef]
9. Bogdan, P.; Kordialik-Bogacka, E. Alternatives to Malt in Brewing. *Trends Food Sci. Technol.* **2017**, *65*, 1–9. [CrossRef]
10. Piddocke, M.P.; Kreisz, S.; Heldt-Hansen, H.P.; Nielsen, K.F.; Olsson, L. Physiological Characterization of Brewer's Yeast in High-Gravity Beer Fermentations with Glucose or Maltose Syrups as Adjuncts. *Appl. Microbiol. Biotechnol.* **2009**, *84*, 453–464. [CrossRef]
11. Asikin, Y.; Wada, K.; Imai, Y.; Kawamoto, Y.; Mizu, M.; Mutsuura, M.; Takahashi, M. Compositions, Taste Characteristics, Volatile Profiles, and Antioxidant Activities of Sweet Sorghum (*Sorghum bicolor* L.) and Sugarcane (*Saccharum officinarum* L.) Syrups. *Food Meas.* **2018**, *12*, 884–891. [CrossRef]
12. Feng, S.; Luo, Z.; Zhang, Y.; Zhong, Z.; Lu, B. Phytochemical Contents and Antioxidant Capacities of Different Parts of Two Sugarcane (*Saccharum officinarum* L.) Cultivars. *Food Chem.* **2014**, *151*, 452–458. [CrossRef] [PubMed]
13. Xu, K.; Guo, M.; Du, J.; Zhang, Z. Cloudy barley Beer Enriched with Okra [*Abelmoschus esculentus* (L.) Moench]: Effects on Volatile Compound and Sensorial Attributes. *Int. J. Food Prop.* **2018**, *21*, 289–300. [CrossRef]
14. Vesely, P.; Lusk, L.; Basarova, G.; Seabrooks, J.; Ryder, D. Analysis of Aldehydes in Beer Using Solid-Phase Microextraction with On-Fiber Derivatization and Gas Chromatography/Mass Spectrometry. *J. Agric. Food Chem.* **2003**, *51*, 6941–6944. [CrossRef] [PubMed]
15. Du, R.; Guo, W.; Shen, Y.; Dai, J.; Zhang, H.; Fu, M.; Wang, X. In Situ Assay of the Reducing Sugars in Hydrophilic Natural Deep Eutectic Solvents by a Modified DNS Method. *J. Mol. Liq.* **2023**, *385*, 122286. [CrossRef]
16. Hahn, C.D.; Lafontaine, S.R.; Pereira, C.B.; Shellhammer, T.H. Evaluation of Nonvolatile Chemistry Affecting Sensory Bitterness Intensity of Highly Hopped Beers. *J. Agric. Food Chem.* **2018**, *66*, 3505–3513. [CrossRef] [PubMed]
17. Castro, L.F.; Affonso, A.D.; Lehman, R.M. Impact of Specialty Malts on Wort and Beer Characteristics. *Fermentation* **2021**, *7*, 137. [CrossRef]
18. Yang, Y.; Chang, L.; Lu, Y.; Yang, L. Analysis of Amino Acids in Foodstuff by Nuclear Magnetic Resonance. In Proceedings of the 2010 3rd International Conference on Biomedical Engineering and Informatics, Yantai, China, 16–18 October 2010; IEEE: Yantai, China, 2010; pp. 750–754.
19. Men, H.; Shi, Y.; Fu, S.; Jiao, Y.; Qiao, Y.; Liu, J. Mining Feature of Data Fusion in the Classification of Beer Flavor Information Using E-Tongue and E-Nose. *Sensors* **2017**, *17*, 1656. [CrossRef] [PubMed]

20. Molina-Calle, M.; Priego-Capote, F.; Luque De Castro, M.D. Headspace−GC–MS Volatile Profile of Black Garlic vs Fresh Garlic: Evolution along Fermentation and Behavior under Heating. *LWT* **2017**, *80*, 98–105. [CrossRef]
21. Gaulejac, N.S.-C.; Provost, C.; Vivas, N. Comparative study of polyphenol scavenging activities assessed by different methods. *J. Agric. Food Chem.* **1998**, *47*, 425–431. [CrossRef]
22. Aguiar, D.; Pereira, A.C.; Marques, J.C. Assessment of the Prediction Power of Forced Ageing Methodology on Lager Beer Aldehyde Evolution during Maritime Transportation. *Molecules* **2023**, *28*, 4201. [CrossRef]
23. Ortiz, R.M. Analysis of Selected Aldehydes in Packaged Beer by Solid-Phase Microextraction (SPME)–Gas Chromatography (GC)–Negative Chemical Ionization Mass Spectrometry (NCIMS). *J. Am. Soc. Brew. Chem.* **2015**, *73*, 266–274. [CrossRef]
24. Ojala, M.; Kotiaho, T.; Siirilä, J.; Sihvonen, M.L. Analysis of aldehydes and ketones from beer as O-(2,3,4,5,6-pentafluorobenzyl)hy droxylamine derivatives. *Talanta* **1994**, *41*, 1297–1309. [CrossRef] [PubMed]
25. Vilela-Moura, A.; Schuller, D.; Mendes-Faia, A.; Silva, R.D.; Chaves, S.R.; Sousa, M.J.; Côrte-Real, M. The Impact of Acetate Metabolism on Yeast Fermentative Performance and Wine Quality: Reduction of Volatile Acidity of Grape Musts and Wines. *Appl. Microbiol. Biotechnol.* **2011**, *89*, 271–280. [CrossRef] [PubMed]
26. Meilgaard, M. Hop Analysis, Cohumulone Factor, and the Bitterness of Beer: Review and Critical Evaluation. *J. Inst. Brew.* **1960**, *66*, 35–50. [CrossRef]
27. Černá, S.; Benešová, K. Stanovení Organických Kyselin ve Speciálních Pivech A Nápojích Na Bázi Piva Pomocí Kapilární Izotachoforézy. *Chem. Listy* **2023**, *117*, 516–521. [CrossRef]
28. Dušek, M.; Olšovsk, J.; Krofta, K.; Jurková, M.; Mikyska, A. Qualitative Determination of β-Acids and Their Transformation Products in Beer and Hop Using HR/AM-LC-MS/MS. *J. Agric. Food Chem.* **2014**, *62*, 7690–7697. [CrossRef] [PubMed]
29. Meneses, F.J.; Henschke, P.A.; Jiranek, V. A Survey of Industrial Strains of Saccharomyces cerevisiae Reveals Numerous Altered Patterns of Maltose and Sucrose Utilisation. *J. Inst. Brew.* **2002**, *108*, 310–321. [CrossRef]
30. Hambraeus, G.; Nyberg, N. Enzymatic Hydrogenation of Trans-2-Nonenal in Barley. *J. Agric. Food Chem.* **2005**, *53*, 8714–8721. [CrossRef]
31. Bushnell, S.E.; Guinard, J.-X.; Bamforth, C.W. Effects of Sulfur Dioxide and Polyvinylpolypyrrolidone on the Flavor Stability of Beer as Measured by Sensory and Chemical Analysis. *J. Am. Soc. Brew. Chem.* **2003**, *61*, 133–141. [CrossRef]
32. Andersen, M.L.; Skibsted, L.H. Electron Spin Resonance Spin Trapping Identification of Radicals Formed during Aerobic Forced Aging of Beer. *J. Agric. Food Chem.* **1998**, *46*, 1272–1275. [CrossRef]
33. Saison, D.; De Schutter, D.P.; Uyttenhove, B.; Delvaux, F.; Delvaux, F.R. Contribution of Staling Compounds to the Aged Flavour of Lager Beer by Studying Their Flavour Thresholds. *Food Chem.* **2009**, *114*, 1206–1215. [CrossRef]
34. Vanderhaegen, B.; Neven, H.; Verachtert, H.; Derdelinckx, G. The Chemistry of Beer Aging—A Critical Review. *Food Chem.* **2006**, *95*, 357–381. [CrossRef]

foods

MDPI

Article

Fractionation of Winemaking Grape Stalks by Subcritical Water Extraction to Obtain Added-Value Products

Irene Maté, Maria Vargas *, Lorena Atarés and Amparo Chiralt

Instituto Universitario de Ingeniería de Alimentos-FoodUPV, Universitat Politècnica de València, 46022 Valencia, Spain; irmagar1@upv.edu.es (I.M.); loathue@tal.upv.es (L.A.); dchiralt@tal.upv.es (A.C.)
* Correspondence: mavarco@tal.upv.es

Abstract: Grape stalks (GSs) from winemaking were submitted to a green process to valorise its lignocellulosic biomass that applied subcritical water extraction (SWE) at 170 °C and 180 °C to obtain active extracts and cellulose-enriched fractions. The sum of the total phenolic content of the soluble extract and the solid residue fractions from the SWE exceeded that of the GS, which suggests the generation of compounds with antioxidant properties through SWE. All SWE fractions showed high antioxidant power. The increased temperature promoted the extraction of polyphenolic compounds, enhancing the antioxidant power of both extracts and solid residues. These solid residue fractions were bleached with alkaline hydrogen peroxide solutions (4 and 8% v/v) to purify cellulose. After two bleaching cycles, no notable delignification progress was observed, as the bleaching yield or whiteness index did not significantly change in the further cycles. The first bleaching cycle led to a significant reduction in the lignin content at both SWE temperatures. The cellulose purity was higher in the samples obtained at 170 °C and bleached with 4% alkaline hydrogen peroxide. SWE at 180 °C led to greater cellulose oxidation during the bleaching step regardless of the hydrogen peroxide concentration.

Keywords: hydrothermal treatment; integral valorisation; lignocellulosic waste; hydrogen peroxide bleaching; phenolic compounds; antioxidant; cellulose fibres

1. Introduction

Grapes are one of the most abundant fruit crops in the world. In 2022, the world's vineyards of approximately 7.3 million ha produced a total of 80.1 million tons of fresh grapes. Of the total volume available, 37.3 million tons were destined for pressing (34.1 million tons to produce wine and 3.2 million tons for the production of musts and juices) [1]. Grape stalks constitute a significant by-product of wine production, representing up to 7% of the raw material [2–4]. The production of 100 L of white or red wine yields approximately 4 kg of grape stalks [2]. Even though grape stalks are not hazardous waste, their high content of organic matter and the seasonal production of grapes represent an environmental problem. The discharge of grape stems into the soil inhibits its germinative properties [3] and may affect groundwater's chemical and biological oxygen demand [2,5]. Grape skins can be used for animal feed, whereas grape stalks are mainly utilised as fertilisers [2]. However, the chemical composition of grape stalks offers interesting prospects for their valorisation by obtaining phenolic compounds, cellulose, or lignin in the context of the circular economy by using sustainable processes with low environmental impact.

The grape stalk is mainly composed of cellulose (12–36%), hemicellulose (14–26%), and lignin (23–34%) [5–8], as well as other non-structural components extractable in water or organic solvents (tannins, other polyphenols, etc.), proteins, waxes, and ashes. Compositional discrepancies in the literature may arise from using different analytical procedures, varying definitions of biomass fractions, and differing origins and types of grape stalks. The valorisation of lignocellulosic materials such as grape stalks is linked to the potential applications of some components, mainly the cellulosic fraction and the extractable phase with antioxidant/antimicrobial properties. The demand for cellulose fibres has been increasing recently and is expected to grow [3,9]. The environmental impact of cellulose extraction from wood pulp, namely forest devastation and its consequential contribution to global warming, have driven research efforts toward utilising more environmentally friendly alternative cellulose sources, such as lignocellulosic agricultural residues [9]. Exploring grape stalk as a cellulose source has been carried out by other authors [3,5,10] but using a non-environmentally friendly process with a great amount of chemicals for pulping and bleaching treatments. Araújo et al. [3] evaluated cellulose extraction from grape stalks using sequenced acid hydrolysis, alkaline hydrolysis, and a single 24 h bleaching step with alkaline hydrogen peroxide, performed on samples of varying particle sizes. Delignification, indirectly evaluated with colour measurements, was more efficient in small particles with low cellulose yields. Other fractionation processes have been tested for the valorisation of grape stalks, aiming for the release of fermentable sugars. In this regard, Atatoprak et al. [10] used an alkaline treatment for delignification, whereas Ping et al. [11] found that dilute sulfuric acid and ethanol organosolv pretreatments resulted in low degrees of delignification due to the high content of lignin and tannins of the feedstock.

Grape stalks also contain high proportions of non-structural functional compounds that have demonstrated favourable effects on human health [12]. These include polyphenols such as tannins, flavonols, and stilbenes [4]. Antioxidants are naturally present in many plant-based food products, such as nuts, oilseeds, vegetables, and fruits. Using low-cost industrial wastes instead of fresh raw materials for phenolic compound extraction could reduce production costs and increase the margin profit of these high-added-value products [12]. As well as other winemaking industry residues, grape stalks are considered a promising rich source for these functional compounds. Grape stalk extracts find, therefore, potential applications in the pharmacy and food industries, such as the shelf-life extension of fatty products, the inhibition of food pathogens, or the formulation of active food packaging [4]. Solvent extraction is the most frequently used procedure for antioxidant recovery [12]. Even though these processes may achieve efficient antioxidant extraction, using hazardous chemicals entails recovery costs and the environmental risks linked to their disposal. In contrast, subcritical water extraction, SWE (also known as hydrothermal treatment or autohydrolysis), using water as a solvent, accomplishes acidic hydrolysis without the use of any external acid because of hydrogen ions produced by the water autoionisation. This process results in the fractionation of lignocellulosic materials by using only hot compressed water, enabling the simultaneous removal of antioxidant water-soluble extractives, hemicelluloses, and soluble lignin. In this way, cellulose and Klason lignin quantitatively remain in the extraction solid residue, constituting a cellulose-enriched fraction that can be more easily purified by a delignification step. Difonzo et al. [13] reported on the potential of grape stalks as a source of antioxidant compounds extracted by this process. Freitas et al. [14] obtained antioxidant and antibacterial extracts from grape stalks (red and white varieties) by using SWE at 160 °C and 180 °C, which allows for active films for food packaging to be obtained. Amendola et al. [6] found that SWE efficiently extracted solid residues from grape stalks enriched in cellulose and lignin with lowered proportions of ashes and hemicellulose. Similar results were reported by Senila et al. [15], who additionally applied a subsequent delignification step with sodium chlorite in an acid medium to improve cellulose purification.

Cellulose pulping and delignification are usually carried out through different chemical treatments of lignocellulosic biomass procedures, such as Kraft pulping, alkaline or acid hydrolysis, ionic liquids, and oxidant delignification treatments such as acidified sodium chlorite [15,16]. All of these implied the use of great amounts of chemicals and water, making the process sustainable difficult. Therefore, reducing chemicals is essential to designing more environmentally friendly and lower-cost processes to purify cellulose. In this sense, the present study proposes the use of hydrogen peroxide in the bleaching process, which allows for operation under mild conditions without forming chlorine-toxic by-products that are recalcitrant for biodegradation. In alkaline media, hydrogen peroxide dissociates to yield hydroperoxyl anions (HOO^-) and superoxide ions (O_2^-), which cleave side chains in lignin and oxidise chromophores. These potent antioxidants produce delignification, yielding low molecular weight water-soluble products, but cellulose depolymerisation could also occur as an undesirable side-effect in achieving efficient lignocellulose fractionation [15,17,18]. In previous studies, hydrogen peroxide treatment in an alkaline medium has been successfully used to obtain cellulose fibres from rice straw [14], almond skins [19] and almond shells [20], *Posidonia oceanica* waste [21] or vineyard pruning residues [17]. However, the combined effect of subcritical water treatment and hydrogen peroxide delignification for obtaining cellulose-enriched fractions from winemaking grape stalks has not been evaluated yet.

This study aimed to evaluate the potential of using subcritical water extraction and hydrogen peroxide alkaline delignification to fractionate winemaking grape stalks, producing phenol-rich aqueous extracts and cellulosic fractions, both valuable in developing food packing materials. Both extracts and solid residues were characterised in terms of their phenolic richness and antioxidant activity. The structural components (cellulose, hemicellulose, and lignin) were analysed in the solid residues of the extraction and at the different steps of the bleaching process to assess the efficacy of the fractionation treatments.

2. Materials and Methods

2.1. Materials

Grape stalk (GS) residues from red grape variety Bobal, origin Utiel-Requena (Valencia, Spain), were supplied by a local wine producer after the winemaking process. Folin–Ciocalteu reagent (2 N), gallic acid, methanol (>99.9 purity), 2,2-Diphenyl-1-picrylhydrazyl (DPPH), sodium hydroxide, glucose, and arabinose were purchased from Sigma-Aldrich (St. Louis, MO, USA). Ethanol (98%), hydrogen peroxide (H_2O_2, 30%), sulphuric acid (H_2SO_4, 98%), and sodium carbonate (Na_2CO_3, 99.5%) were obtained from Panreac Quimica S.L.U (Castellar del Vallés, Barcelona, Spain). Phosphorous pentoxide (P_2O_5, 98.2%) was obtained from VWR Chemicals (Leuven, Belgium). D(+)-Xylose was supplied by Merck KGaA (Darmstadt, Germany).

2.2. Subcritical Water Extraction of Grape Stalks

The grape stalk raw material was dried at 105 °C for 24 h and ground with a mill using a 1 mm diameter mesh (Model Zyklon SM 300 stainless steel, Retsch, Haan, Germany). The milled GS was sieved with a 0.55 mm mesh and stored at 0% RH until further use.

The milled and sieved GS was subjected to subcritical water extraction (SWE) in a pressure reactor (Model 1-TAP-CE, 5 L capacity, Amar Equipment PVT. Ltd., Mumbai, India) at a solids/water mass ratio of 1:10. The extraction was carried out for 30 min, and two temperatures were tested, 170 °C and 180 °C. These temperatures were selected based on previous studies that report a higher hydrolysis of hemicellulose near 180 °C [14]. Subsequently, the extraction residues were left to decant for 3 h to separate the liquid extract and the solid residue. The liquid extracts were filtered with a qualitative filter (Filter-lab S.A., Barcelona, Spain, pore size < 0.5 mm), frozen at −40 °C and subsequently freeze-dried (Telstar, model Lyoquest-55; Barcelona, Spain) at −45 °C, 6 mbar, and 72 h. The lyophilised extracts (E-170 and E-180) were stored at −20 °C until further use.

The solid extracts obtained at either temperature (R-170 and R-180) were washed with distilled water, filtered with a qualitative filter (pore size < 0.5 mm, Filter-lab S.A., Barcelona, Spain,), and dried at 50 °C for 24 h. The dried mass of the extracts and solid residues were used to analyse mass balance in the SWE process and quantify the yield of each fraction.

2.3. Characterisation of Phenolic Compounds and Antioxidant Activity of SWE Fractions

Total phenolic content (TPC) and antioxidant activity with the DPPH assay were determined in the raw material (GS), in the soluble extracts (E-170 and E-180), and in the insoluble fractions (R-170 and R-180) obtained from SWE. For raw material and insoluble residues (GS, R-170, and R-180), an ultrasound extraction of phenols in methanol-water mixtures (40:60 *v*/*v*) at 1:10 S/L mass ratio and 30 °C was performed before the quantification test. To this aim, a high-intensity ultrasonic homogeniser was used (Vibra CellTM VCX750, Sonics & Material, Inc., Newtown, CT, USA), applying a frequency of 20 kHz, a power of 750 W and a sonication amplitude of 40% for 30 min.

TPC was determined using the Folin–Ciocalteu method in triplicate, according to the procedure described by Freitas et al. [22]. The assay consisted of mixing 0.5 mL of extract with 6 mL of distilled water and 0.50 mL of Folin reagent. After one minute, 1.5 mL of a 20% Na_2CO_3 solution was added and made up to 10 mL with distilled water. It was then shaken and kept for 2 h in the dark at room temperature. After that time, absorbance was measured at 750 nm using a UV–Vis spectrophotometer (model Evolution 201, Thermo Scientific, Waltham, MA, USA). Finally, TPC was determined using a gallic acid calibration curve (R^2 = 0.9991), and the results were expressed as g gallic acid equivalents (g GAE) per 100 g of sample solids. The results were also expressed as g GAE per 100 g of grape stalk to evaluate phenolic balance in the fractions.

The ability of GS and SWE fractions to scavenge DPPH free radicals was determined using the 2,2-Diphenyl-1-pikryl-hydroxyl (DPPH) method, with some modifications [23]. Each sample extract was mixed with a methanolic solution of DPPH at different ratios. The reaction was left to complete in the dark at room temperature, and the absorbance was measured at 515 nm. The initial and final DPPH concentration was obtained from a calibration curve fitted by linear regression (R^2 = 0.9992). The EC50 was determined, which is defined as the ratio of sample to DPPH that is required to reduce the initial concentration of DPPH to 50% once the stability of the reaction has been reached. Thus, the lower the EC50 values, the greater the antioxidant activity of the tested sample. The results were expressed in mg sample per mg DPPH and mg GS per mg DPPH.

2.4. Cellulose Purification from the Insoluble SWE Fractions

Aiming to purify cellulose fibres from the insoluble SWE fractions (extraction residues), the solid extracts were bleached with hydrogen peroxide solutions at a 1:30 S/L mass ratio at pH 12. Two different hydrogen peroxide concentrations were tested, namely 4 and 8% *v*/*v*. Four consecutive 1 h bleaching cycles at 40 °C were performed, and after filtering and washing with abundant distilled water, the samples obtained in each cycle were air-dried at room temperature for three days, and the mass yield was determined in each step. The bleached residues were coded as BR-170-4, BR-170-8, BR-180-4, and BR-180-8.

The colour coordinates L* (lightness), a* (red-green), and b* (yellow-blue) of each bleached fraction were measured with a CM-3600d spectrocolorimeter (Minolta Co., Tokyo, Japan). The effectiveness of the bleaching treatment was evaluated through the mass yield and the whiteness index (WI) calculated with Equation (1):

$$\mathbf{WI} = \mathbf{100} - \sqrt{(\mathbf{100} - \mathbf{L^*})^2 + \mathbf{a^{*2}} + \mathbf{b^{*2}}} \quad (1)$$

Additionally, the morpho-geometric characteristics of the fibres after different process steps were qualitatively evaluated using a field emission scanning electron microscope (ULTRATM 55, Zeiss, Oxford Instruments, UK). The conditioned samples (P_2O_5 at 25 ± 2 °C for one week) were covered with a gold layer (EM MED020 sputter coater, Leica Biosystems, Barcelona, Spain), and the images were taken at 2.0 kV acceleration voltage.

2.4.1. Analysis of Structural Components in the Lignocellulosic Fractions

The content of cellulose, hemicellulose and lignin of GS, the extraction solid residues (R-170 and R-180), and the bleached materials (BR-170-4, BR-170-8, BR-180-4, and BR-180-8) after different bleaching cycle were determined. The extractive contents in ethanol and water were previously determined in duplicate in non-bleached samples according to the NREL standard procedure (NREL/TP-510-42619-2008) [24]. After eliminating the non-structural components, structural carbohydrates (cellulose and hemicellulose) and lignin were quantified according to the NREL standard procedure (NREL/TP-510-42618-2008) [25]. This method involves acid hydrolysis with sulphuric acid in two steps and a filtration step to separate the soluble and insoluble hydrolysed fractions. The insoluble hydrolysed fraction was washed with abundant water and dehydrated at 105 °C for two days to quantify the acid-insoluble fraction (AIF) gravimetrically. The insoluble lignin was determined from the AIF by subtracting the ash content of AIF determined by incineration at 550 °C. The free sugars resulting from the hydrolysis of structural carbohydrates (glucose, xylose, and arabinose) were determined in the soluble hydrolysed fraction.

Free sugars (glucose, xylose, and arabinose) were quantified by HPLC (Agilent Technologies, model 1120 Compact LC, Frankfurt, Germany). A HILIC Luna Omega Sugars HILIC column (150 × 4.6 mm, three μm) and an evaporative light scattering detector (ELSD Agilent Technologies 1200 Series, Waldbronn, Germany) were used. The mobile phase consisted of water: acetonitrile (25:75) in isocratic mode at a 0.8 mL min-1 flow rate. The detector conditions were 40 °C, 3.0 bar N_2 pressure, and a gain of 3. The software used for data acquisition was ChemStation (version LTS 01.11, Agilent Technologies, Waldbronn, Germany). Results were analysed using the Origin program (version OriginPro 2021, OriginLab Corporation, Northampton, MA, USA) by applying the Gaussian model for peak area determination. The hemicellulose content was calculated from the sum of the xylose and arabinose concentrations, and the cellulose content was obtained from the glucose concentration by applying the corresponding correction factors related to the water molecular weight, as described by Sluiter et al. [24]. The cellulose content would be overestimated because glucose units are also present in the hemicellulose chains, whereas hemicellulose content would be underestimated.

The ash contents in the different samples were quantified as the mass percentage of residue remaining after incineration at 550 °C for 24 h (NREL/TP-510-42622-2008) [25].

2.4.2. Thermal Analyses (TGA) of the Lignocellulosic Fractions

The thermal behaviour of the different samples was analysed with a thermogravimetric analyser under a nitrogen atmosphere (TGA 1 Stare System analyser, Mettler-Toledo, Greifensee, Switzerland). Before the tests, the samples were placed in a desiccator with P_2O_5 for two weeks to eliminate moisture. The TGA test consisted of heating from 25 °C to 900 °C at 10 °C min^{-1} under a constant nitrogen flow (10 mL). All samples were run in duplicate. The results were evaluated using STARe software (version V12.00a, Mettler-Toledo, Inc., Greifensee, Switzerland).

2.4.3. FTIR Analyses of the Lignocellulosic Fractions

An FTIR spectrophotometer (Vertex 80, Bruker AXS GmbH, Karlsruhe, Germany) was used to evaluate the vibrational patterns of the functional groups present in the samples (GS, R-180, R-170, and bleached samples). FTIR spectra were obtained at a 6 cm^{-1} resolution, in the wavelength range 4000–650 cm^{-1}, and 128 scans were performed for each spectrum. The measurements were performed in duplicate.

2.4.4. X-Ray Diffraction Analysis (XRD) of the Lignocellulosic Fractions

An X-ray diffractometer (AXS/D8 Advance, Bruker, Karlsruhe, Germany) was used to analyse the X-ray diffraction pattern spectra of the GS and the different cellulosic fractions (R-170, R-180, BR-170-4-4, BR-170-8-4, BR-180-4-4, and BR-180-8-4). The analyses were performed using Kα-Cu radiation (λ: 1.542 Å), 40 kV, 40 mA, a step size of 2.0° per minute, and a scan angle of 2θ between 5° and 40°. Once conditioned at 53% RH and 25 °C, the samples were appropriately compacted on the sample holder. The data were obtained using XRD Commander software (version 8.61, Bruker AXS/D8 Advance, GmbH, Karlsruhe, Germany) and processed with DIFFRAC.EVA and DRXWin (Windows, version 2.3) software. The crystallinity index (CI%) was determined from the spectra using Equation (2) [26], where the maximum intensity of 200 lattice diffraction (I_{200}, crystalline peak) and the diffraction intensity at $2\theta = 18°$ ($I_{2\theta}$ 18°, valley of the amorphous phase) are related.

$$CI(\%) = \frac{(I_{200} - I_{2\theta\,18°})}{I_{200}}100 \quad (2)$$

2.5. Statistical Analysis

The experimental data were submitted to analyses of variance (ANOVA) with a 95% confidence level using Statgraphics version 19-X64. Comparisons were performed using 95% Fisher's LSD intervals.

3. Results and Discussion

3.1. Yields of the SWE Process and Composition of the Grape Stalk and Extraction Residues

Figure 1 shows the flow chart of the process applied to revalorise grape stalks, where the values of mass yields (Y: dry out-flow mass to dry in-flow mass) of the different fractions obtained in the SWE process steps at both 170 °C and 180 °C, are specified.

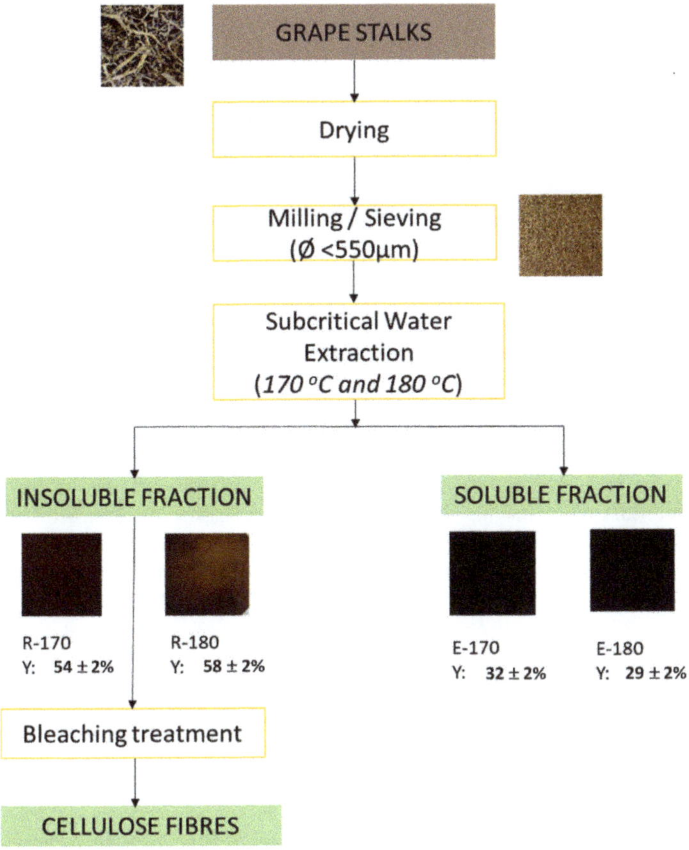

Figure 1. Flow chart of the process applied to valorise grape stalks. The Y-values represent the mass yield (dry out-flow mass with respect to the dry in-flow mass) of the different fractions obtained in the subcritical water extraction process at 170 °C and 180 °C. Mean values ± standard deviation. Insoluble residues (R-170, R-180) and extracts (E-170, E-180).

The extraction step with subcritical water at 170 °C and 180 °C showed a solid extraction yield of 32% and 29%, respectively, in the range of those reported by Amendola et al. [6].

This yield indicated a high water solubilisation of the GS components under the conditions applied. The dry, solid, insoluble residues yield represented 54% and 58% for R-170 and R-180, respectively, in the range of those found in previous studies. In this sense, Senila et al. [15] performed the autohydrolysis of vine-shoots waste of different grape varieties at 165 °C and 180 °C and reported solid yields ranging between 55% and 70%. In the present study, no significant differences were observed between the yields in the lignocellulosic insoluble residue at 170 °C and 180 °C. Still, the R-170 sample could have more cellulose since more extraction of non-cellulosic components occurred at 170 °C.

The sums of the yields of soluble fractions (E) and the insoluble fractions (R) were 86% and 87% at 170 °C and 180 °C, respectively, which corresponds to a mass loss of 13–14%. This could be attributed to the experimental variability and the partial degradation of the organic matter caused by hydrolysis (e.g., loss of formic or acetic acids resulting from hydrolysis of methoxy or ethoxy groups in some molecules) at these temperatures. In fact, a slight pressure increase was detected in the reactor during the extraction processes (up to 1.5 bars above the initial pressure), which can be attributed to the volatile formation in the closed reactor.

The contents of extractives and structural components of the GS and the SWE solid, insoluble residues (R-170 and R-180) obtained at 170 °C and 180 °C are shown in Table 1.

Table 1. Extractives, structural components, and ashes of winemaking grape stalks (GS) and the subcritical water extraction insoluble residues obtained at 170 °C (R-170) and 180 °C (R-180). Mean values ± standard deviation.

Sample	Extractives in Water (g/100 g Sample)	Cellulose (g/100 g Sample)	Hemicellulose (g/100 g Sample)	Lignin (g/100 g Sample)	Ashes (g/100 g Sample)
GS	52.2 ± 0.0^{a}	22 ± 4^{a}	7 ± 4^{a}	14.7 ± 1.3^{a}	7.6 ± 0.1^{a}
R-170	25.9 ± 0.0^{b}	32 ± 4^{b}	1.6 ± 0.1^{b}	41.3 ± 1.6^{b}	2.9 ± 0.1^{b}
R-180	21.2 ± 0.1^{b}	24.2 ± 1.8^{b}	0.4 ± 0.1^{b}	40 ± 0.2^{b}	2.9 ± 0.1^{b}

Different letters (a, b) in the same column indicate significant differences among samples ($p < 0.05$).

In the GS samples, more than 50% of GS solids were water-extractable in Soxhlet, whereas no ethanol-soluble compounds were quantitatively extracted. These water-soluble compounds must be mainly tannins and free sugars, as reported by other authors for other GS grape cultivars [1,2,5]. In the SWE extraction residues (R-170 and R-180), the extractive amount was lower, due to the previous extraction of some of these constituents during the SWE step. SWE at 170 and 180 °C resulted in the removal of an important fraction of hemicellulose from the grape stalk, as has been previously described for different lignocellulosic materials [19,27,28]. A high hydrolysis degree of the hemicellulose at 160 and 180 °C was also reported in GS by Freitas et al. [14]. The lignin and cellulose contents significantly increased in R-170 and R-180 samples due to the significant removal of water extractives and hemicellulose during the SWE step. In contrast, the ash content was significantly reduced without a significant effect of temperature ($p > 0.05$).

The cellulose content in the GS was approximately 22%, in the range previously reported for the cellulose content GS (12 to 36%, depending on the grape variety and origin) [1,2,6,7,29]. After the SWE step, the grape stalk residues exhibited an increased cellulosic content, thus reflecting the cellulose enrichment provoked by the process. This has also been found in other lignocellulosic materials such as rice straw [14,30], defatted almond skin [19] or the waste of the Posidonia oceanica leaves [21]. When the cellulose content of R-170 and R-180 samples was referred to per mass unit of GS, taking the process yield into account, it can be noted that only 14–17 g cellulose/g GS was obtained. This could suggest a cellulose loss during the SWE process by the autohydrolysis process. Nevertheless, the difference could also be attributed to the overestimated value of the cellulose content of the raw GS, where more hemicellulose is present and more hemicellulosic glucose would be released during the hydrolysis step with sulphuric acid.

3.2. Total Phenol Content and Antioxidant Properties of the GS and SWE Fractions

The total phenolic content (TPC) and antioxidant activity (EC_{50} values) of the GS and the different SWE fractions obtained at 170 °C and 180 °C are reported in Table 2. These results are shown per mass unit of the sample solids and per mass unit of GS, which enables the analysis of the phenolic mass balance in the corresponding fractions.

Table 2. Total phenolic content (TPC) and EC50 of the grape stalk (GS) and the subcritical water extraction fractions obtained at 170 °C and 180 °C: Insoluble residues (R-170, R-180) and extracts (E-170, E-180). Mean values ± standard deviation.

Sample	TPC (g GAE/100 g Sample Solids)	TPC (g GAE/100 g GS)	EC 50 (mg Sample Solids/mg DPPH)	EC 50 (mg GS/mg DPPH)
GS	9.3 ± 0.6^{4}	3.8 ± 0.2^{1}	0.64 ± 0.07^{1}	27 ± 3^{1}
R-170	$17.8 \pm 0.3^{b,2}$	$1.3 \pm 0.0^{b,5}$	$0.38 \pm 0.05^{a,2}$	$2.8 \pm 0.4^{a,4}$
R-180	$19.7 \pm 0.3^{a,1}$	$1.6 \pm 0.0^{a,4}$	$0.38 \pm 0.01^{a,2,3}$	$3.11 \pm 0.07^{a,4}$
E-170	$8.3 \pm 0.0^{b,5}$	$2.7 \pm 0.0^{b,3}$	$0.42 \pm 0.00^{a,2,3}$	$13.33 \pm 0.13^{a,2}$
E-180	$11.1 \pm 0.1^{a,3}$	$3.2 \pm 0.0^{a,2}$	$0.30 \pm 0.01^{b,2}$	$8.9 \pm 0.4^{b,3}$

Different superscripts indicate significant differences among all samples in each column (1–5) or due to temperature (a, b) for the same subcritical water extraction fraction ($p < 0.05$).

Subcritical water extraction allowed for obtaining phenolic-rich extracts with more than 70% of the TPC of GS, whereas a part of these compounds also remained in the solid fraction. The highest temperature provided the extract E-180 with greater phenol concentration, according to a greater hydrolysis degree of the biomass and release of bonded phenols. At a determined temperature, the sum of the TPC of both E and R samples from SWE exceeded the TPC value of the initial GS material, mainly at 180 °C. This suggests the generation of some antioxidant compounds that are reactive with the Folin reactant at high extraction temperatures. This could be attributed to the release of lignin-derived polyphenol antioxidants, as has been previously observed by other authors [31] or to the formation of Maillard and caramelisation compounds at high temperatures, as reported by different authors due to the reaction of free sugars [28]. The neo-formed brown compounds, such as 5-hydroxymethylfurfural (HMF), exhibit antioxidant properties [28] and can be sensitive to the Folin reactant. This implies the enhancement of the antioxidant capacity of the biomass fractions, but the potential toxicity of the neo-formed compounds must be analysed.

The TPC found for the extracts was significantly higher than others previously reported for the soluble fraction of grape stalk obtained by SWE [6,13,14]. They were also higher than TPCs obtained from grape stalks using solvent extraction procedures for antioxidant compound recovery. In this sense, Spigno et al. [12] extracted polyphenols from the grape stalk using ethanol and ethylacetate as solvents at varying temperatures and time, reporting less than 0.4 g GAE per 100 g of GS, which is significantly lower than the values obtained in the present study.

The values of EC_{50} supported the high antioxidant power of all obtained GS fractions in the range of pure, potent antioxidants, like ascorbic acid or tocopherol (0.12 and 0.26 mg compound.mg^{-1} DPPH, respectively [23]. Freitas et al. [32] found 0.6 and 1.1 mg extract mg^{-1} DPPH for the EC_{50} values of SWE extracts from GSs of red and white grape varieties, respectively, which is in the range of the obtained values. This demonstrates the great antioxidant power of the grape stalk extracts obtained by SWE. The variability in the phenolic extraction efficiency can be attributed to different factors, such as the biomass variety and origin and the sample pretreatments (drying and milling), which may affect the solvent penetration and the polyphenol recovery, depending on the temperature applied in the SWE [33,34]. Cheng et al. [34] reviewed the optimal temperature/time conditions for extracting polyphenols from different plant matrices and found that the optimal values ranged from 120 to 225 °C. Increased temperature within a determined range seemed to promote the extraction of polyphenolic compounds from the grape residue feedstock. Gullón et al. [35] found the same positive effect of temperature on TPC and DPPH radical scavenging capacity of SWE extracts of dry vine shoots between 180 °C and 215 °C.

The phenolic richness of the SWE extraction residues (R-170 and R-180) is also remarkable since these retained approximately 30–40% of the total phenols of GS. The corresponding EC_{50} values also reveal that these residues have as strong antioxidant power as the extracts. The EC_{50} values referred to per mass unit of GS showed that their antioxidant capacity is approximately 10 times higher than that of the raw GS. This fact suggests that even after the extraction process, the solid SWE residues could represent an attractive source of antioxidant compounds potentially applicable in various fields. For instance, these residues could be used as fillers of biodegradable polymeric matrices to obtain lower-cost, reinforced, and potentially active food packaging materials.

3.3. Obtaining Cellulosic Fibres from SWE Insoluble Residues

The cellulose-enriched extraction residues (R-170 and R-180) obtained from SWE at 170 °C and 180 °C were submitted to four successive bleaching steps with alkalinised H_2O_2 to promote cellulose delignification. Figure 2a shows the bleaching yield (BY, g of bleached sample per 100 g of SWE residue) after one to four bleaching cycles together with the corresponding whiteness index (WI) as indicators of the delignification progress. Figure 2b shows the appearance of the samples after the successive process steps.

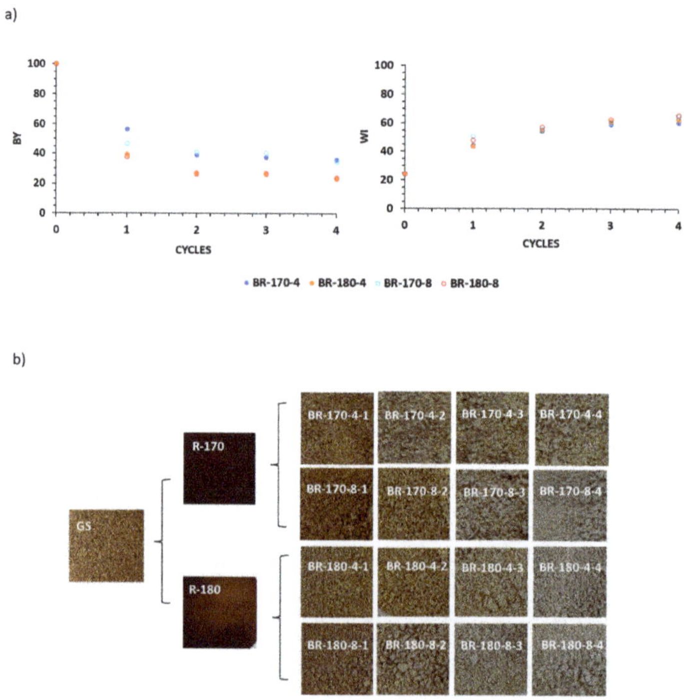

Figure 2. (**a**) Bleaching yield (BY) and whiteness index (WI) obtained with the different one-hour treatment cycles with alkaline hydrogen peroxide at 4 and 8% v/v. The LSD intervals are 2.22 for BY and 0.03 for WI. (**b**) Images of the raw material, subcritical water extraction (SWE) residues at 170 °C (R-170) and 180 °C (R-180), and bleached residues (BR) at the different steps of the bleaching process. Bleached residues are coded as the BR temperature of SWE concentration (% v/v) of alkaline hydrogen peroxide number of 1 h bleaching cycles.

Bleaching treatments allowed for the elimination of both coloured and non-coloured compounds of the cellulosic residue that are sensitive to the action of oxidants, while their oxidised forms are solubilised in the bleaching medium, thus reducing the mass of the solid residue. Bleaching yields after the fourth cycle ranged between 25% and 40%, which implies that a great part of the components of the SWE residues were susceptible to alkaline oxidation and, therefore, solubilised in the liquid medium. Bleaching yields were higher for the R-170 samples than for R-180, which indicates a more significant elimination of non-cellulosic components in the residue obtained at 170 °C. The concentration of hydrogen peroxide did not significantly affect bleaching yield in R-180 samples, whereas the 8% concentration of H_2O_2 was more effective in R-170 samples, but only during the first bleaching cycle. In this sense, the minute change in the blanching yield after the second cycle is remarkable, which suggests that no further compound elimination occurred after this cycle.

Concerning the whiteness index (WI), which allows for controlling the removal of coloured compounds, relatively low WI values were observed for the R-170 and R-180 samples without significant differences between them (Figure 2a,b). The severe conditions used for SWE promote caramelisation or Maillard reactions in the GS substrate, as previously observed by other authors in terms of rice straw [14], which resulted in deeply coloured extraction residues. The development of WI throughout the bleaching cycles revealed very similar behaviour of both R-170 and R-180 samples with 4 or 8% H_2O_2. Only slight differences could be appreciated after the first cycle, where treatments with 8% H_2O_2 provided the samples with a slightly higher WI value. It is remarkable that even after four bleaching cycles, the samples did not show WI values greater than 60, which indicates the recalcitrance of GS-coloured components to the peroxide treatment. This can be attributed to the high tannin content of GS, which makes delignification difficult. Senila et al. [15] reported that the high proportion of tannins in the vine-shoot waste, structurally associated with the lignin fraction, through the formation of alkali–aryl condensed structures, makes the separation difficult. The persistence of coloured compounds in bleached cellulosic material obtained from grape stalks was also observed in previous studies [3].

Structural components were also analysed in the samples R-170 and R-180 after one and two bleaching cycles, with 4% and 8% hydrogen peroxide solutions, since no progress in the cellulose purification after two cycles can be deduced from the bleaching yield development (Figure 2a). Table 3 shows the content of cellulose, lignin, and ashes of the different lignocellulosic residues, referred per mass unit of GS to evaluate the mass balances better, and the global yield of the fraction submitted to the corresponding process steps. Hemicellulose was not reflected in Table 3 since no quantitative amounts of xylose and arabinose could be detected in the bleached samples. Therefore, the small hemicellulose amount in the extraction residues R-170 and R-180 (Table 2) was practically eliminated after the first alkaline bleaching cycle. Likewise, approximately 65–75% of lignin from GS was removed during the two bleaching cycles without significant effect on the hydrogen peroxide concentration. After bleaching, the R-180 sample had slightly lower lignin content than R-170, which suggests a better aptitude for bleaching of the sample extracted at the highest temperature, as previously observed for other lignocellulosic residues submitted to a previous SWE step [14,20]. The combined treatments provoked a remarkable reduction of ashes in the cellulosic fractions due to their progressive lixiviation in the liquid phases promoted by the weakening of the matrix and the increased exposure area during the different treatments. Figure 3 shows the contents (g/100 g sample) of cellulose and lignin reached after the two bleaching cycles in comparison to those of the non-bleached residues (R-170 and R-180).

The results indicated that alkaline bleaching with hydrogen peroxide was not effective in removing the total lignin content of the SWE residues (R-170 and R180). However, a significant reduction of lignin content was observed in both materials, even at the first bleaching cycle, without a remarkable effect of hydrogen peroxide concentration. The final lignin content in the bleached samples (more than 20%wt.) was higher than those previously reported for other lignocellulosic matrices, such as rice straw [14] or almond shells [20] by applying a similar combined process (SWE plus alkaline bleaching with hydrogen peroxide). This can be mainly attributed to the high lignin content of GS and the particular structural association of lignin with other macromolecular components of grape stalks [2]. Ping et al. [11] also achieved low degrees of delignification of grape stalks using dilute sulphuric acid and ethanol organosolv treatments. These authors also stated that the valorisation of grape stalks as a cellulose source is challenging due to their high content of lignin and tannins.

Table 3. Composition of grape stalk (GS), subcritical water extraction residues (SWE) at 170 °C (R-170) and 180 °C (R-180), and the corresponding bleached residues (BR), as referred to the initial GS basis, together with the global mas yield of each fraction. Mean values ± standard deviation. Bleached residues are coded as the BR temperature of SWE concentration (% v/v) of alkaline hydrogen peroxide number of 1 h bleaching cycles.

Sample	Cellulose (g/100 g GS)	Klason Lignin (g/100 g GS)	Ash (g/100 g GS)	Global Yield (g Fraction/100g GS)
GS	22.3 ± 4^{e}	14.4 ± 1.3^{e}	7.6 ± 0.1^{e}	-
R-170	17.1 ± 1.9^{bcd}	11.0 ± 0.9^{abc}	0.8 ± 0.0^{abc}	54 ± 2^{d}
BR-170-4-1	18.9 ± 0.3^{bcd}	6.0 ± 0.0^{abc}	0.5 ± 0.0^{abc}	30 ± 2^{c}
BR-170-4-2	15.0 ± 2.5^{bcd}	4.4 ± 0.5^{abc}	0.3 ± 0.1^{abc}	21 ± 3^{b}
BR-170-8-1	14.3 ± 1.1^{bcd}	7.3 ± 0.8^{d}	0.6 ± 0.0^{d}	25 ± 2^{b}
BR-170-8-2	13.0 ± 1.5^{bcd}	5.4 ± 0.5^{abc}	0.3 ± 0.0^{abc}	22 ± 2^{b}
R-180	14 ± 1.0^{bcd}	12 ± 0.1^{abc}	0.8 ± 0.1^{abc}	58 ± 2^{d}
BR-180-4-1	12.4 ± 1.4^{bcd}	6.8 ± 0.1^{abc}	0.3 ± 0.0^{abc}	23 ± 2^{b}
BR-180-4-2	9.2 ± 1.1^{ab}	3.5 ± 0.1^{ab}	0.2 ± 0.0^{ab}	15 ± 2^{a}
BR-180-8-1	12 ± 2.0^{bcd}	5.7 ± 0.7^{abc}	0.3 ± 0.0^{abc}	22 ± 2^{b}
BR-180-8-2	10.3 ± 2.0^{cd}	3.1 ± 0.5^{cd}	0.3 ± 0.1^{cd}	16 ± 2^{a}

[a–e] Different letters in the same column indicate significant differences among samples ($p < 0.05$).

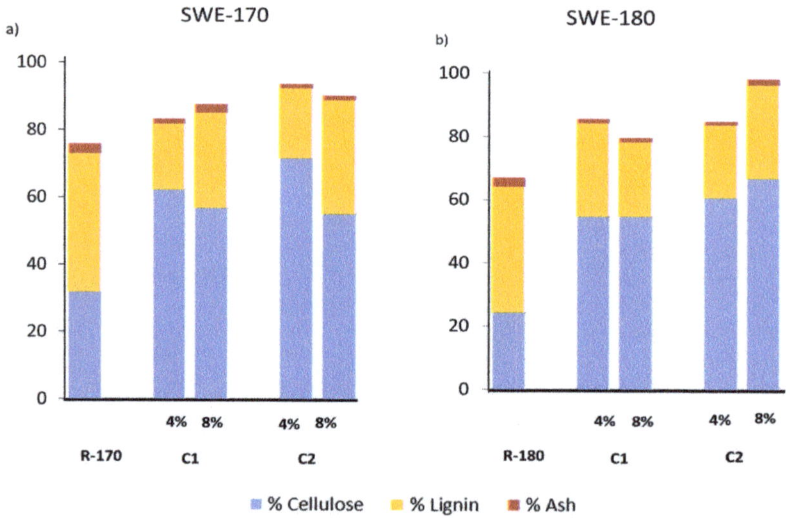

Figure 3. Cellulose, lignin, and ash content of residues after subcritical water extraction (SWE) at 170 °C (R-170) and 180 °C (R-180) and after two bleaching cycles (C1 and C2) with hydrogen peroxide solutions at 4 and 8% v/v. (**a**) Results for SWE at 170 °C (LSD intervals: 13.04 for cellulose, 8.10 for lignin, and 2.49 for ash). (**b**) Results for SWE at 180 °C (LSD intervals: 23.95 for cellulose, 11.63 for lignin, and 0.88 for ash).

Concerning the cellulose yield obtained in the different treatments, expressed as the mass of cellulose recovered with respect to that of raw GS, the overestimated concentration of this component in GS samples by the applied quantification method should be considered. A more realistic cellulose content would be determined in the SWE residues where hemicellulose is present in a very small proportion. Thus, considering 14 g/100 g of GS (as determined in the R-180 sample with the lowest hemicellulose content), a practically total recovery of cellulose from the R-170 samples was achieved after the two bleaching cycles. In contrast, approximately 25% of the GS cellulose was lost during the bleaching treatments of R-180 samples. This loss could be attributed to the cellulose oxidation by the oxygen radicals through the formation of alpha hydroxyalkyl radicals, giving rise to chain scission and carbonyl compounds [36]. The greater progress of this cellulose degradation in the R-180 sample could be attributed to the greater exposure area of this sample, where a higher extraction ratio was promoted by SWE, which facilitates the access of radicals to the cellulose points.

3.4. Characterisation of the Cellulose Fibres

The properties of the different lignocellulosic fractions were studied in terms of their morphological characteristics, FTIR analysis, X-ray diffraction patterns, and thermal degradation. Figure 4 shows representative FESEM images of the two-bleaching-cycle fibres from R-170 and the R-180 samples in comparison with the non-bleached fibres (R-180 and GS samples).

Untreated GS fibres showed uneven surfaces, where the fibrils are bound together by hemicellulose and lignin, which is in line with previous observations [37]. The SWE insoluble residues showed smoother surfaces than untreated GS, which agrees with the removal of different surface compounds during the SWE treatment. The structural changes caused by SWE are linked to the compositional modifications caused by the process, such as the quantitative extraction of hemicellulose and partial removal of lignin, causing relevant morphological changes in the fibre structure. Bleached samples still showed an increase in the fibre surface smoothness and some defibrillation without appreciable differences due to the different SWE temperatures and hydrogen peroxide concentrations. This morphological change must also be related to the progressive removal of the residual hemicellulose and lignin, giving rise to the typical morphology of more purified cellulose fibres previously described by other authors [37,38]. Thus, the bleached fibres achieved less cohesive entanglement fibrils in the bundles, with clear voids previously filled with other non-cellulosic constituents.

The FTIR spectra of the non-bleached (GS, R-170, and R-180) and bleached fibres are shown in Figure 5. The bands agreed with those reported in previous studies for GS materials [14,39]. A strong absorbance band due to OH^- stretching at 3600–3000 cm^{-1} was observed, which is associated with the presence of this moiety in cellulose, hemicellulose, and lignin. The two peaks at approximately 2900 cm^{-1} correspond to the asymmetric and symmetric vibration of C-H, respectively. A strong band at 1033 cm^{-1} is found due to C-O stretching and bending. Characteristic lignin peaks appear at 1606, 1509, and 1423 cm^{-1}, assigned to aromatic skeletal vibrations of syringil and guayacil aromatic rings [8,40]. Likewise, the band at 1730 cm^{-1} is attributed at C=O in ester groups, present in phenolic and uronic acids, and aromatic compounds in lignin and tannins. The band corresponding to the β-linkage of cellulose appears at 898 cm^{-1} [41].

After SWE treatment, the carbonyl vibration band practically disappears from the FTIR spectra. In contrast, the vibration band of cellulosic β-linkage was enhanced, according to cellulose enrichment of the samples with the hemicellulose removal and other sugars and phenols. After bleaching (Figure 5), the typical vibration bands of lignin aromatic rings still had a notable intensity in the FTIR spectra, along with the remaining amount of these compounds in the fibres, as deduced from the compositional analysis (Figure 3).

In lignocellulosic natural fibres, cellulose shows a semicrystalline structure, unlike hemicellulose and lignin, which are amorphous and have no effect on total crystallinity [38].

The X-ray diffraction spectra of GS, the SWE-insoluble fractions (R-170 and R-180), and the bleached residues (BR) coming out from the second bleaching at either 4% or 8% hydrogen peroxide (BR-170-4, BR-170-8, BR-180-4, and BR180-8) are shown in Figure 6, where the diffraction peaks characteristic of the type I cellulose crystalline lattice at 2θ = 15°, 16°, 22°, and 34° have been indicated [42].

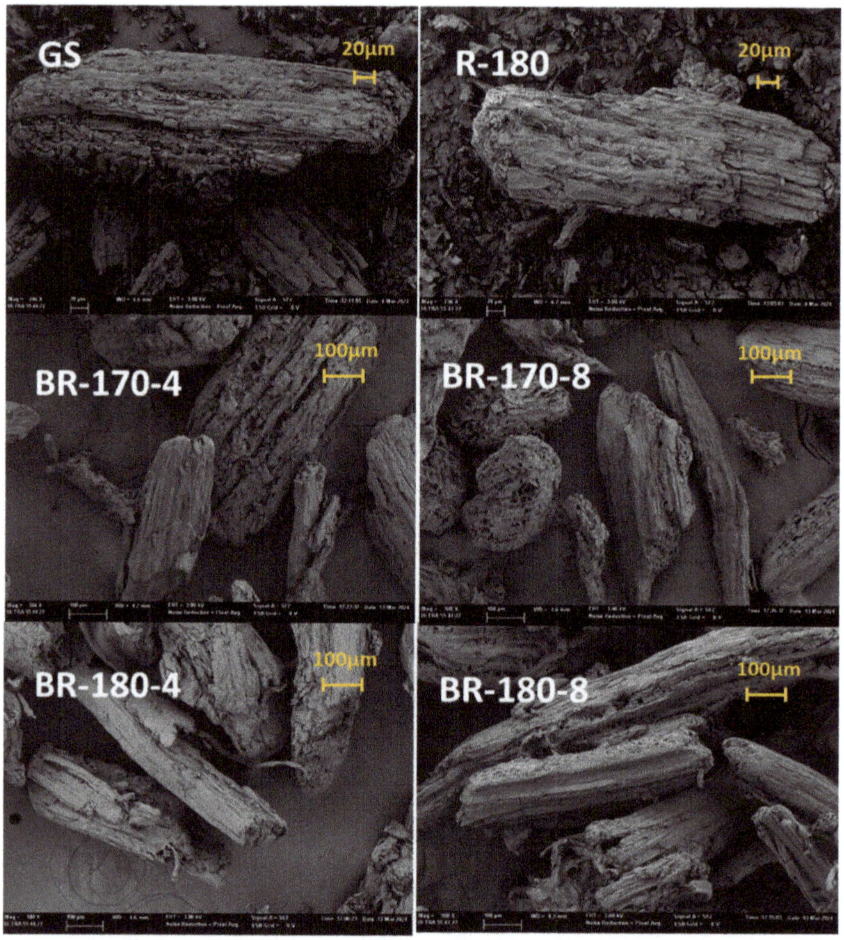

Figure 4. FESEM images of cellulose fibres after two bleaching cycles with 4% or 8% (*v*/*v*) alkaline hydrogen peroxide solutions (BR-170-4, BR-170-8, BR-18-4, and BR-180-8), compared with non-bleached fibres from subcritical water extraction at 180 °C (R-180) and grape stalk (GS) samples.

GS presented an X-ray pattern similar to previous studies, with two low-intensity peaks at around 15–16° and 22° [38,43] and a low crystallinity index. The extraction of non-cellulosic compounds during the SWE treatments only implied a slight increase in the crystallinity index of the material while promoting the intensity of the peaks at 15°. In contrast, the bleaching step had a noticeable effect on crystallinity. The progression of cellulose purification with the partial elimination of the amorphous fractions entailed a remarkable increase in the crystallinity index. At the same time, the diffraction peak at 22° became narrower and more defined, according to the typical cellulose X-ray diffraction pattern. No relevant differences in the crystallinity of the different bleached samples, with CI ranging between 54–58%, were found. These values are lower than those reported by

Prozil et al. [2], who reported extremely high values (75.4%) compared with those found in purified cellulose from other biomass (e.g., 55–65% in wood). As the final lignin content was high in every bleached sample, a low crystallinity degree is expected in the fibres where the final cellulose content ranged between 60–70%, depending on the SWE residue and bleaching agent concentration. However, the BR-180 samples contained approximately 65% cellulose, regardless of the H_2O_2 concentration used, and the final cellulose contents of BR-170 samples were approximately 60 and 70%, respectively, for samples treated with H_2O_2 at 8 and 4%.

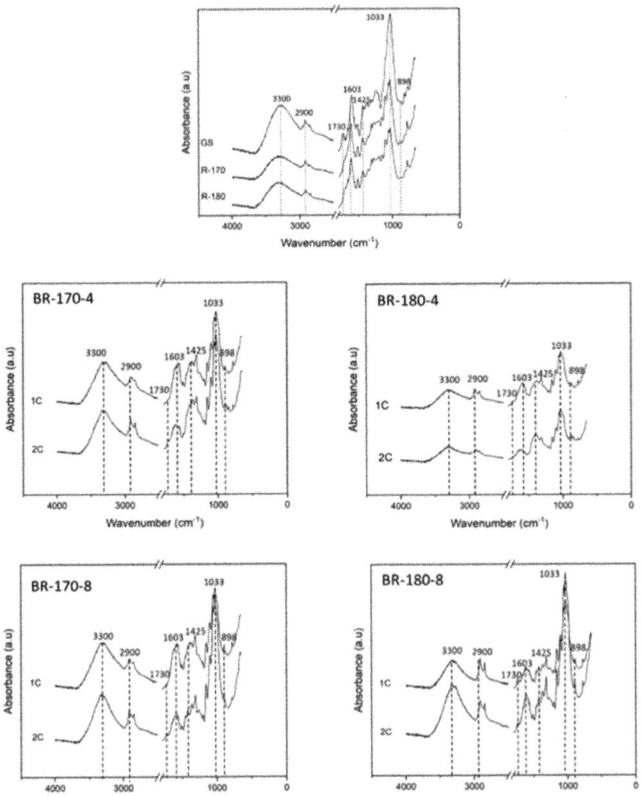

Figure 5. FTIR spectra of grape stalk (GS), non-bleached samples (R-170 and R-180) and samples bleached with 4 or 8% v/v alkaline hydrogen peroxide solutions (BR-170-4, BR-170-8, BR-180-4, and BR-180-8) after one and two bleaching cycles (1C–2C).

Figure 7 shows the thermal degradation behaviour of the non-bleached (GS, R-170, and R-180) samples and one-and-two-cycles-bleached fibres through the TGA curves and their corresponding derivative curves. The SWE treatment improved thermal stability due to the removal of low molecular compounds and a great part of hemicellulose, which degrade at lower temperatures. Thermal decomposition of GS occurred in several mass loss steps. The first event at approximately 100 °C corresponded to the loss of bonded water, as observed in previous studies [37,44]. The mass loss at approximately 300 °C was attributed to cellulose degradation, which occurred at a lower temperature than the usually reported for this pure polymer due to the lack of purity. The peak at approximately 200 °C in GS samples can be attributed to the degradation of hemicellulose, which is not observed quantitatively in the samples after SWE treatment due to hemicellulose removal. The peak degradation temperature of pure hemicellulose has been reported to be approximately 270 °C [39], but structural differences and the presence of other compounds can shift this temperature to lower values. After SWE, the removal of different low molecular compounds and hemicellulose was reflected in the displacement of the peak temperature, which was mainly attributed to the cellulose degradation towards a higher temperature, in accordance with the higher cellulose purity. The decomposition of pure lignin has been reported to occur gradually over a wide range of temperatures [39]. The observed peak between 400 °C and 600 °C in GS and the R-170 and R-180 samples can be assigned to the present lignin in the samples, whose content increased in the R-170 and R-180 residues due to the removal of other SWE extractables. SWE affected the composition of the lignin fraction, modifying its thermal stability and shifting the peak temperature towards lower temperatures. Condensed tannins in the GS could also contribute to this degradation step [38].

In the bleached samples, a sharper weight loss step associated with cellulose degradation was observed, which was in line with higher cellulose richness. However, the onset and peak temperature were shifted to lower values: 276–278 °C in the first bleaching cycle and 285 °C in the R-170 sample after the second cycle. Additionally, at T > 325 °C, a thermal resistant mass was observed in almost all the samples, which was supposed to be approximately 50 and 40% of the samples in the first and second bleaching cycles, respectively. The latter could be associated with partially degraded compounds by the oxidative treatment (including lignin), which become more thermally stable. In the two-cycles-bleached R-170 samples, this thermally stable mass was absent, and only a second degradation peak associated with the residual lignin could be observed. Differences in the peak degradation temperature of cellulose from R-170 and R-180 samples also point to the greater oxidation/depolymerisation of cellulose in samples SWE-treated at 180 °C than at 170 °C, as commented on above due to the greater cellulose exposure to the action of radicals.

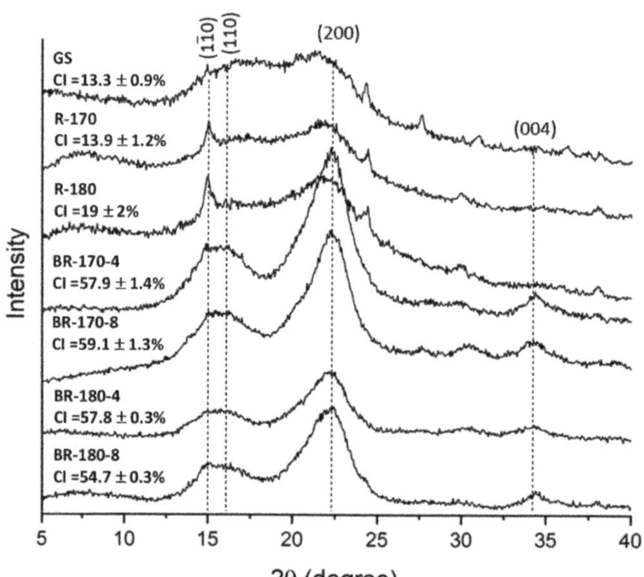

Figure 6. XRD spectra and crystallinity index (CI) of the grape stalk (GS), non-bleached samples (R-170 and R-180) and the two-cycles-bleached samples (BR) under different conditions (4 or 8% *v*/*v* alkaline hydrogen peroxide solutions).

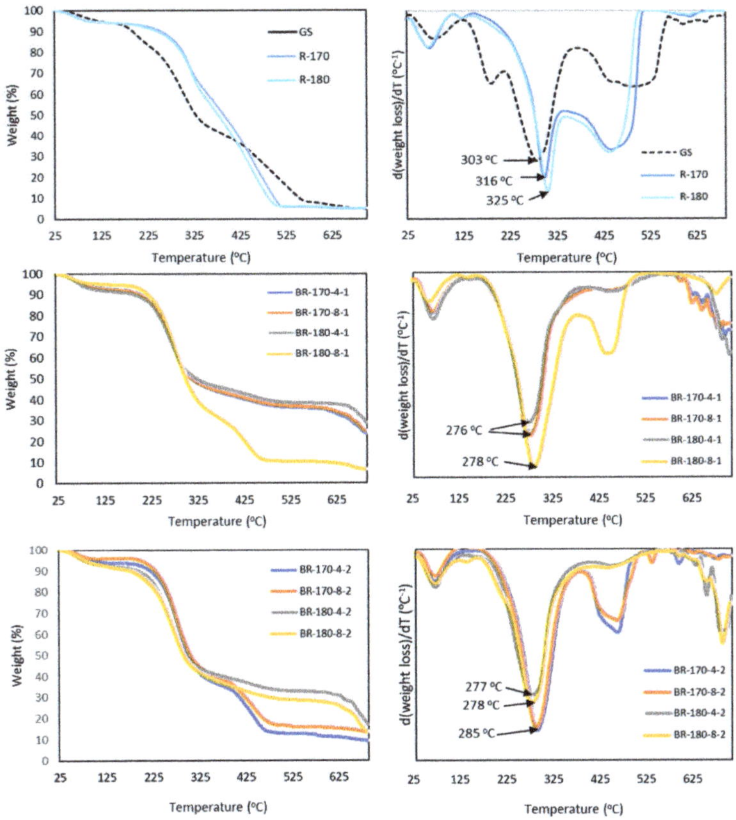

Figure 7. Thermogravimetric analysis (TGA) and DGTA curves of the grape stalk (GS), non-bleached subcritical water extraction residues (R-170 and R-180) and the bleached residues (BR) after one and two bleaching cycles with 4% or 8% (*v*/*v*) alkaline hydrogen peroxide solutions. Bleached residues are coded as BR temperature of subcritical water extraction concentration (%) of hydrogen peroxide-number of 1 h bleaching cycles.

4. Conclusions

Applying subcritical water extraction to grape stalk allows for obtaining phenolic-rich extracts and solid residues with high antioxidant capacity, which can be used in different industrial fields such as pharmaceutical, cosmetics, and food industries, including food packaging. The subcritical water extraction process partially removed lignin, and most of the hemicellulose from the solid fractions became enriched in cellulose. The increase in subcritical water extraction temperature promoted the extraction of polyphenolic compounds in the extracts and the neoformation of antioxidant brown compounds, enhancing the antioxidant power of both extracts and solid residues. Thus, subcritical water extraction is useful for designing sustainable grape stalk fractionation processes, obtaining soluble or dispersible products with high antioxidant activity.

Moreover, the recovery of cellulose from the insoluble fractions, obtained by subcritical water extraction at 170 or 180 °C, was possible by applying a green delignification process with alkaline hydrogen peroxide. The first bleaching cycle led to a maximum reduction in the lignin content of the insoluble fractions of 75% at both subcritical water extraction temperatures. After two bleaching cycles, the mass yield and the whiteness index did not significantly change, which indicated that no notable delignification progress occurred in the samples. However, the final lignin content in the bleached samples was still higher than 25%. The cellulose purity was higher in samples treated at 170 °C and bleached with 4% hydrogen peroxide than in those treated at 180 °C when greater cellulose oxidation during the bleaching step seems to be promoted regardless of the oxidant concentration. Therefore, despite the cellulose richness of grape stalk, its recovery and purification using sustainable processes need to be optimised in further studies.

Author Contributions: Conceptualisation, M.V., L.A. and A.C.; methodology, I.M., M.V., L.A. and A.C.; investigation, I.M., M.V. and L.A.; resources, A.C.; data curation, I.M., M.V. and L.A.; writing—original draft preparation I.M., M.V. and L.A.; writing—review and editing, I.M., M.V., L.A. and A.C.; supervision, M.V. and L.A.; project administration, A.C.; funding acquisition, A.C. All authors have read and agreed to the published version of the manuscript.

Funding: This research was funded by the AGROALNEXT program, grant number AGROALNEXT/2022/026 and was supported by MCIN with funding from the European Union NextGenerationEU (PRTR-C17.I1) and the project CIPROM/2021/071 (Generalitat Valenciana).

Institutional Review Board Statement: Not applicable.

Informed Consent Statement: Not applicable.

Data Availability Statement: The original contributions presented in the study are included in the article, further inquiries can be directed to the corresponding author.

Conflicts of Interest: The authors declare no conflicts of interest. The funders had no role in the design of the study; in the collection, analyses, or interpretation of data; in the writing of the manuscript; or in the decision to publish the results.

References

1. International Organisation of Vine and Wine 2022. Annual Assessment of the World Vine and Wine Sector in 2022. Available online: https://www.oiv.int/sites/default/files/documents/OIV_Annual_Assessment-2023.pdf (accessed on 23 October 2024).
2. Prozil, S.O.; Evtuguin, D.V.; Lopes, L.P.C. Chemical composition of grape stalks of *Vitis vinifera* L. from red grape pomaces. *Ind. Crops Prod.* **2012**, *35*, 178–184. [CrossRef]
3. Araújo, L.; Machado, A.R.; Pintado, M.; Vieira, E.; Moreira, P. Toward a circular bioeconomy: Extracting cellulose from grape stalks. *Eng. Proc.* **2023**, *37*, 86. [CrossRef]
4. Troilo, M.; Difonzo, G.; Paradiso, V.M.; Summo, C.; Caponio, F. Bioactive Compounds from Vine Shoots, Grape Stalks, and Wine Lees: Their Potential Use in Agro-Food Chains. *Foods* **2021**, *10*, 342. [CrossRef]
5. Spigno, G.; Pizzorno, T.; De Faveri, D.M. Cellulose and hemicelluloses recovery from grape stalks. *Bioresour. Technol.* **2008**, *99*, 4329–4337. [CrossRef]
6. Amendola, D.; De Faveri, D.M.; Egües, I.; Serrano, L.; Labidi, J.; Spigno, G. Autohydrolysis and organosolv process for recovery of hemicelluloses, phenolic compounds and lignin from grape stalks. *Bioresour. Technol.* **2012**, *107*, 267–274. [CrossRef] [PubMed]
7. Lorenzo, M.; Moldes, D.; Couto, S.R.; Sanroman, A. Improving laccase production by employing different lignocellulosic wastes in submerged cultures of Trametes versicolor. *Bioresour. Technol.* **2002**, *82*, 109–113. [CrossRef] [PubMed]
8. Prozil, S.O.; Evtuguin, D.V.; Silva, A.M.; Lopes, L.P. Structural characterisation of lignin from grape stalks (*Vitis vinifera* L.). *J. Agric. Food Chem.* **2014**, *62*, 5420–5428. [CrossRef]
9. Vallejo, M.; Cordeiro, R.; Dias PA, N.; Moura, C.; Henriques, M.; Seabra, I.J.; Malça, C.M.; Morouço, P. Recovery and evaluation of cellulose from agroindustrial residues of corn, grape, pomegranate, strawberry-tree fruit and fava. *Bioresour. Bioprocess.* **2021**, *8*, 25. [CrossRef]
10. Atatoprak, T.; Amorim, M.M.; Ribeiro, T.; Pintado, M.; Madureira, A.R. Grape stalk valorisation for fermentation purposes. *Food Chem. Mol. Sci.* **2022**, *4*, 100067. [CrossRef]
11. Ping, L.; Brosse, N.; Sannigrahi, P.; Ragauskas, A. Evaluation of grape stalks as a bioresource. *Ind. Crops Prod.* **2011**, *33*, 200–204. [CrossRef]
12. Spigno, G.; De Faveri, D.M. Antioxidants from grape stalks and marc: Influence of extraction procedure on yield, purity and antioxidant power of the extracts. *J. Food Eng.* **2007**, *78*, 793–801. [CrossRef]
13. Difonzo, G.; Troilo, M.; Casiello, M.; D'Accolti, L.; Caponio, F. Autohydrolysis Application on Vine Shoots and Grape Stalks to Obtain Extracts Enriched in Xylo-Oligosaccharides and Phenolic Compounds. *Molecules* **2023**, *28*, 3760. [CrossRef] [PubMed]
14. Freitas, P.A.; Santana, L.G.; González-Martínez, C.; Chiralt, A. Combining subcritical water extraction and bleaching with hydrogen peroxide to obtain cellulose fibres from rice straw. *Carbohydr. Polym. Technol. Appl.* **2024**, *7*, 100491. [CrossRef]
15. Senila, L.; Kovacs, E.; Scurtu, D.A.; Cadar, O.; Becze, A.; Senila, M.; Roman, C. Bioethanol production from vineyard waste by autohydrolysis pretreatment and chlorite delignification via simultaneous saccharification and fermentation. *Molecules* **2020**, *25*, 2606. [CrossRef] [PubMed]

16. Li, J.; Feng, P.; Xiu, H.; Zhang, M.; Li, J.; Du, M.; Zhang, X.; Kozliak, E.; Ji, Y. Wheat straw components fractionation, with efficient delignification, by hydrothermal treatment followed by facilitated ethanol extraction. *Bioresour. Technol.* **2020**, *316*, 123882. [CrossRef]
17. Argun, H.; Onaran, G. Delignification of vineyard pruning residues by alkaline peroxide treatment. *Ind. Crops Prod.* **2015**, *74*, 697–702. [CrossRef]
18. Mittal, A.; Katahira, R.; Donohoe, B.S.; Black, B.A.; Pattathil, S.; Stringer, J.M.; Beckham, G.T. Alkaline Peroxide Delignification of Corn Stover. *ACS Sustain. Chem. Eng.* **2017**, *5*, 6310–6321. [CrossRef]
19. Freitas, P.A.; Martín-Pérez, L.; Gil-Guillén, I.; González-Martínez, C.; Chiralt, A. Subcritical Water Extraction for Valorisation of Almond Skin from Almond Industrial Processing. *Foods* **2023**, *12*, 3759. [CrossRef]
20. Gil-Guillén, I.; Freitas, P.A.; González-Martínez, C.; Chiralt, A. Obtaining Cellulose Fibers from Almond Shell by Combining Subcritical Water Extraction and Bleaching with Hydrogen Peroxide. *Molecules* **2024**, *29*, 3284. [CrossRef]
21. Camarena-Bononad, P.; Freitas, P.A.; Chiralt, A.; Vargas, M. Subcritical water extraction for recovering cellulose fibres from *Posidonia oceanica* waste. *Carbohydr. Polym. Technol. Appl.* **2024**, *8*, 100550. [CrossRef]
22. Freitas, P.A.V.; González-Martínez, C.; Chiralt, A. Application of Ultrasound Pretreatment for Enhancing Extraction of Bioactive Compounds from Rice Straw. *Foods* **2020**, *9*, 1657. [CrossRef] [PubMed]
23. Brand-Williams, W.; Cuvelier, M.; Berset, C. Use of a free radical method to evaluate antioxidant activity. *LWT* **1995**, *28*, 25–30. [CrossRef]
24. Sluiter, A.D.; Ruiz, R.; Scarlata, C.; Templeton, D.W. Determination of Extractives in Biomass; Laboratory Analytical Procedure (LAP); Issue Date 17 July 2005. 2008. Available online: https://www.nrel.gov/docs/gen/fy08/42619.pdf (accessed on 23 October 2024).
25. Sluiter, A.; Hames, B.; Ruiz, R.; Scarlata, C.; Sluiter, J.; Templeton, D.; Crocker, D. Determination of Structural Carbohydrates and Lignin in Biomass: Laboratory Analytical Procedure (LAP). (Revised July 2011). 2008. Available online: https://www.nrel.gov/docs/gen/fy13/42618.pdf (accessed on 23 October 2024).
26. Segal, L.; Creely, J.; Martin, A.; Conrad, C. An Empirical Method for Estimating the Degree of Crystallinity of Native Cellulose Using the X-Ray Diffractometer. *Text. Res. J.* **1959**, *29*, 786–794. [CrossRef]
27. Ong, E.S.; Cheong, J.S.H.; Goh, D. Pressurised hot water extraction of bioactive or marker compounds in botanicals and medicinal plant materials. *J. Chromatogr. A* **2006**, *1112*, 92–102. [CrossRef]
28. Plaza, M.; Amigo-Benavent, M.; Del Castillo, M.D.; Ibáñez, E.; Herrero, M. Facts about the formation of new antioxidants in natural samples after subcritical water extraction. *Food Res. Int.* **2010**, *43*, 2341–2348. [CrossRef]
29. Egüés, I.; Serrano, L.; Amendola, D.; De Faveri, D.; Spigno, G.; Labidi, J. Fermentable sugars recovery from grape stalks for bioethanol production. *Renew. Energy* **2013**, *60*, 553–558. [CrossRef]
30. Freitas, P.A.; González-Martínez, C.; Chiralt, A. Influence of the cellulose purification process on the properties of aerogels obtained from rice straw. *Carbohydr. Polym.* **2023**, *312*, 120805. [CrossRef]
31. Zhang, J.; Wen, C.; Zhang, H.; Duan, Y.; Ma, H. Recent advances in the extraction of bioactive compounds with subcritical water: A review. *Trends Food Sci. Technol.* **2020**, *95*, 183–195. [CrossRef]
32. Freitas, P.A.V.; Meyer, S.; Hernández-García, E.; Rebaque, D.; Vilaplana, F.; Chiralt, A. Antioxidant and antimicrobial extracts from grape stalks obtained with subcritical water. Potential use in active food packaging development. *Food Chem.* **2024**, *451*, 139526. [CrossRef]
33. García-Pérez, J.V.; Cárcel, J.A.; Benedito, J.; Blasco, M.; Mulet, A. Extraction of antioxidant compounds from grape stalk dried at different conditions. *Defect Diffus. Forum* **2009**, *283*, 604–609. [CrossRef]
34. Cheng, Y.; Xue, F.; Yu, S.; Du, S.; Yang, Y. Subcritical Water Extraction of Natural Products. *Molecules* **2021**, *26*, 4004. [CrossRef] [PubMed]
35. Gullón, B.; Eibes, G.; Moreira, M.T.; Dávila, I.; Labidi, J.; Gullón, P. Antioxidant and antimicrobial activities of extracts obtained from the refining of autohydrolysis liquors of vine shoots. *Ind. Crops Prod.* **2017**, *107*, 105–113. [CrossRef]
36. Łojewska, J.; Miśkowiec, P.; Łojewski, T.; Proniewicz, L. Cellulose oxidative and hydrolytic degradation: In situ FTIR approach. *Polym. Degrad. Stab.* **2005**, *88*, 512–520. [CrossRef]
37. Borsoi, C.; Júnior, M.A.D.; Beltrami, L.V.R.; Hansen, B.; Zattera, A.J.; Catto, A.L. Effects of alkaline treatment and kinetic analysis of agroindustrial residues from grape stalks and yerba mate fibers. *J. Therm. Anal. Calorim.* **2019**, *139*, 3275–3286. [CrossRef]
38. Engel, J.B.; Luchese, C.L.; Tessaro, I.C. Insights on the properties of physically and chemically treated grape stalks. *Sustain. Mater. Technol.* **2022**, *34*, e00506. [CrossRef]
39. Yang, H.; Yan, R.; Chen, H.; Lee, D.H.; Zheng, C. Characteristics of hemicellulose, cellulose and lignin pyrolysis. *Fuel* **2007**, *86*, 1781–1788. [CrossRef]
40. Pujol, D.; Liu, C.; Fiol, N.; Olivella, M.À.; Gominho, J.; Villaescusa, I.; Pereira, H. Chemical characterisation of different granulometric fractions of grape stalks waste. *Ind. Crops Prod.* **2013**, *50*, 494–500. [CrossRef]
41. Abidi, N.; Cabrales, L.; Haigler, C.H. Changes in the cell wall and cellulose content of developing cotton fibers investigated by FTIR spectroscopy. *Carbohydr. Polym.* **2014**, *100*, 9–16. [CrossRef]
42. Chen, X.; Yu, J.; Zhang, Z.; Lu, C. Study on structure and thermal stability properties of cellulose fibers from rice straw. *Carbohydr. Polym.* **2011**, *85*, 245–250. [CrossRef]

43. Engel, J.B.; Ambrosi, A.; Tessaro, I.C. Development of biodegradable starch-based foams incorporated with grape stalks for food packaging. *Carbohydr. Polym.* **2019**, *225*, 115234. [CrossRef]
44. Dorez, G.; Ferry, L.; Sonnier, R.; Taguet, A.; Lopez-Cuesta, J. Effect of cellulose, hemicellulose and lignin contents on pyrolysis and combustion of natural fibers. *J. Anal. Appl. Pyrolysis* **2014**, *107*, 323–331. [CrossRef]

foods

MDPI

Article

Impact of Thermophysical and Biological Pretreatments on Antioxidant Properties and Phenolic Profile of Broccoli Stem Products

Claudia Bas-Bellver, Cristina Barrera and Lucía Seguí *

Institute of Food Engineering—FoodUPV, Universitat Politècnica de València, Camino de Vera, s/n, 46022 Valencia, Spain; clbabel@upv.es (C.B.-B.); mcbarpu@tal.upv.es (C.B.)
* Correspondence: lusegil@upvnet.upv.es

Abstract: Fruit and vegetable industrialisation is a major contributor to food waste; thus, its integral transformation into functional powders has gained attention. Pretreatments can be incorporated into valorisation processes to generate structural or biochemical changes that improve powders' characteristics. This study deepens into the impact of biological (fermentation, FERM) and thermophysical (autoclaving, AUTO; microwaves, MW; ultrasound, US; and pasteurisation, PAST) pretreatments, combined with dehydration (hot air-drying, HAD; or freeze-drying, FD) on the characteristics of powdered products obtained from broccoli stems. The impact of pretreatments on physicochemical (moisture, water activity, total soluble solids) and antioxidant properties (phenols, flavonoids, antioxidant capacity by ABTS and DPPH) on residue and powdered products was studied, together with their impact on plant tissue structure (Cryo-SEM) and the powders' phenolic profile (HPLC). Probiotic viability was also determined on the fermented samples. The pretreatments applied, particularly the ultrasound, improved the antioxidant properties of the broccoli stems compared to the unpretreated samples, in line with microscopic observations. Dehydration did also improve the antioxidant attributes of the broccoli wastes, especially drying at 60 °C. However, pretreatments combined with dehydration did not generally lead to an improvement in the antioxidant properties of the powders. Probiotic properties were preserved in the freeze-dried products (>10^7 CFU/g). In conclusion, pretreatments may be applied to enhance the antioxidant attributes of broccoli wastes, but not necessarily that of dried powdered products.

Keywords: broccoli stem; waste valorisation; dehydrated powders; pretreatments; fermentation; ultrasounds; microwaves; antioxidants; functional ingredient

1. Introduction

The food industry generates a large amount of waste, a current issue related to the scarcity of natural resources [1], with waste being of major relevance in the horticulture industry [2]. It has been estimated that around 45% of total fruit and vegetable production is lost each year during the various stages of the food chain, from fields to processing and consumption [3].

For many years, the continuous production of waste has been encouraged by a linear economy design of food systems, increasing the discharge of waste to the environment, and ending its passage through the food chain. With the 2030 Agenda for Sustainable Development, it is proposed that the linear economy is to be replaced by a circular economy design, which comprises all activities related to the reintroduction of waste into production, distribution, and consumption processes [1], contributing to the achievement of the Sustainable Development Goals (SDGs).

One of the most widely grown cruciferous vegetables in the world is broccoli (*Brassica oleracea* var. *italica*) [4], with a global production higher than 26 million tonnes, together with cauliflower, in 2022 [5], which is directly related to a disproportionate increase in waste generation. Commonly, only the broccoli florets are used for food purposes, so in the fresh-cut or frozen industry, 47% of broccoli is classified as waste (stem and leaf) [4], despite its richness in minerals, vitamins, bioactive compounds (such as glucosinolates, phenolic compounds, and carotenoids), and antioxidant properties that can prevent oxidative stress that contributes to pathogenesis [6–8].

Fresh stems have a high moisture content, which promotes spoilage [3,9], for which broccoli residues need to be processed into products with improved stability and shelf life. One of the approaches to valorise these waste products is the production of powdered functional ingredients through processes based on disruption, dehydration, and milling [10]. Vegetable powders may have applications as colouring or flavouring agents, as well as in fortifying foods and improving their nutritional value [7,11–13].

Hot air-drying (HAD) is one of the most widely used dehydration techniques to increase the stability of foods and allow their proper preservation [14]. However, the characteristics of the resulting product depend on the process variables [15,16]. Another method used in product preservation is freeze-drying (FD), which is gaining relevance in food stabilisation as it leads to high-quality products due to the low temperature and vacuum conditions applied [17]. The response of vegetal material to drying can be modified by the application of pretreatments, which may improve water diffusion through the tissue and reduce drying times [18]. Additionally, pretreatments cause tissue disruption, which facilitates the extraction of bioactive compounds and the release of enzymes that promote biochemical reactions, generating more active forms and thereby increasing the bioactivity of certain compounds [19]. Pretreatments can also contribute to the release of active compounds which are naturally bound to cellular structures [20]. One of the simplest pretreatments that can be applied to facilitate water transfer and release compounds of interest is physical tissue disruption. The impact of grinding intensity, as well as freeze–thaw pretreatment, have already been evaluated in previous studies [10,11]. In the present work, other pretreatments which have the potential to modify plant structures are investigated.

Biological treatments include enzymatic and microbial treatments. Vegetables' fermentation with lactic acid bacteria is a common practice in the food industry. Fermentation may modify the vegetable matrix and release antioxidant compounds. Lactic acid bacteria (LAB) fermentation allows the transformation of polyphenols and other bioactive compounds into more bioactive forms through the production of certain enzymes [21–23]. Among LAB, *Lactiplantibacillus plantarum*, a QPS-status microorganism, has been proposed for broccoli fermentation due to its ability to grow in this plant matrix [23,24].

Thermophysical pretreatments include classical and emerging technologies. In this case, the structure of the plant tissue might be modified due to a high temperature or pressure, as in the case of autoclave or pasteurisation treatments, or due to the response of the interaction of the plant material with mechanical (ultrasounds) or electromagnetic waves (microwaves). The interaction with microwaves generates thermal and non-thermal effects which may cause structural changes due to the vapour explosion generated in overheating points (hotspots). The interaction of the plant matrix with electromagnetic waves accelerates physicochemical reactions by heating, which, together with structural modifications, makes it possible to reduce the drying time, improving the nutritional profile of the products [25,26]. Ultrasound is an emerging technology increasingly applied in food product and process research which, combined with drying, can reduce energy expenditure and result in final products with improved properties [27,28]. It is known that ultrasound pretreatment shortens the drying time for both HAD and FD [29] due to the so-called 'sponge effect', which enhances the intracellular water transfer to the surface [27], as well as the dispersion of intracellular components [30]. Pasteurisation and autoclaving are thermal pretreatments widely used in the food processing industry [31]. On the one hand, autoclaving is performed under high pressure and temperature conditions, which have been linked to cell wall breakage due to cellulose solubilisation [32]. Thus, damage to cell walls can promote the extraction of bioactive compounds and increase their bioaccessibility [31]. Pasteurisation has been proposed for the inactivation of undesired enzymes, as well as to preserve the nutritional content of food products [33]; furthermore, the use of this pretreatment is also justified by the need of reducing microbial loads prior to fermentation.

The aim of this work was to evaluate the impact of different pretreatments (microwaves, ultrasounds, autoclave, pasteurisation, and fermentation with *L. plantarum*) and drying techniques (HAD and FD) on the physicochemical, antioxidant, and structural properties of broccoli ground stems, as well as to evaluate their convenience as pretreatments to obtain sustainable broccoli powdered ingredients with improved functional properties.

2. Materials and Methods

2.1. Raw Material

Broccoli (*Brassica oleracea* var. *italica*) heads were purchased from a local supermarket in Valencia (Valencia, Spain) and the stems, which corresponds with the IV range or frozen bags waste, were manually separated using a knife. The fresh broccoli stems were washed with a 1% (v/v) sodium hypochlorite in water solution. After that, the stems were disrupted in a food processor (Thermomix®, Vorwerk, Madrid, Spain) for 8 s at 10,000 rpm [34] before undergoing the processes described below.

2.2. Inoculum Preparation

Lactiplantibacillus plantarum spp. CECT 749 (Colección Española de Cultivos Tipo, Valencia, Spain) was used to inoculate the broccoli waste. This microorganism was selected on the basis of its potential probiotic effect [35], its ability to degrade different polysaccharides, and its metabolic diversity and adaptability [36].

The freeze-dried strain was reactivated in Man, Rogosa, and Sharpe (MRS) broth (Scharlab, Barcelona, Spain) and incubated at 37 °C for 24 h (Incugidit, PSelecta, Barcelona, Spain), as described in [34]. The starter culture obtained contained 9.4 ± 0.2 log CFU/mL (according to plate count measurements). All materials used for microbiological analyses were conveniently sterilised in an autoclave at 120 °C for 2 h (Systec GmbH VB-40, Linden, Germany).

2.3. Powder Manufacturing

2.3.1. Preliminary Fermentation and Drying Study

A preliminary study was carried out to evaluate the effect of fermentation and drying on the properties of powdered products. Powders were obtained by HAD at 50, 60, or 70 °C or FD, from fermented and unfermented ground broccoli stems (Figure 1). To reduce the variability due to the sample origin, all powders were obtained from the same batch. The broccoli stems were ground and mixed, distributed as explained next, and processed simultaneously.

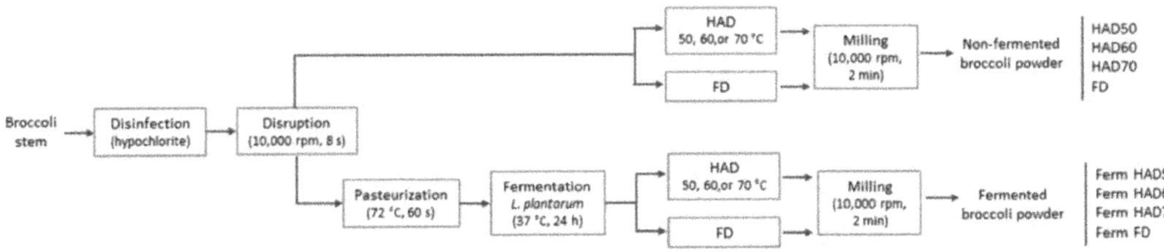

Figure 1. Process flow diagram for the integral transformation of broccoli stems into non-fermented and fermented powdered products. HAD: hot air-drying; FD: freeze-drying; Ferm: fermentation.

To proceed with fermentation, ground broccoli residue was pasteurised in a hot water bath to reduce the initial microbial load. For this aim, the sample was distributed in glass jars containing 200 g of the ground residue, which were immersed in a hot water bath at

82 °C, until reaching a temperature of 72 °C in the geometric centre, and maintained for 1 min. A total of 2 mL of the prepared inoculum (~10^9 CFU/mL) was added to each glass jar containing 200 g of broccoli and incubated at 37 °C for 24 h. All the material used in this process was previously sterilised in an autoclave (Systec GmbH VB-40, Linden, Germany) at 120 °C for 2 h.

HAD and FD were used as dehydration techniques. In this preliminary study, HAD was carried out on a laboratory scale, in three temperature-controlled bench tray dryers (Gastroback, Natural Plus 46600, Hollenstedt, Germany), each one at a corresponding temperature (50, 60, or 70 °C). The sample was prepared in 1 cm-thick layers and placed on the drier-perforated trays with a load of 200 g/tray. HAD was carried out with a constant air temperature of 50, 60, or 70 °C until a a_w value below 0.3 was reached, to guarantee the stability of the dehydrated residue [37]. For FD, fermented and unfermented residues were distributed in 1 cm-thick layers in aluminium trays. The samples were first deep-frozen at −40 °C in a CVN-40/105 freezer (Matek, Barcelona, Spain) and next freeze-dried for 48 h (LyoQuest-55 freeze-dryer, Terrasa, Spain) at −45 °C (condenser temperature) and at a sub-atmospheric pressure (0.1 mbar). After dehydration, either by HAD or FD, the dried residue was milled in a Thermomix® food processor (10,000 rpm for 2 min in 30 s intervals) [34] to yield a fine powder. The powders were then stored in twist-off glass jars in a light-free environment at room temperature until analysis.

2.3.2. Impact of Thermophysical and Biological Pretreatments on Ground Broccoli Stems and Powdered Products

Thermophysical and biological pretreatments were applied to the residue prior to dehydration (Figure 2). Again, all the raw material used in this series of experiments came from one batch, and was ground and processed simultaneously, to reduce variability due to the sample origin. Thus, the broccoli ground waste was distributed into sterile twist-off glass jars (200 g per jar) and subjected to the corresponding pretreatment, which were the following: autoclave at 120 °C for 5 min (Systec GmbH VB-40, Linden, Germany), microwave oven at 4 W/g for 5 min (Samsung GW72N, Samsung Electronics, Suwon, Republic of Korea), ultrasound at 40 kHz for 10 min (Ultrasons-H, Selecta, Barcelona, Spain), and fermentation with *Lactiplantibacillus plantarum* for 24 h at 37 °C (Incugidit, PSelecta, Barcelona, Spain) after pasteurisation. In addition, fresh and pasteurised residues were used as controls. The selection of pretreatment conditions was based on the available literature [20], along with unpublished results from our own laboratory. Pasteurisation and fermentation processes were carried out as previously described (Section 2.3.1).

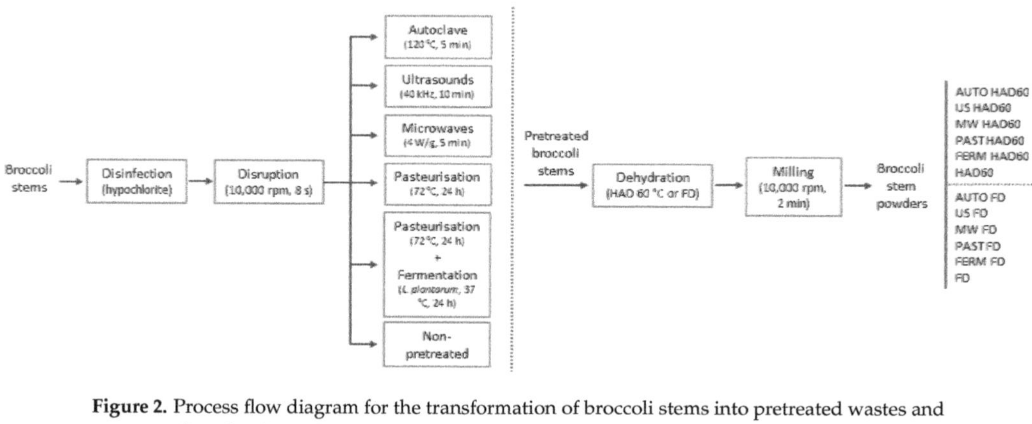

Figure 2. Process flow diagram for the transformation of broccoli stems into pretreated wastes and corresponding dried powders. HAD: hot air-drying; FD: freeze-drying; Ferm: fermentation.

Fresh and pretreated samples were then dehydrated by HAD or FD. In this case, all HAD samples were dried under the same conditions and simultaneously to ensure the homogeneity of the treatment. Therefore, the HAD of the fresh and pretreated samples was carried out in a pilot plant convective transverse flow tray dryer (CLW 750 TOP+, Pol-Eko-Aparatura SPJ, Katowice, Poland) with air at 60 °C and an air velocity of 2 m/s for 10 h. The drying time and temperature were decided based on the results of preliminary test. Samples were spread in 1 cm-thick layers on perforated dryer trays with a load of 200 g/tray (2 trays/pretreatment). The FD, milling, and storage of powders were performed as explained earlier (Section 2.3.1).

2.4. *Characterisation of Intermediate and Final Broccoli Stem Products*

2.4.1. Microbial Counts

The viability of *Lactiplantibacillus plantarum* cells was determined by serial dilution (10^{-1} to 10^{-8}) with sterile buffered peptone water (Scharlab, Barcelona, Spain) and subsequent surface seeding on MRS agar (Scharlab, Barcelona, Spain), followed by incubation at 37 °C for 24–48 h. The first dilution (10^{-1}) was obtained by homogenising in stomacher equipment (Interscience, BagMixer® 400 model, St Nom, France) for 2 min 3 g of solid sample (fresh residue or dehydrated powder) with 27 mL of sterile peptone water. After incubation, colonies present on the plates were counted.

2.4.2. Physicochemical Properties: Water Activity, Moisture Content, and Soluble Solids

Water activity (a_w) was obtained with an Aqualab® 4TE dew point hygrometer at 25 °C (Decagon Devices Inc., Pullman, Washington, DC, USA). Moisture content (x_w) was obtained as described in the official method of the AOAC 934.06 [38], based on the weight loss before and after drying in a vacuum oven (Vaciotem-T, JP Selecta, Barcelona, Spain) at 60 °C and 200 mbar to a constant weight. The mass fraction of total soluble solids (x_{ss}) was calculated from the moisture content and the measurement of the Brix degrees read at 20 °C in a thermostatic Abbe refractometer NAR-3T (Atago, Tokyo, Japan). In the powders, measurements were obtained from an aqueous extract (1:10 (*w*/*v*) ratio).

2.4.3. Antioxidant Properties: Total Phenols and Flavonoids and Antiradical Activity

Antioxidant compounds were extracted by mixing 4 g of undried sample or 0.5 g of powdered product with 10 mL of 80% (*v*/*v*) methanol/water solution, followed by shaking (WY-100 horizontal shaker, COMECTA, Barcelona, Spain) for 1 h and further centrifugation for 5 min at 10,000 rpm in an Eppendorf centrifuge (5804/5804R, Eppendorf SE, Hamburg, Germany).

Total phenols were measured by the Folin–Ciocalteu method [39,40]. Thus, 0.125 mL of the extract was mixed with 0.5 mL of bidistilled water and 0.125 mL of the Folin–Ciocalteu reagent (Scharlab S.L., Barcelona, Spain) and let react in darkness for 6 min. Then, 1.25 mL of a 7% (*w*/*v*) sodium carbonate solution was added together with 1 mL of bidistilled water. Absorbance was measured at 760 nm, after 90 min in darkness, in a Cary 60 UV/Vis spectrophotometer. Results were given as mg of gallic acid equivalents (GAE) per g of dry matter, using gallic acid as the standard (purity $\geq$ 98%, Sigma-Aldrich, St Louis, MO, USA).

The total flavonoid content was determined following the modified aluminium chloride method [41]. A total of 1.5 mL of the extract and 1.5 mL of a 2% (*w*/*v*) aluminium chloride solution (Thermo Fisher Scientific Inc., Waltham, MA, USA) were mixed and reacted for 10 min in darkness. Absorbance was measured at 368 nm in a Cary 60 UV/Vis spectrophotometer. Results were given as mg of quercetin equivalents (QE) per g of dry matter, using quercetin as the standard (purity $\geq$ 95%, Sigma-Aldrich, St Louis, MO, USA).

The DPPH (1,1 diphenyl-2-picryl hydrazyl) and ABTS (2,20-azobis-3-ethyl benzothiazolin-6-sulphonic acid) methods were applied for determining the antioxidant activity of the samples. For the former [42], 0.1 mL of the extract was mixed with 2.9 mL of a 0.1 mM DPPH solution in methanol (Merck KGaA and affiliates, Darmstadt, Germany) and let react for 60 min in darkness. Then, absorbance was measured at 575 nm in a Cary 60 UV/Vis spectrophotometer (Agilent Technologies, Santa Clara, CA, USA). Results were expressed as mg of trolox equivalent (TE) per gram of dry matter. As for ABTS [43], 0.1 mL of the extract was mixed with 2.9 mL of an ABTS+ (VWR International LLC, Radnor, PA, USA) solution in phosphate buffer with an absorbance of 0.70 ± 0.02 at 734 nm. After 7 min of reacting, absorbance was measured at 734 nm in a Cary 60 UV/Vis spectrophotometer. Results were given as mg of trolox equivalent (TE) per g of dry matter, using trolox as standard (purity $\geq$ 97%, Sigma-Aldrich, St Louis, MO, USA).

2.4.4. Phenolic Constituents by High Performance Liquid Chromatography (HPLC)

The phenolic profile of the powdered products was also determined. For this aim, the method proposed by Caprioli et al. [44] and Giusti et al. [45] was used, with some modifications. The methanolic extracts of the powders were prepared using an 80% (v/v) HPLC-grade methanol (Scharlab S.L., Barcelona, Spain) solution in double distilled water as a solvent at an extraction ratio of 1:10 (w/v). The mixture was shaken for 1 h in a horizontal shaker, kept in the dark, and centrifuged at 10,000 rpm for 5 min. The supernatants were collected and filtered through 0.45 µm PTFE filter (Scharlab S.L., Barcelona, Spain) and the resulting extract was analysed by HPLC. Two extracts per powder were prepared.

The equipment used for the HPLC analyses was HPLC 1200 Series Rapid Resolution equipment coupled to a diode detector (Agilent, Palo Alto, CA, USA). Separations were carried out on a Kinetex column (250 × 4.6 mm I.D., 5 µm; Phenomenex; Torrance, CA, USA) at 31 °C and using 1% formic acid (Scharlab S.L., Barcelona, Spain) as mobile phase A and acetonitrile (Scharlab S.L., Barcelona, Spain) as mobile phase B. The gradient used was as follows: 0 min, 90% A; 25 min, 40% A; 26 min, 20% A; held for 30 min; 35 min, 90% A, with holding for up to for 40 min. The injection volume was 10 µL and the flow rate was 0.5 mL/min. The phenolic compounds were identified by their retention times and spectra, as compared to HPLC reference standards (Extrasynthese, Genay, France), at a wavelength of 280 nm. The compounds were quantified by standard curves and results were given as mg/100 g_{dm}.

2.5. Cryo-Scanning Electron Microscopy (Cryo-SEM)

The impact of the pretreatments on the microstructure of the ground broccoli stems was examined by cryo-SEM. For that purpose, a ZEISS ULTRA55 microscope (Carl Zeiss AG, Oberkochen, Germany) (Microscopy Service of the Universitat Politècnica de València), equipped with an external cryo-freezing chamber (OXFORD mod. CT-1500, Oxford, UK), was used. For the sample preparation, a small fragment of a fresh or fresh and pretreated ground broccoli stem was placed on the slide and fixed with a mixture of two components, colloidal graphite dispersion in water (G303 Colloidal Graphite AQUADAC, Agar Scientific, Stansted, UK) and a tissue fixative (Tissue-Tek AutoTEC® a120, Sakura, Barcelona, Spain), to ensure proper fixation for subsequent fracture. The sample was placed in a cryo-chamber and frozen with liquid nitrogen (−210 °C) under vacuum conditions. The sample was fractured inside the observation chamber to expose the inner tissue microstructure. Subsequently, the sample was sublimated for 20 min at −85 °C with a pressure of 10^{-5} mmHg, after which a platinum coating was applied to provide the electron beam with a suitable reflecting surface. The coating or quorum sputtering was performed for 15 s. Finally, the samples were observed through the screen of a computer equipped with ZEISS SmartSEM software https://www.zeiss.com/microscopy/de/produkte/software/zeiss-smartsem.html (accessed on 12 September 2024) (Carl Zeiss AG, Oberkochen, Germany), at −150 °C and 10–20 kV. During tissue observation, micrographs were taken and saved at appropriate magnifications to allow sample comparison.

2.6. Statistical Analysis

Analytical determinations were performed in the powdered products obtained. Measurements were conducted at least in triplicate. Statistical analysis was carried out with

Statgraphics Centurion XVIII software (version 17.1.04) (StatPoint Technologies, Inc., Warrenton, VA, USA). Analyses of variance (one-way ANOVA and multifactorial ANOVA) were performed with a confidence level of 95% ($p < 0.05$), after checking the data normality. A multiple range test using Fisher's LSD method was performed to discriminate among the means and identify homogeneous groups.

3. Results and Discussion

3.1. Impact of Drying on Fermented and Non-Fermented Broccoli Stems

The results of the moisture content (x_w), water activity (a_w), and total soluble solids content (x_{ss}) of the fresh, fermented, and dehydrated broccoli stems are shown in Table 1.

Table 1. Water content (x_w), water activity (a_w), and total soluble solids content (x_{ss}) of non-dehydrated samples (fresh, fermented) and dehydrated (by hot air-drying at 50, 60, or 70 °C (HAD) or freeze-drying (FD)) powders obtained from fermented (Ferm HAD, Ferm FD) and unfermented (HAD, FD) broccoli stems. Mean ± standard deviation of three replicates.

Treatment	x_w (g_w/g_{total})	a_w	x_{ss} (g_{ss}/g_{dm})
FRESH	0.9163 ± 0.0011^{f}	0.9932 ± 0.0014^{g}	0.69 ± 0.04^{de}
FERM	0.9128 ± 0.0007^{f}	0.9920 ± 0.0012^{g}	0.654 ± 0.008^{bcd}
HAD 50	0.054 ± 0.005^{a}	0.23 ± 0.02^{c}	0.600 ± 0.016^{a}
HAD 60	0.054 ± 0.002^{a}	0.215 ± 0.011^{bc}	0.599 ± 0.016^{a}
HAD 70	0.055 ± 0.003^{a}	0.249 ± 0.006^{d}	0.61 ± 0.00^{ab}
Ferm HAD 50	0.093 ± 0.012^{c}	0.200 ± 0.006^{ab}	0.673 ± 0.017^{cde}
Ferm HAD 60	0.1075 ± 0.0014^{de}	0.193 ± 0.012^{a}	0.637 ± 0.017^{abc}
Ferm HAD 70	0.0991 ± 0.0010^{cd}	0.2110 ± 0.0016^{bc}	0.64 ± 0.00^{abc}
FD	0.075 ± 0.011^{b}	0.280 ± 0.013^{e}	0.72 ± 0.00^{e}
Ferm FD	0.115 ± 0.006^{e}	0.330 ± 0.013^{f}	0.70 ± 0.04^{de}

a–g Different superscript letters in the same column indicate statistically significant differences at the 95% confidence level (p-value < 0.05), according to the multiple range test.

The moisture content and a_w values for the fresh and fermented samples were within the range obtained in previous studies [34] and implied a risk of spoilage, for which dehydration is justified. No significant differences were found between the fermented and non-fermented samples regarding moisture content and water activity. As observed, the soluble solids content remained constant after fermentation. This behaviour has also been observed in mixed mango and carrot juice fermented with *L. plantarum* [46], likely due to microorganisms' action on more complex polysaccharides. During fermentation, bacteria degrade simple sugars but also use glycosidases and glycosyl hydrolases enzymes, among others, to breakdown polysaccharides from the plant matrix into sugar monomers that are more easily metabolised [46,47]. As expected, dehydration caused a significant reduction in the a_w and x_w (p-value < 0.05) to values which ensure a prolonged shelf life [48]. Fermentation caused a reduction in the drying times, since the processing time needed to reduce the a_w to safe values was as follows: 10 h at 50 °C, 7 h at 60 °C, and 5 h at 70 °C; this was in contrast to the drying of non-fermented broccoli residue which was completed in 12 h at 50 °C, 10 h at 60 °C, and 6 h at 70 °C. The fermented powders exhibited a higher moisture content than their non-fermented counterparts, a difference which could be explained by their structural breakdown and decompartmentalisation due to microbial action. However, this increase in moisture content was not observed in the water activity values, which were similar or slightly lower than in the non-fermented samples. These results suggest that microbial growth and metabolism modifies the way in which water interacts with the matrix, so fermented powders contain a higher proportion of bound water.

The soluble solids content of the powders was in the range of that reported in previous studies [10]. As evidenced in the literature, dehydration and subsequent milling may contribute to soluble solids' increase due to the release of soluble compounds from broken cells and the breakage of fibres into simpler compounds [11,49,50]. This was observed in the FD powders, both fermented and non-fermented, which exhibited higher x_{ss} values (p-value < 0.05) than the HAD powders, a result which can be explained by the higher efficiency of milling in the FD samples due to the fragility and porous structure which is characteristic of FD products [14]. Thus, FD facilitates milling, resulting in a smaller particle size and promoting fibres' breakdown and soluble compounds' release. On the contrary, HAD can cause crusting phenomena, also known as case-hardening, which generates rubbery cores that make subsequent milling more difficult, resulting in powders with a lower soluble solids content [10,51,52].

Figure 3 shows the antioxidant properties values of the fresh and fermented stems before and after dehydration by the two techniques tested. Values are shown for the total phenol content (mg GAE/g_{dm}), total flavonoid content (mg QE/g_{dm}), and antioxidant capacity (mg TE/g_{dm}) by the ABTS and DPPH radical methods. Among the non-dehydrated samples, only the total flavonoids content exhibited statistically significant differences, with a decrease in the fermented samples. Similar results were reported by Septembre-Malaterre et al. [47], who found a slight decrease in antioxidant properties (total phenol, flavonoid content, and DPPH scavenging activity) after the fermentation of cabbage with *L. plantarum* at 37 °C for 24 h. Similarly, Kiczorowski et al. [22] obtained a lower content of total phenols in fermented broccoli compared to an unfermented control. However, these results differ from those obtained in other studies. Li et al. [21] reported that *L. plantarum* improved the antioxidant capacity of apple juice, due to the consumption of glucose molecules available in phenolic compounds, thus generating metabolites with more hydroxyl groups or less steric hindrance. Also, Kwaw et al. [53] studied the impact of *L. plantarum* on the antioxidant activity of mulberry juice, finding that during fermentation there was a release of soluble phenolic compounds from the plant cell walls, increasing the antioxidant activity [54]. These disparities have already been discussed in the literature. According to Knez et al. [55], although one of the fundamental aspects of fermentation is the increase in antioxidant potential due to the release of phenols and other antioxidants from the plant matrix, the methodology for analysing antioxidant potential is not yet standardised, and there may be variations between the results of different studies. In addition, the fermentation of plant foods is a complex combination of factors in which interactions between microbiological, enzymatic, chemical, and biochemical reactions and physical processes occur [56], adding variability to determination methods and making it difficult to standardise procedures.

Antioxidant properties significantly increased for samples after the HAD and subsequent milling compared to their corresponding controls, but not for the FD samples. The multifactorial ANOVA considering the factors of fermentation and dehydration, and the interaction between them, revealed that both factors and their interaction were significant for the four antioxidant parameters analysed (p-value < 0.05) except for flavonoids, in which the interaction was not significant (p-value = 0.3450). For some drying conditions such as HAD there were statistically significant differences between the fermented and non-fermented samples, whereas these differences were not statistically significant when FD was applied, particularly for the total flavonoids and antioxidant capacities. The best results were obtained for the HAD60 powders, which exhibited higher values for all the antioxidant parameters analysed. However, increasing the temperature to 70 °C had a negative impact on antioxidant properties, as compared to lower temperatures. Similar results were reported by other authors in kiwifruit slices, in which the total phenol content increased significantly compared to a fresh control when drying at 60 °C [57]. The improvement of the antioxidant properties of vegetables after HAD has been reported by several authors [58–61]. Improved antioxidant properties can be explained by the formation of new antioxidant compounds as a result of biochemical reactions, such as Maillard reactions, due to exposure to high temperature [56,62] or isomerisations to more active forms [63], together with the reduction in certain pro-oxidant enzymes capable of degrading antioxidant compounds [15,16,51].

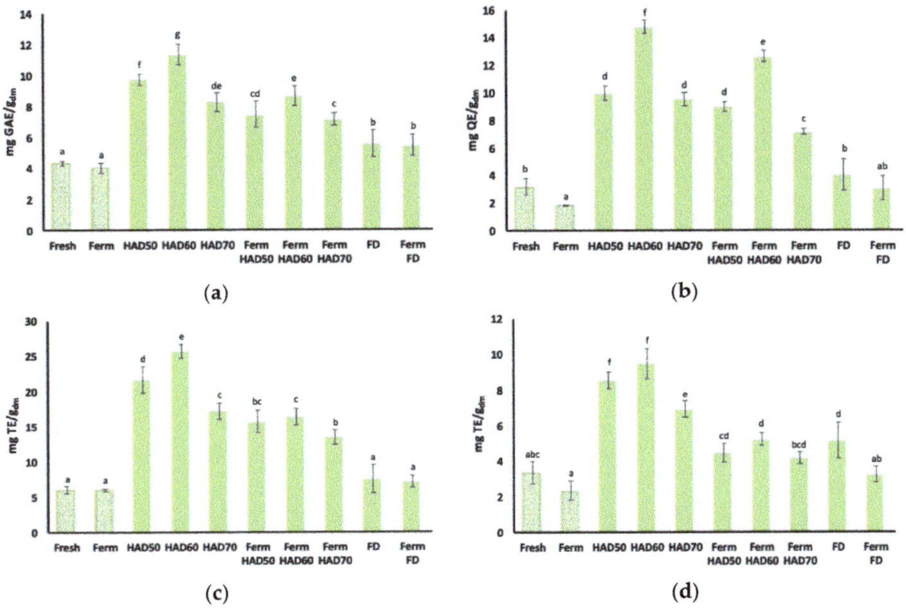

Figure 3. Antioxidant properties of fresh and fermented broccoli stems and their respective powders obtained by different dehydration techniques: (**a**) Total phenolic content expressed as mg gallic acid equivalent (GAE) per gram of dry matter. (**b**) Total flavonoid content expressed in mg quercetin equivalent (QE) per gram of dry matter. (**c**,**d**) ABTS and DPPH antioxidant capacity, respectively, expressed in mg trolox equivalent (TE) per gram of dry matter. HAD: hot air-drying at 50, 60, or 70 °C; FD: freeze-drying; Ferm: fermentation. Filled columns represent dehydrated samples, and dotted columns represent non-dried ones. Error bars correspond to standard deviation of six replicates from two replicas (3 replicates/replica). $^{a–g}$ Different superscript letters indicate statistically significant differences at 95% confidence level (p-value < 0.05), according to multiple range test.

On the other hand, the Ferm HAD powders showed a decrease in antioxidant parameters, as compared to the dried samples which had not been fermented prior to dehydration (HAD). Differences between the Ferm HAD and HAD powders were more significant than differences between the fermented and non-fermented fresh residue. One possible explanation could be related to reducing sugars' consumption by microorganisms [64], which could decrease Maillard reactions' incidence. The decreased antioxidant properties of the Ferm HAD powders could also be attributed to the action of bacterial enzymes, which would have released antioxidant compounds from the structures [65], which would then become more exposed to oxygen and high temperatures during HAD. Differences between the fermented and non-fermented samples after HAD were statistically significant in all cases, but less remarkable for the flavonoids than for the other antioxidant parameters analysed.

The FD powders generally presented decreased antioxidant properties as compared to the HAD ones (Figure 3). This could be explained by a higher exposure to oxidative conditions once the atmospheric pressure is restored, due to the higher porosity of FD products [65]. The high porosity of FD products allows easy access to oxygen, which may lead to higher levels of oxidation or the degradation of bioactive compounds [66,67]. These phenomena would explain the lower content of antioxidant compounds in the powders obtained by FD compared to the HAD ones. Dalmau et al. [68] reported similar results in apples subjected to FD as compared to HAD at 60 °C. Also, Rudy et al. [69] observed a decrease in total phenol content in FD blueberries. Similarly, previous studies carried out in the same laboratory on residues of various vegetables confirmed that HAD generally results in products with higher antioxidant capacity than FD ones. However, the presence of specific bioactive compounds such as carotenoids or sulforaphane is usually higher in FD products [34,70].

3.2. Probiotic Properties of Broccoli Waste Powders

Fermentation can be explored as a pretreatment for modifying the physicochemical and antioxidant properties of a residue and the corresponding powder but also for obtaining probiotic products when probiotic strains are used as starters, on the condition that microbial viability is preserved. Figure 4 shows the results for *L. plantarum* counts in freshly inoculated broccoli stems, after 24 h of fermentation, and in the corresponding powders obtained by HAD at 50, 60, and 70 °C and FD.

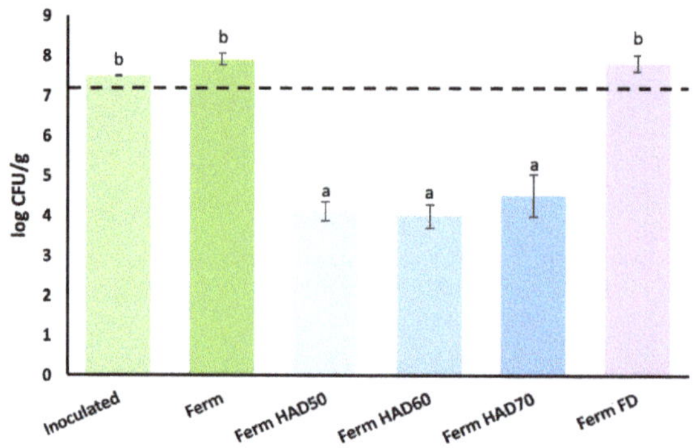

Figure 4. Viable counts in freshly inoculated broccoli residue after 24 h of fermentation with *L. plantarum* and in powders obtained from the fermented stems dried by hot air-drying (HAD) at 50, 60, and 70 °C and freeze-drying (FD). Error bars represent the standard deviation of four replicates from two replicas (two replicates/replica). ^{a,b} Different superscript letters indicate statistically significant differences at the 95% confidence level (p-value < 0.05), according to the multiple range test.

To be considered probiotic, a product must contain around 10^6–10^8 CFU/g viable cells [71,72], with 10^7 CFU/g generally being the minimum probiotic concentration found in commercial products [73].

The broccoli residue freshly inoculated with the starter culture had a viable cells concentration higher than 10^7 CFU/g. After 24 h of fermentation, the fermented broccoli waste had a slightly increased concentration, confirming broccoli waste as a favourable plant matrix for *L. plantarum* growth, as reported by other authors [23,24].

The results evidenced a statistically significant impact of the drying technique on microbial viability (p-value < 0.05). Whereas FD allowed the preservation of microorganisms, HAD significantly reduced microbial counts below the minimum considered probiotic, regardless of the temperature applied.

These results are consistent with previous lab results [16] and other authors' research [74]. Thus, FD is confirmed as a technique for preparing high-value stable probiotic cultures [75], as well as a modern and innovative method of drying and processing primary agricultural products [76]. However, this technique presents high equipment and operating costs and requires specialised staff [77]. Another option would be optimising HAD conditions to improve cell viability, by proposing a progressive decreasing temperature during drying, for instance.

After this first part of the study, and based on the antioxidant properties of the dried powders, it was decided to proceed with HAD at 60 °C for the second part of the research, in which pretreatments were investigated prior to HAD and FD.

3.3. Impact of Thermophysical and Biological Treatments on Broccoli Stem Products

3.3.1. Physicochemical, Antioxidant, and Microstructural Properties of Broccoli Wastes as Affected by Thermophysical and Biological Pretreatments

The impact of the pretreatments applied on the moisture content, water activity, and soluble solids content of the broccoli wastes are shown in Table 2. As observed, the moisture content was quite similar for all samples, but slightly lower for the AUTO and MW (p-value < 0.05). According to the literature, autoclaving can reduce moisture content due to evaporation caused by high temperatures [78]; similarly, the MW treatment may heat up the product due to electromagnetic energy conversion into heat and evaporate part of the liquid water [79]. Similar results were obtained in a previous study on camellia seed samples subjected to microwave pretreatment at 640 W per 500 g of a sample for 5 and 8 min, where a significant reduction in moisture content was obtained [80].

Table 2. Effect of thermophysical and biological pretreatments on the moisture content (x_w), water activity (a_w), and total soluble solids content (x_{ss}) of broccoli stems. PAST: pasteurisation; AUTO: autoclaving; MW: microwaves; FERM: fermentation; US: ultrasounds. Mean ± standard deviation of three replicates.

Treatment	x_w (g_w/g_{total})	a_w	x_{ss} (g_{ss}/g_{dm})
FRESH	0.934 ± 0.004^{c}	0.9910 ± 0.0009^{ab}	0.66 ± 0.04^{b}
PAST	0.934 ± 0.002^{c}	0.9940 ± 0.0013^{c}	$0.65 \pm 5\ 0.018^{b}$
AUTO	0.928 ± 0.004^{b}	0.9931 ± 0.0009^{bc}	0.66 ± 0.00^{b}
MW	0.913 ± 0.004^{a}	0.9947 ± 0.0009^{c}	0.678 ± 0.006^{c}
FERM	0.938 ± 0.003^{c}	0.990 ± 0.003^{a}	0.55 ± 0.08^{a}
US	0.93 ± 0.00^{c}	0.9948 ± 0.0014^{c}	0.669 ± 0.011^{b}

a–c Different superscript letters in the same column indicate statistically significant differences at the 95% confidence level (p-value < 0.05), according to the multiple range test.

As for a_w, the fresh and pretreated samples exhibited values higher than 0.99, which imply a high susceptibility to spoilage due to enzymatic action or microbial growth [81,82].

The PAST, AUTO, and US treatments did not significantly affect the soluble solids content in the ground broccoli stems. In contrast, the application of MW implied a significant increase, whereas fermentation also reduced the soluble solids content significantly (p-value < 0.05). Other authors have also reported soluble solids' increase after MW treatment [83,84], which could be related to water evaporation and subsequent solutes' concentration due to microwave heating [85]. In addition, the interaction of biological materials with microwaves generates thermal and non-thermal effects which may cause structural changes due to the vapour explosions generated in overheating points (hotspots) as a consequence of the heterogeneous heating [86]. These points can undergo self-explosion phenomena that cause ruptures and structural modifications such as the depolymerisation of cellulose and solubilisation of lignin, releasing simpler sugars among other constituents [86,87]. Regarding FERM, the decrease in soluble sugars may be due to microorganisms using them as an energy source [88]. Similarly, Peng et al. [89] confirmed that after fermenting apple juices of different cultivars with *Lactobacillus* spp., there was a decrease in sucrose, lactose, and glucose content due to their consumption by probiotics. However, this result differs from those obtained in the first part of this study.

The antioxidant properties of the pretreated broccoli wastes are shown in Table 3, where values are given for the total phenol content (mg GAE/g_{dm}), total flavonoid content (mg QE/g_{dm}), and antioxidant capacity measured by the ABTS and DPPH methods (mg TE/g_{dm}).

Table 3. Effect of different pretreatments on total phenol content (mg GAE/g_{dm}), total flavonoid content (mg QE/g_{dm}), and antioxidant capacity by ABTS and DPPH (mg TE/g_{dm}) of fresh broccoli stems. PAST: pasteurisation; AUTO: autoclaving; MW: microwaves; FERM: fermentation; US: ultrasounds. Mean ± standard deviation of six replicates from two replicas (3 replicates/replica).

Treatment	Total Phenols (mg GAE/g_{dm})	Total Flavonoids (mg QE/g_{dm})	ABTS (mg TE/g_{dm})	DPPH (mg TE/g_{dm})
FRESH	4.2 ± 0.2 ^{a}	2.05 ± 0.17 ^{a}	5.8 ± 0.2 ^{a}	1.01 ± 0.10 ^{a}
PAST	6.1 ± 0.004 ^{c}	2.58 ± 0.18 ^{b}	9.5 ± 0.4 ^{c}	3.5 ± 0.7 ^{b}
AUTO	4.05 ± 0.11 ^{a}	4.1 ± 0.2 ^{c}	7.56 ± 0.14 ^{b}	1.9 ± 0.2 ^{a}
MW	4.8 ± 0.3 ^{ab}	2.02 ± 0.11 ^{a}	7.6 ± 0.3 ^{b}	3.87 ± 0.14 ^{b}
FERM	5.3 ± 0.3 ^{bc}	2.9 ± 0.3 ^{b}	7.9 ± 0.4 ^{b}	1.43 ± 0.07 ^{a}
US	5.79 ± 0.18 ^{c}	4.88 ± 0.12 ^{d}	9.2 ± 0.7 ^{c}	3.7 ± 0.2 ^{b}

$^{a–c}$ Different superscript letters in the same column indicate statistically significant differences at the 95% confidence level (p-value < 0.05), according to the multiple range test.

Pretreatments generally improved the antioxidant properties as compared to the fresh broccoli stems (p-value < 0.05). The pasteurisation and ultrasound pretreatments resulted in the highest values. US promote cavitation phenomena which cause plant tissues' breakage and microchannels' formation [27,28], which may lead to the release of antioxidant compounds bound to complex structures in the plant matrix. Different examples in the literature report US efficiency for increasing the antioxidant properties of vegetable matrices [90–93]. Regarding the increase in antioxidant properties after pasteurisation, similar results were reported by Urquieta-Herrero et al. [94], who obtained changunga pulp enriched in phenolic compounds after pasteurisation. In a different study about fruit juices, pasteurisation pretreatment increased the antioxidant capacity measured by the ABTS and DPPH methods [9]. This increase in antioxidant properties after pasteurisation pretreatment may be due to the temperature-induced disruption and permeabilisation of the plant matrix [95]. Also, during heat pretreatment with pasteurisation, biochemical reactions may occur, resulting in forms of antioxidant compounds with higher bioactivity [96].

Fermentation also caused an improvement in the antioxidant properties, except for the ability to scavenge the DPPH free radical. Other studies have reported an increase in antioxidant properties after fermentation with LAB in broccoli samples [34,97] or in other products such as fermented loquat juice, where higher phenols and flavonoids contents were obtained [98]. This increase could be due to the LAB, which promote the hydrolysis of complex molecules, like polyphenols and other bioactive compounds, into free and simple forms with higher bioactivities [21,23,98,99]. This is achieved through the production of certain enzymes, such as glycosidases, tannases, and esterases, which convert phenolic esters into aglycones and phenolic acids with greater antioxidant activities. Moreover, the disruption of protein–polyphenol complexes by LAB proteases may further increase phenolic content [100,101] and the LAB's production of new antioxidant compounds [102]. Fermentation also promotes the breakdown of sugars, vitamins, and other compounds present in plant matrices [54], thus releasing phenolic constituents that would otherwise remain bound to the plant matrix. On the contrary, other authors have reported opposite results, such as those obtained in the present research in the preliminary study of fermentation and drying. Another example was found in the lactic acid fermentation of apple juice, where Wu et al. [103] reported a dramatic decrease in total phenols and flavonoids. All the previous information confirms that this pretreatment shows variability in its efficacy and could be influenced by interactions between various microbiological, enzymatic, chemical, and biochemical reactions and physical processes [56].

The MW pretreatment also caused an improvement in some antioxidant properties, which were statistically significant for the DPPH and ABTS radicals' scavenging activities. As previously stated, the increase in antioxidant compounds may be due to the vapour explosions generated at the points where there is overheating, also known as hotspots [86,104], which occur because of the direct interaction of the food or plant material with microwaves. These hotspot explosions may release simpler phenolic compounds, which otherwise remain bound to the structure, in a more complex organisation. MW treatment has shown efficiency in lignocellulosic biomass, such as sorghum grains [105], where it was shown that the antioxidant capacity of microwave-treated samples increased significantly. The impact of MW on phenolics content and antioxidant properties may also be due to improved extractability. Álvarez et al. [106] reported that MW pretreatments intensified phenolics extraction in apple pomace and boosted anthocyanin product richness. During MW treatment, moisture is heated, leading to vaporisation and increased pressure within the vacuole; consequently, the porous cell wall ruptures, releasing phenolics from the solid and facilitating their extraction [107]. The AUTO pretreatment was the least efficient in improving the antioxidant properties of the ground broccoli stems, maybe due to the thermal decomposition of antioxidant compounds because of the temperature effect [108]. Nevertheless, the total flavonoids and ABTS antiradical activity improved with respect to the control sample in the autoclaved samples, suggesting that this thermophysical treatment may also release certain antioxidant compounds due to the high pressure and heat reached due to their effect on the tissue structure. Other authors have found a positive effect of AUTO in the phenolic content of nuts, an increase which was attributed to higher extraction yields, the formation of Maillard reaction products, and the possible release of some bound phenolic compounds due to processing conditions [109]. Nevertheless, the results were variable among nuts (pistachio, cashew, chestnut) and improved significantly when harsher autoclave conditions were applied. In addition, improvement was more significant for specific phenolic compounds and less remarkable for the antioxidant properties measured.

Microscopical images of the pretreated broccoli residue were obtained to verify the impact of the pretreatments on the plant tissue structure. Micrographs of selected pretreatments were obtained by scanning electron microscopy at low temperatures (cryo-SEM). Non-pretreated (a), fermented (b), ultrasonicated (c), and autoclaved (d) samples are shown in Figure 5.

The non-pretreated ground broccoli (Figure 5a) shows classic parenchymatic tissue in which large rounded cells with cell walls can be identified. The intercellular spaces or pores appear empty, with no reticulum due to liquid phase release, which is a sign of tissue integrity. In contrast, in the micrograph corresponding to the ultrasonicated residue (Figure 5b), it can be seen that some intercellular spaces or pores are occupied by fluid coming from inside the cells, which evidences cell walls' and membranes' permeabilisation due to cavitation phenomena [110]. Kumar et al. [111] confirmed that the generation and collapse of cavitation bubbles induces turbulence within the fluid, resulting in cell walls' and membranes' rupture releasing active compounds of interest. Also, we can observe (Figure 5b) the appearance of larger intercellular spaces and membrane separations from the corresponding cell walls due to mechanical vibration [112]. These structural changes, together with increased porosity, may facilitate the extraction of components from plant matrices [113]. These microscopic observations are consistent with the improvement in the antioxidant properties observed for the US pretreated samples.

Figure 5c shows a micrograph of the fermented tissue, where a significant loss of cell integrity and compartmentalisation of cell structures can be observed. While some cells maintain a defined shape, others show manifest signs of decompartmentalisation, such as an irregular shape and unstructured and degraded walls, suggesting a certain depolymerisation of the cell walls. Additionally, the liquid phase is present throughout all the tissue, indicating a greater loss of compartmentalisation and release of cellular compounds. This breakdown of the plant matrix is due to the ability of lactic acid bacteria, such as *L. plantarum*, to metabolise and transform complex and indigestible proteins, cellulose, and other substances into simpler ones [54]. These microscopic observations are consistent with the improved antioxidant properties obtained for the fermented residue, since the observed decompartmentalisation and depolymerisation contribute to the release of compounds with antioxidant activity (Table 3).

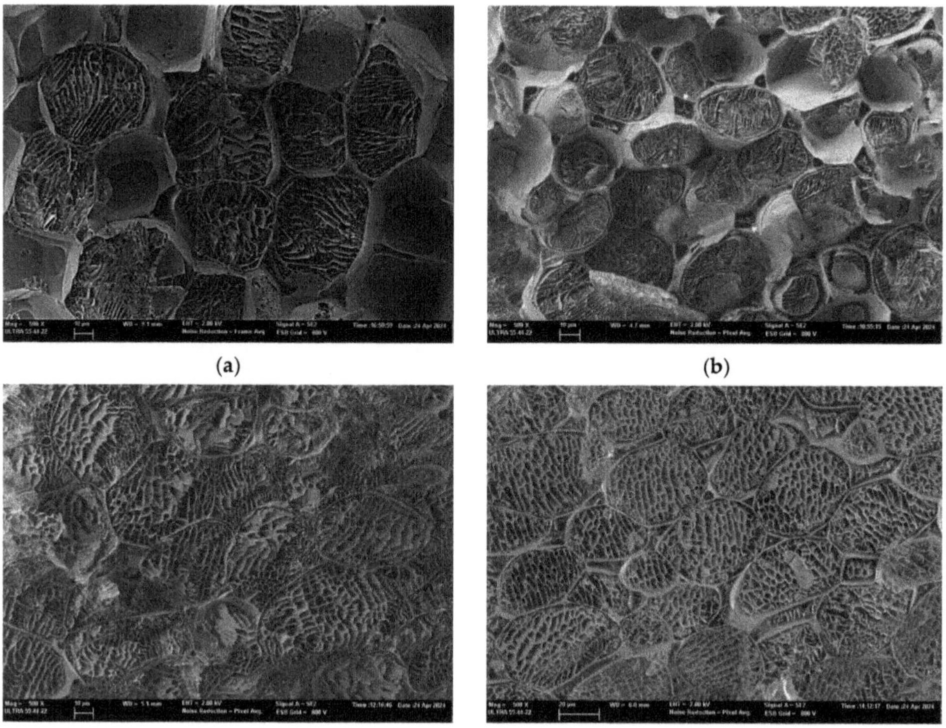

Figure 5. Micrographs of the pretreated residue obtained by low-temperature scanning electron microscopy (cryo-SEM) at 500× magnification (bar = 10 microns). (**a**) Ground broccoli residue; (**b**) ultrasonically pretreated broccoli residue; (**c**) fermented broccoli residue; (**d**) autoclaved broccoli residue. Arrows in (**b**) indicate intercellular spaces.

Finally, in the case of the residue which underwent autoclave pretreatment (Figure 5d), the differences with respect to the non-pretreated control (Figure 5a) are not as evident, which agrees with the less significant improvement in the antioxidant properties observed in this case. However, this high temperature and pressure treatment would also have caused some permeabilisation of membranes, as reticulated spaces are observed inside the pores and intercellular spaces, indicating the presence of the liquid phase both inside and outside the cells. In addition, the image obtained suggests less rigidity in the cell walls, which is consistent with some degradation or solubilisation of the structures.

3.3.2. Impact of Pretreatments on the Properties of Powdered Broccoli Stem Products

Table 4 shows the physicochemical properties of the powders obtained by the HAD and FD of the fresh and pretreated broccoli stems.

As expected, the powders showed significantly lower a_w and x_w values (p-value < 0.05) than their non-dehydrated counterparts (Table 2). These values fall within the range considered adequate to ensure the stability of such products (a_w between 0.20 and 0.35; x_w < 0.1 g_w/g_{total}), effectively preventing the growth of spoilage bacteria and extending the shelf life of the resulting powders [48,114]. FD resulted in powders with generally lower a_w values (p-value < 0.05), probably due to an enhanced efficiency of water removal achieved through the prior freezing of the ground residue, followed by sublimation and desorption [115], thus leading to the formation of porous channels as the ice sublimates.

Table 4. Values for moisture content (x_w), water activity (a_w), and total soluble solids content (x_{ss}) of powders obtained by hot air-drying or freeze-drying of fresh and pretreated broccoli stems. PAST: pasteurisation; AUTO: autoclaving; MW: microwaves; FERM: fermentation; US: ultrasounds. HAD: hot air-drying; FD: freeze-drying. Mean ± standard deviation of three replicates.

Treatment	x_w (g_w/g_{total})	a_w	x_{ss} (g_{ss}/g_{dm})
HAD60	0.059 ± 0.002^{de}	0.243 ± 0.005^{b}	0.670 ± 0.016^{bcdef}
PAST HAD60	0.068 ± 0.002^{f}	0.307 ± 0.005^{h}	0.655 ± 0.016^{abcd}
AUTO HAD60	0.060 ± 0.003^{e}	0.305 ± 0.005^{h}	0.64 ± 0.02^{abc}
MW HAD60	0.0493 ± 0.0011^{c}	0.268 ± 0.006^{de}	0.630 ± 0.011^{a}
FERM HAD60	0.115 ± 0.003^{g}	0.302 ± 0.004^{gh}	0.631 ± 0.011^{a}
US HAD60	0.0520 ± 0.0014^{cd}	0.26 ± 0.008^{cd}	0.688 ± 0.016^{def}
FD	0.033 ± 0.005^{a}	0.261 ± 0.008^{cd}	0.678 ± 0.011^{cdef}
PAST FD	0.059 ± 0.004^{de}	0.2495 ± 0.0017^{bc}	0.663 ± 0.016^{abcde}
AUTO FD	0.046 ± 0.004^{bc}	0.281 ± 0.006^{ef}	0.68 ± 0.02^{def}
MW FD	0.042 ± 0.003^{b}	0.260 ± 0.007^{cd}	0.70 ± 0.02^{f}
FERM FD	0.057 ± 0.010^{de}	0.208 ± 0.005^{a}	0.691 ± 0.005^{ef}
US FD	0.048 ± 0.002^{bc}	0.290 ± 0.008^{fg}	0.64 ± 0.03^{ab}

$^{a–h}$ Different superscript letters in the same column indicate statistically significant differences at the 95% confidence level (p-value < 0.05), according to the multiple range test.

Statistically significant differences in moisture content values (p-value < 0.05) were observed, with higher values in the HAD powders, indicating differences in water removal mechanisms depending on the drying technique used. During convective drying, water is transferred from inside the product to the product–air interface and then removed from the surface in the vapour state, leading to tissue shrinkage that can limit further moisture transfer [52,116]. This process results in the formation of a surface layer with increased resistance, which results in a dry surface, while the inner side remains moist [52,77]. This phenomenon is more pronounced as the drying rate increases [11,117]. As evidenced in previous studies, fermentation may accelerate surface water removal and contribute to case-hardening incidence [34]. The major impact of drying on fermented residues is related to structural changes, including the breakdown of the plant cell wall and pore formation [34,117]. This could explain why the moisture content of powders obtained through fermentation followed by hot air-drying at 60 °C (FERM HAD60) was the highest.

Among the HAD60 samples, the US HAD60 and MW HAD60 ones reached lower moisture contents compared to the non-pretreated powders, as well as significantly lower a_w values (p-value < 0.05) than those subjected to other pretreatments. The moisture content of the US HAD60 powders was similar to the respective FD powders. Increased water availability in the liquid phase of the tissue, or reduced mass transfer resistance in the tissue due to membrane and cell wall permeabilisation caused by cavitation, could explain this. Additionally, the US HAD60 powders had significantly lower a_w values (p-value < 0.05) than the US FD powders, further confirming, along with the moisture data, the contribution of the pretreatment to drying efficiency. Most of the HAD powders exhibited lower x_{ss} values than their respective FD ones. This was confirmed by the multifactorial ANOVA considering the pretreatment and dehydration techniques as factors, since the dehydration technique resulted in being significant (p-value < 0.05). This could be due to the FD process, which facilitates extraction due to ice crystals' formation, which causes disruption [115,118]. In addition, FD produces a more porous structure [14], which facilitates milling, and generally yields a smaller particle size.

The antioxidant properties of the powdered products obtained are shown in Figure 6, where the total phenols (mg GAE/g_{dm}), total flavonoids (mg QE/g_{dm}), and overall antioxidant activity measured by the ABTS and DPPH methods (mg TE/g_{dm}) are shown.

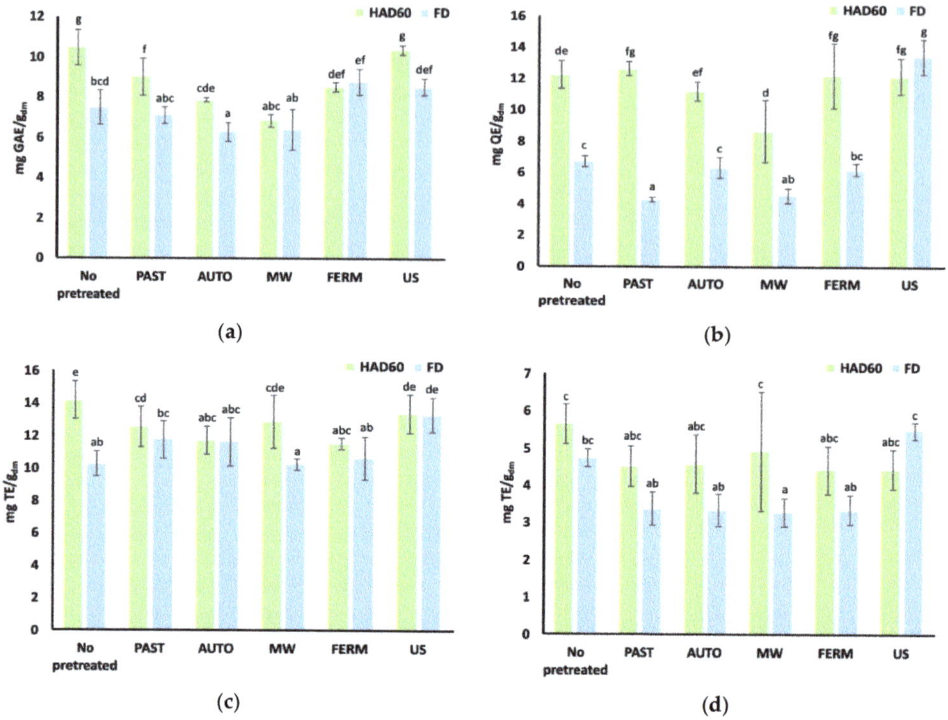

Figure 6. Antioxidant properties of pretreated and non-pretreated powders obtained by different dehydration techniques: (**a**) Total phenol content expressed as mg gallic acid equivalent (GAE) per gram of dry matter. (**b**) Total flavonoid content expressed as mg quercetin equivalent (QE) per gram of dry matter. (**c**,**d**) ABTS and DPPH antioxidant capacity, respectively, expressed in mg trolox equivalent (TE) per gram of dry matter. PAST: pasteurisation; AUTO: autoclaving; MW: microwaves; FERM: fermentation; US: ultrasounds. Error bars represent standard deviation of six replicates from two replicas (3 replicates/replica). $^{a\text{–}g}$ Different superscripts indicate statistically significant differences at 95% confidence level (p-value < 0.05), according to multiple range test.

Dehydration caused an improvement in the antioxidant properties of the broccoli residue (p-value < 0.05), as compared to the non-dehydrated stems (Table 3). This increase might be attributed to the various processing steps, including grinding, pretreatment, drying, and milling, which likely contributed to the breakdown of the plant matrix, facilitating the release of bioactive compounds or triggering other biochemical changes that improved the antioxidant properties [119]. Notably, the antioxidant properties were generally higher in powders obtained by HAD at 60 °C than by FD. This difference may be due to Maillard reactions, which occur during HAD [120], leading to the formation of antioxidant compounds. Such reactions are unlikely during FD because of the low temperatures involved. The role of Maillard reactions in enhancing antioxidant compounds was studied by Somjai et al. [121], who observed increased antioxidant capacity in Chinese lemon peels dried at 60 °C, as measured by ABTS and DPPH methods. In addition to Maillard reactions, other biochemical reactions favoured by heat could generate more active forms of antioxidant compounds; trans–cis isomerisation [63], for instance, inhibits pro-oxidant enzymes' activity [122] or activates enzymes that hydrolyse compounds into more active forms, such as in the hydrolysis of glucosinolates into isothiocyanates by myrosinase action [123]. These factors collectively could explain the observed increase in the antioxidant properties.

Multifactor ANOVA analysis confirmed that both the pretreatment and the drying method significantly influenced (p-value < 0.05) the antioxidant properties of the powders. Likewise, the interaction between the pretreatment used and the type of dehydration method was also significant (p-value < 0.05) for all the antioxidant properties analysed. This reveals that the impact of the drying technique on antioxidant properties varies depending on the pretreatment applied, which is likely influenced by the different conditions that occur in each of the dehydration techniques [124].

Regarding total phenol content (Figure 6a), none of the pretreatments improved total phenol levels in the products subjected to HAD. However, among the FD powders, pretreatments with US, and especially FERM, led to significant improvements. This suggests that the antioxidant compounds released or generated during fermentation and ultrasound pretreatment may be more susceptible to degradation by HAD or are better extracted due to the physical characteristics of FD powders [17,54,111]. In both cases, these pretreatments, along with pasteurisation, contributed to higher phenol content in both the HAD and FD powders. Conversely, the pasteurised residue, which initially had the highest total phenol content before dehydration, resulted in powders with phenolic levels comparable to those without pretreatment.

In terms of flavonoids, the HAD powders showed a significant improvement (p-value < 0.05) compared to non-pretreated powders in all cases except for the MW and AUTO pretreatments, where no significant differences were observed (p-value > 0.05). Similarly, Vargas et al. [125] found that the HAD of broccoli, kale, and spinach led to a temperature-dependent release of flavonoids bound in the plant matrix. This release may also be favoured by pretreatments such as FERM and US, which cause direct damage to the plant matrix [54,111], explaining the increase in flavonoids for these pretreatments (Figure 6b). Similarly, the FD powders were significantly (p-value < 0.05) enriched in flavonoids when undergoing US pretreatment, likely due to the increased damage to the plant tissue architecture, as previously mentioned.

In terms of antioxidant capacity, different results were obtained depending on the method used. No improvement was obtained according to the DPPH radical method (Figure 6d). However, the ABTS radical method showed enhanced antioxidant capacity in the powders pretreated with US and further FD (Figure 6c). This increase in the extraction of antioxidant compounds could be attributed to the cavitation effect induced in the broccoli residue by high ultrasound frequencies, as well as by the formation of channels in the plant matrix after the sublimation of ice crystals [110,118]. These findings confirm that the US pretreatment is the one that allowed us to obtain FD powders with better antioxidant properties.

With regard to powders obtained from the broccoli residues subjected to a fermentation pretratement, viable cell counts were obtained for the freshly inoculated broccoli stems, after 24 h of fermentation with *L. plantarum*, and following dehydration by HAD and FD. As mentioned in the preliminary study, a product must contain a viable cell concentration of approximately 10^6–10^7 CFU/g to be considered probiotic [71,72]. For the freshly inoculated broccoli stem, the count was at a concentration of 10^7 CFU/g, so it could be considered potentially probiotic. After 24 h of fermentation, there was a significant increase (p-value < 0.05) in the viable cell count, confirming that broccoli is a suitable plant matrix for the growth of probiotic microorganisms [23,80].

After subjecting the fermented broccoli stems to HAD in the pilot plant's convective dryer, and unlike in preliminary studies using benchtop dryers, no viable counts were observed in the seeded plates (counts < 10^4 CFU/g). This result is likely due to the negative impact of the drying temperature or to an excessively prolonged treatment. Previous studies have shown that convective drying at temperatures above 40 °C can reduce the viability of probiotic bacteria [126], but other research on similar waste products demonstrated that viable microorganisms can be preserved in powders dried at 50, 60, and 70 °C [34].

Therefore, it becomes crucial to optimise the drying time to not extend the duration of the falling drying rate period, during which the product's temperature can exceed the probiotic tolerance threshold. In the present study, all pretreated products were dried under the same temperature and time conditions for comparison purposes. This decision could have been detrimental for the viability of the probiotics, in the light of the results obtained for the microorganism viability after drying. In contrast, the FD powders exhibited fewer viable cells after dehydration (p-value < 0.05), but the powders still maintained a potentially probiotic count of 10^6 CFU/g. FD is generally preferred for preserving microbial viability due to its use of low temperatures and absence of oxygen, making it a suitable technique for producing high-value, stable probiotic cultures [75].

Table 5 displays the powders' phenolic profiles obtained from chromatographic analyses (HPLC). The content of specific phenolic compounds may vary depending on the pretreatment and the dehydration method used. After comparing with standard absorption spectra and retention times, none of the following phenolics were identified in the powder extracts: 4-Hydroxibezoic acid, epicatechin, vanillin, apigenin-7-glucoside, trans-cinnamic acid, naringenin, and kaempherol.

To facilitate the results' interpretation, phenolic constituents were gathered in groups so that the total hydroxycinnamic acids, total hydroxybenzoic acids, and total flavonoids were presented, besides the total phenolic content. The HAD powders not subjected to pretreatments showed higher phenolic contents than the FD ones, which is consistent with the antioxidant properties exhibited (Figure 6). However, when pretreatments were applied, the values in the FD powders were generally higher than in their HAD counterparts. On the one hand, this could be explained by the facilitated extraction of bioactive compounds from FD products due to their characteristic porous structure, coupled with the effect of the pretreatments on the cellular tissue, which would have favoured the release of phenolic compounds from the vegetal matrix. On the other hand, release of phenolic constituents during pretreatments followed by HAD could have resulted in an increased exposure of these compounds to high temperatures and oxidative conditions, thus reducing their concentration in the pretreated HAD samples.

In the fermented samples (both FD and HAD), a decrease in the glycosides analysed, together with an increase in aglycone forms, was registered, a result which is in accordance with Lee et al. [127], who reported the conversion of flavonoid glycosides to flavonols in silkworm thorn leaves due to fermentation with *Lactobacillus plantarum*. On the other hand, phenolic acids' degradation by lactic acid bacteria is an important mechanism for the detoxification of these compounds. According to the literature, *Lactobacillus* spp. exhibit the strain-specific metabolism of phenolic acids including hydroxybenzoic acids, hydroxycinnamic acids, and hydroxycinnamic acid derivatives; particularly, Filannino et al. [128] evidenced that the metabolism of phenolic acids by *L. plantarum* is strain-specific. In this work, fermented samples were amongst the ones which presented higher phenolic acids content.

As for flavonoids, rutin was only identified in fermented powders (FERM HAD60 and FERM FD), exhibiting a significantly higher value for the HAD powder than for the FD one (4.89 ± 0.03 vs. 1.7 ± 0.4 mg/100 g_{dm}, respectively). The fact that this compound was only present in the fermented samples suggests that *L. plantarum* might be involved in the release of this compound. On the other hand, quercetin was only found in the non-pretreated powders (HAD60 and FD), thus suggesting that pretreatments might have a negative impact on it.

Table 5. Phenolic content (mg/100 g_{dm}) of powders obtained by hot air-drying or freeze-drying from fresh and pretreated broccoli stems. PAST: pasteurisation; AUTO: autoclaving; MW: microwaves; FERM: fermentation; US: ultrasounds. HAD: hot air-drying; FD: freeze-drying. Mean ± standard deviation of three replicates. n.d. not detected. $^{a–g}$ Different superscripts indicate statistically significant differences at 95% confidence level (p-value < 0.05), according to multiple range test.

	Phenolic Compounds (mg/100 g_{dm})																			
	Hydroxycinamic Acids							Hydroxybenzoic Acids			Flavonoids									
	Sinapic Acid	Caffeic Acid	*p*-Coumaric Acid	Ferulic Acid	4-*O*-Caffeoyl-Quinic	*Trans*-Cinnamic Acid	Total	Gallic Acid	4-Hydroxibezoic Acid	Total	Epicatechin	Quercitin 3-Glucoside	Rutin	Quercitrin	Naringenin	Apigenin-7-Glucoside	Quercetin	Kaempherol	Total	Total
HAD60	0.928 ± 0.014 ^{f}	0.609 ± 0.006 ^{a}	0.406 ± 0.002 ^{a}	0.91 ± 0.11 ^{c}	2.1 ± 0.2 ^{c}	n.d.	5.0 ± 0.3 ^{cd}	3.13 ± 0.08 ^{a}	n.d.	3.13 ± 0.08 ^{a}	n.d.	2.5 ± 0.5 ^{d}	n.d.	0.71 ± 0.08 ^{b}	n.d.	n.d.	1.640 ± 0.008 ^{b}	n.d.	4.9 ± 0.6 ^{e}	13.03 ± 0.23 ^{f}
PAST HAD60	0.64 ± 0.02 ^{ab}	3.25 ± 0.13 ^{de}	0.570 ± 0.016 ^{cd}	0.75 ± 0.04 ^{b}	0.46 ± 0.03 ^{a}	n.d.	5.7 ± 0.2 ^{de}	5.3 ± 1.7 ^{a}	n.d.	5.3 ± 1.7 ^{a}	n.d.	1.28 ± 0.09 ^{abc}	n.d.	n.d.	n.d.	n.d.	n.d.	n.d.	1.28 ± 0.09 ^{ab}	12.3 ± 1.5 ^{f}
AUTO HAD60	0.79 ± 0.03 ^{bcdef}	1.4 ± 0.3 ^{b}	n.d.	n.d.	1.148 ± 0.016 ^{b}	n.d.	3.0 ±0.5 ^{a}	n.d.	n.d.	n.d.	n.d.	1.8 ± 0.2 ^{cd}	n.d.	n.d.	n.d.	n.d.	n.d.	n.d.	1.8 ± 0.2 ^{bc}	4.9 ± 0.3 ^{a}
MW HAD60	0.74 ± 0.08 ^{bcde}	1.64 ± 0.07 ^{bc}	0.53 ± 0.05 ^{c}	n.d.	0.78 ± 0.02 ^{ab}	n.d.	3.64 ± 0.14 ^{ab}	n.d.	n.d.	n.d.	n.d.	1.58 ± 0.05 ^{bc}	n.d.	0.523 ± 0.013 ^{a}	n.d.	n.d.	n.d.	n.d.	2.10 ± 0.03 ^{c}	5.74 ± 0.16 ^{ab}
FERM HAD60	0.95 ± 0.02 ^{f}	0.59 ± 0.02 ^{a}	n.d.	1.03 ± 0.06 ^{d}	0.55 ± 0.02 ^{a}	n.d.	3.12 ± 0.11 ^{a}	n.d.	n.d.	n.d.	n.d.	1.624 ± 0.016 ^{bc}	4.89 ± 0.03 ^{b}	0.674 ± 0.015 ^{b}	n.d.	n.d.	n.d.	n.d.	7.18 ± 0.02 ^{f}	10.30 ± 0.14 ^{de}
US HAD60	0.68 ± 0.05 ^{abcd}	3.30 ± 0.17 ^{e}	0.466 ± 0.013 ^{b}	n.d.	0.556 ± 0.016 ^{a}	n.d.	4.8 ± 0.3 ^{c}	n.d.	n.d.	n.d.	n.d.	0.73 ± 0.05 ^{a}	n.d.	n.d.	n.d.	n.d.	n.d.	n.d.	0.73 ± 0.05 ^{a}	5.6 ± 0.3 ^{a}
FD	0.56 ± 0.03 ^{a}	0.634 ± 0.005 ^{a}	n.d.	0.600 ± 0.006 ^{a}	4.59 ± 0.17 ^{d}	n.d.	6.38 ± 0.16 ^{e}	n.d.	n.d.	n.d.	n.d.	0.8252 ± 0.0016 ^{a}	n.d.	1.05 ± 0.10 ^{c}	n.d.	n.d.	1.613 ± 0.002 ^{a}	n.d.	3.40 ± 0.10 ^{d}	9.9 ± 0.2 ^{d}
PAST FD	0.7 ± 0.3 ^{bcde}	2.0 ± 0.4 ^{c}	n.d.	0.60 ± 0.03 ^{a}	n.d.	n.d.	3.4 ± 0.7 ^{ab}	10 ± 3 ^{b}	n.d.	10 ± 3 ^{b}	n.d.	1.04 ± 0.03 ^{ab}	n.d.	0.53 ± 0.09 ^{a}	n.d.	n.d.	n.d.	n.d.	1.58 ± 0.05 ^{bc}	15 ± 3 ^{ef}
AUTO FD	0.676 ± 0.012 ^{abc}	0.834 ± 0.006 ^{a}	n.d.	1.615 ± 0.016 ^{f}	1.172 ± 0.014 ^{b}	n.d.	4.29 ± 0.04 ^{bc}	n.d.	n.d.	n.d.	n.d.	3.4 ± 1.2 ^{e}	n.d.	n.d.	n.d.	n.d.	n.d.	n.d.	3.4 ± 1.2 ^{d}	7.3 ± 1.4 ^{bc}
MW FD	0.88 ± 0.04 ^{ef}	2.9 ± 0.5 ^{d}	0.58 ± 0.02 ^{d}	1.29 ± 0.04 ^{e}	1.10 ± 0.05 ^{b}	n.d.	6.1 ± 0.8 ^{e}	n.d.	n.d.	n.d.	n.d.	1.60 ± 0.12 ^{bc}	n.d.	0.434 ± 0.013 ^{a}	n.d.	n.d.	n.d.	n.d.	2.03 ± 0.11 ^{c}	8.1 ± 0.8 ^{c}
FERM FD	0.84 ± 0.10 ^{def}	1.91 ± 0.03 ^{c}	0.409 ± 0.005 ^{a}	0.881 ± 0.006 ^{c}	8.5 ± 0.9 ^{e}	n.d.	13.0 ± 0.9 ^{g}	n.d.	n.d.	n.d.	n.d.	1.85 ± 0.05 ^{cd}	1.7 ± 0.4 ^{a}	0.48 ± 0.05 ^{a}	n.d.	n.d.	n.d.	n.d.	3.82 ± 0.08 ^{d}	16.8 ± 0.8 ^{g}
US FD	0.80 ± 0.02 ^{cdef}	4.92 ± 0.06 ^{e}	0.609 ± 0.004 ^{d}	1.65 ± 0.05 ^{f}	0.83 ± 0.05 ^{ab}	n.d.	8.6 ± 0.4 ^{f}	n.d.	n.d.	n.d.	n.d.	1.70 ± 0.06 ^{bc}	n.d.	2.015 ± 0.013 ^{d}	n.d.	n.d.	n.d.	n.d.	3.72 ± 0.07 ^{d}	12.3 ± 0.4 ^{f}

Hydroxycinnamic acids act as powerful antioxidants in dried broccoli [129,130]. As observed in Table 5, the value of individual hydroxycinnamic acids varied significantly among the different pretreatments. However, powders obtained by the FD method presented higher contents of these compounds than their HAD counterparts, except for PAST-pretreated samples. Since hydroxycinnamic acids are heat-sensitive, FD conditions, i.e., lower temperatures and reduced oxygen, could have prevented the degradation found in the HAD samples. This fact has been previously reported in other studies about dried broccoli, in which hydroxycinnamic acids (such as ferulic acid, caffeic acid, or coumaric acid) were affected by the heat and oxidative conditions of convective drying [125,129]. Finally, gallic acid was the only hydroxybenzoic acid identified in the broccoli powders. It was present in the unpretreated HAD samples and in powders obtained from the pasteurised samples, both with FD or HAD.

4. Conclusions

Biological and thermophysical pretreatments have an impact on tissue structure, thus releasing bioactive compounds which contribute to antioxidant properties. This study on biological and thermophysical pretreatments applied to fresh ground tissue confirmed that these pretreatments can successfully contribute to improving the antioxidant properties of ground broccoli stems. Particularly, ultrasound application exhibited the most remarkable increase, whereas autoclaving and microwaving led to less evident improvements. Changes in the antioxidant properties of the broccoli wastes were related to the changes observed in the microstructure, such as permeabilisation of membranes and cell walls, and loss of cell compartmentalisation. Treatments such as ultrasounds, microwaves, or fermentation could also have led to the release of simpler phenolic constituents initially bound to the structure as part of more complex forms. In contrast, the thermal degradation of some bioactive constituents could also have occurred in treatments such as the autoclave or microwave treatments. Interestingly, dehydration applied to the pretreated ground broccoli stems improved their antioxidant properties, especially in the HAD ones, but not as much as in powders obtained from the non-pretreated broccoli wastes. The powders obtained from the ultrasonicated ground broccoli stems were the only pretreated powders which exhibited improved antioxidant properties, and this was obtained only for some antioxidant parameters. In general, the HAD powders showed better antioxidant properties than the freeze-dried ones. The probiotic properties of the powders were not maintained when scaling up to the pilot plant air drier, although this could be attributed to the excessive duration of the treatment. In contrast, the freeze-dried products maintained their probiotic properties.

In conclusion, thermophysical and biological pretreatments might be proposed to enhance the antioxidant attributes of broccoli wastes but not necessarily that of dried powdered products. Further research should focus on the drying kinetics and duration of this stage after pretreatments to better adjust drying parameters.

Once refined, these processes could be implemented at an industrial scale to support sustainable practises and circular economy principles. By transforming the entire broccoli stem into functional powders, this study contributes to reducing food waste and promotes the reintroduction of these residues into the food supply chain as valuable ingredients. Being rich in antioxidant compounds, these products have market potential for food fortification. The powders could be effectively incorporated into a variety of processed products such as baked goods, pasta, sauces, meat products or analogues, and smoothies, among others, offering nutritional enhancement while supporting sustainable production practises.

Author Contributions: Conceptualisation, L.S. and C.B.; methodology, L.S. and C.B.; formal analysis, C.B.-B.; investigation, C.B.-B.; resources, L.S. and C.B.; data curation, L.S., C.B. and C.B.-B.; writing—original draft preparation, C.B.-B., C.B. and L.S; writing—review and editing, C.B.-B., C.B. and L.S.; visualisation, C.B.-B., C.B. and L.S.; supervision, L.S. and C.B.; project administration, L.S.; funding acquisition, L.S., C.B. and C.B.-B. All authors have read and agreed to the published version of the manuscript.

Funding: This research was funded by Universitat Politècnica de València under the programme PAID-10-23 (Pretramientos termofísicos y biológicos para la mejora de la funcionalidad de productos en polvo obtenidos a partir de residuos de hortalizas, y valoración de su aplicación en el sector agroalimentario).

Institutional Review Board Statement: Not applicable.

Informed Consent Statement: Not applicable.

Data Availability Statement: The original contributions presented in the study are included in the article, further inquiries can be directed to the corresponding author.

Acknowledgments: The authors would like to thank the bachelor student Marta Muñoz Ibáñez and master student María Belén Romero Ramírez for their instrumental role and contribution to the experiment presented in this paper.

Conflicts of Interest: The authors declare no conflicts of interest.

References

1. Rajković, M.B.; Popović, M.D.; Milinčić, D.; Zdravković, M. Circular economy in food industry. *Zaštita Mater.* **2020**, *61*, 229–250. [CrossRef]
2. FAO; WHO. *Sustainable Healthy Diets—Guiding Principles*; Food and Agriculture Organization of the United Nations and World Health Organization: Rome, Italy, 2019. [CrossRef]
3. Sharma, P.; Gaur, V.K.; Sirohi, R.; Varjani, S.; Hyoun Kim, S.; Wong, J.W.C. Sustainable processing of food waste for production of bio-based products for circular bioeconomy. *Bioresour. Technol.* **2021**, *325*, 124684. [CrossRef] [PubMed]
4. Li, H.; Xia, Y.; Liu, H.-Y.; Guo, H.; He, X.-Q.; Liu, Y.; Wu, D.-T.; Mai, Y.-H.; Li, H.-B.; Zou, L.; et al. Nutritional values, beneficial effects, and food applications of broccoli (*Brassica oleracea* var. *italica Plenck*). *Trends Food Sci. Technol.* **2022**, *119*, 288–308. [CrossRef]
5. FAOSTAT. 2022. Available online: https://www.fao.org/faostat/en/#data/QCL (accessed on 12 September 2024).
6. Kaparapu, J.; Pragada, P.M.; Narasimha, M.; Geddada, R. Fruits and Vegetables and its Nutritional Benefits. In *Functional Foods and Nutraceuticals*; Springer: Cham Switzerland, 2020; pp. 241–260.
7. Ganesh, K.S.; Sridhar, A.; Vishali, S. Utilization of fruit and vegetable waste to produce value-added products: Conventional utilization and emerging opportunities-A review. *Chemosphere* **2022**, *287*, 132221. [CrossRef] [PubMed]
8. Núñez-Gómez, V.; González-Barrio, R.; Baenas, N.; Moreno, D.A.; Periago, M.J. Dietary-Fibre-Rich Fractions Isolated from Broccoli Stalks as a Potential Functional Ingredient with Phenolic Compounds and Glucosinolates. *Int. J. Mol. Sci.* **2022**, *23*, 13309. [CrossRef] [PubMed]
9. Mandha, J.; Shumoy, H.; Matemu, A.O.; Raes, K. Characterization of fruit juices and effect of pasteurization and storage conditions on their microbial, physicochemical, and nutritional quality. *Food Biosci.* **2023**, *51*, 102335. [CrossRef]
10. Bas-Bellver, C.; Barrera, C.; Betoret, N.; Seguí, L. Impact of Disruption and Drying Conditions on Physicochemical, Functional and Antioxidant Properties of Powdered Ingredients Obtained from Brassica Vegetable By-Products. *Foods* **2022**, *11*, 3663. [CrossRef]
11. Bas-Bellver, C.; Barrera, C.; Betoret, N.; Seguí, L. Turning Agri-Food Cooperative Vegetable Residues into Functional Powdered Ingredients for the Food Industry. *Sustainability* **2020**, *12*, 1284. [CrossRef]
12. Bas-Bellver, C.; Barrera, C.; Betoret, N.; Seguí, L.; Harasym, J. IV-Range Carrot Waste Flour Enhances Nutritional and Functional Properties of Rice-Based Gluten-Free Muffins. *Foods* **2024**, *13*, 1312. [CrossRef]
13. Singh, L.; Kaur, S.; Aggarwal, P. Techno and bio functional characterization of industrial potato waste for formulation of phytonutrients rich snack product. *Food Biosci.* **2022**, *49*, 101824. [CrossRef]
14. Yao, J.; Chen, W.; Fan, K. Novel Efficient Physical Technologies for Enhancing Freeze Drying of Fruits and Vegetables: A Review. *Foods* **2023**, *12*, 4321. [CrossRef] [PubMed]
15. Miletić, N.; Mitrović, O.; Popović, B.; Nedović, V.; Zlatković, B.; Kandić, M. Polyphenolic Content and Antioxidant Capacity in Fruits of Plum (*Prunus domestica* L.) Cultivars "Valjevka" and "Mildora" as Influenced by Air Drying. *J. Food Qual.* **2013**, *36*, 229–237. [CrossRef]
16. Bas-Bellver, C.; Barrera, C.; Betoret, N.; Seguí, L. Physicochemical, Technological and Functional Properties of Upcycled Vegetable Waste Ingredients as Affected by Processing and Storage. *Plant Foods Hum. Nutr.* **2023**, *78*, 710–719. [CrossRef] [PubMed]
17. Xu, Y.; Xiao, Y.; Lagnika, C.; Li, D.; Liu, C.; Jiang, N.; Song, J.; Zhang, M. A comparative evaluation of nutritional properties, antioxidant capacity and physical characteristics of cabbage (*Brassica oleracea* var. *Capitate* var L.) subjected to different drying methods. *Food Chem.* **2020**, *309*, 124935. [CrossRef]
18. Bassey, E.J.; Cheng, J.H.; Sun, D.W. Novel nonthermal and thermal pretreatments for enhancing drying performance and improving quality of fruits and vegetables. *Trends Food Sci. Technol.* **2021**, *112*, 137–148. [CrossRef]
19. Mohammed, H.H.; Tola, Y.B.; Taye, A.H.; Abdisa, Z.K. Effect of pretreatments and solar tunnel dryer zones on functional properties, proximate composition, and bioactive components of pumpkin (*Cucurbita maxima*) pulp powder. *Heliyon* **2022**, *8*, e10747. [CrossRef]

20. Ummat, V.; Sivagnanam, S.P.; Rajauria, G.; O'Donnell, C.; Tiwari, B.K. Advances in pre-treatment techniques and green extraction technologies for bioactives from seaweeds. *Trends Food Sci. Technol.* **2021**, *110*, 90–106. [CrossRef]
21. Li, Z.; Teng, J.; Lyu, Y.; Hu, X.; Zhao, Y.; Wang, M. Enhanced Antioxidant Activity for Apple Juice Fermented with Lactobacillus plantarum ATCC14917. *Molecules* **2018**, *24*, 51. [CrossRef]
22. Kiczorowski, P.; Kiczorowska, B.; Samolińska, W.; Szmigielski, M.; Winiarska-Mieczan, A. Effect of fermentation of chosen vegetables on the nutrient, mineral, and biocomponent profile in human and animal nutrition. *Sci. Rep.* **2022**, *12*, 13422. [CrossRef]
23. Zdziobek, P.; Jodłowski, G.S.; Strzelec, E.A. Biopreservation and Bioactivation Juice from Waste Broccoli with Lactiplantibacillus plantarum. *Molecules* **2023**, *28*, 4594. [CrossRef]
24. Ye, J.H.; Huang, L.Y.; Terefe, N.S.; Augustin, M.A. Fermentation-based biotransformation of glucosinolates, phenolics and sugars in retorted broccoli puree by lactic acid bacteria. *Food Chem.* **2019**, *286*, 616–623. [CrossRef] [PubMed]
25. Salehi, F.; Inanloodoghouz, M.; Ghazvineh, S. Influence of microwave pretreatment on the total phenolics, antioxidant activity, moisture diffusivity, and rehydration rate of dried sweet cherry. *Food Sci. Nutr.* **2023**, *11*, 7870–7876. [CrossRef] [PubMed]
26. Md Salim, N.S.; Garièpy, Y.; Raghavan, V. Effects of Processing on Quality Attributes of Osmo-Dried Broccoli Stalk Slices. *Food Bioproc. Technol.* **2019**, *12*, 1174–1184. [CrossRef]
27. Pandiselvam, R.; Aydar, A.Y.; Kutlu, N.; Aslam, R.; Sahni, P.; Mitharwal, S.; Gavahian, M.; Kumar, M.; Raposo, A.; Yoo, S.; et al. Individual and interactive effect of ultrasound pre-treatment on drying kinetics and biochemical qualities of food: A critical review. *Ultrason. Sonochem* **2023**, *92*, 106261. [CrossRef]
28. Zhu, X.; Das, R.S.; Bhavya, M.L.; Garcia-Vaquero, M.; Tiwari, B.K. Acoustic cavitation for agri-food applications: Mechanism of action, design of new systems, challenges and strategies for scale-up. *Ultrason Sonochem.* **2024**, *105*, 106850. [CrossRef]
29. Llavata, B.; García-Pérez, J.V.; Simal, S.; Cárcel, J.A. Innovative pre-treatments to enhance food drying: A current review. *Curr. Opin. Food Sci.* **2020**, *35*, 20–26. [CrossRef]
30. Salehi, F. Physico-chemical properties of fruit and vegetable juices as affected by ultrasound: A review. *Int. J. Food Prop.* **2020**, *23*, 1748–1765. [CrossRef]
31. Barba, F.J.; Mariutti, L.R.B.; Bragagnolo, N.; Mercadante, A.Z.; Barbosa-Cánovas, G.V.; Orlien, V. Bioaccessibility of bioactive compounds from fruits and vegetables after thermal and nonthermal processing. *Trends Food Sci. Technol.* **2017**, *67*, 195–206. [CrossRef]
32. Taheri, M.E.; Salimi, E.; Saragas, K.; Novakovic, J.; Barampouti, E.M.; Mai, S.; Malamis, D.; Moustakas, K.; Loizidou, M. Effect of pretreatment techniques on enzymatic hydrolysis of food waste. *Biomass Convers. Biorefin.* **2021**, *11*, 219–226. [CrossRef]
33. Aamir, M.; Ovissipour, M.; Sablani, S.S.; Rasco, B. Predicting the Quality of Pasteurized Vegetables Using Kinetic Models: A Review. *Int. J. Food Sci.* **2013**, *2013*, 271271. [CrossRef]
34. Bas-Bellver, C.; Barrera, C.; Betoret, N.; Seguí, L. Impact of Fermentation Pretreatment on Drying Behaviour and Antioxidant Attributes of Broccoli Waste Powdered Ingredients. *Foods* **2023**, *12*, 3526. [CrossRef] [PubMed]
35. Liu, Y.W.; Liong, M.T.; Tsai, Y.C. New perspectives of *Lactobacillus plantarum* as a probiotic: The gut-heart-brain axis. *J. Microbiol.* **2018**, *56*, 601–613. [CrossRef] [PubMed]
36. Seddik, H.A.; Bendali, F.; Gancel, F.; Fliss, I.; Spano, G.; Drider, D. Lactobacillus plantarum and Its Probiotic and Food Potentialities. *Probiotics Antimicrob. Proteins* **2017**, *9*, 111–122. [CrossRef]
37. Jiang, H.; Zhang, M.; Adhikari, B. Fruits and vegetable powders. In *Handbook of Food Powders*; Elsevier: Amsterdam, The Netherlands, 2024; pp. 423–436.
38. Clifford, P.A. Report on Moisture in Dried Fruit. *J. AOAC Int.* **1934**, *17*, 215–228. [CrossRef]
39. Singleton, V.L.; Orthofer, R.; Lamuela-Raventós, R.M. Analysis of total phenols and other oxidation substrates and antioxidants by means of folin-ciocalteu reagent. *Methods Enzym.* **1999**, *299*, 152–178.
40. Wolfe, K.; Wu, X.; Liu, R.H. Antioxidant Activity of Apple Peels. *J. Agric. Food Chem.* **2003**, *51*, 609–614. [CrossRef] [PubMed]
41. Luximon-Ramma, A.; Bahorun, T.; Soobrattee, M.A.; Aruoma, O.I. Antioxidant Activities of Phenolic, Proanthocyanidin, and Flavonoid Components in Extracts of *Cassia fistula*. *J. Agric. Food Chem.* **2002**, *50*, 5042–5047. [CrossRef]
42. Brand-Williams, W.; Cuvelier, M.E.; Berset, C. Use of a free radical method to evaluate antioxidant activity. *LWT Food Sci. Technol.* **1995**, *28*, 25–30. [CrossRef]
43. Re, R.; Pellegrini, N.; Proteggente, A.; Pannala, A.; Yang, M.; Rice-Evans, C. Antioxidant activity applying an improved ABTS radical cation decolorization assay. *Free Radic. Biol. Med.* **1999**, *26*, 1231–1237. [CrossRef]
44. Caprioli, G.; Nzekoue, F.K.; Giusti, F.; Vittori, S.; Sagratini, G. Optimization of an extraction method for the simultaneous quantification of sixteen polyphenols in thirty-one pulse samples by using HPLC-MS/MS dynamic-MRM triple quadrupole. *Food Chem.* **2018**, *266*, 490–497. [CrossRef]
45. Giusti, F.; Capuano, E.; Sagratini, G.; Pellegrini, N. A comprehensive investigation of the behaviour of phenolic compounds in legumes during domestic cooking and in vitro digestion. *Food Chem.* **2019**, *285*, 458–467. [CrossRef] [PubMed]
46. de Oliveira, P.M.; de Leite Júnior, B.R.C.; Martins, E.M.F.; Martins, M.L.; Vieira, É.N.R.; de Barros, F.A.R.; Cristianini, M.; de Almeida Costa, N.; Ramos, A.M. Mango and carrot mixed juice: A new matrix for the vehicle of probiotic lactobacilli. *J. Food Sci. Technol.* **2021**, *58*, 98–109. [CrossRef] [PubMed]
47. Septembre-Malaterre, A.; Remize, F.; Poucheret, P. Fruits and vegetables, as a source of nutritional compounds and phytochemicals: Changes in bioactive compounds during lactic fermentation. *Food Res. Int.* **2018**, *104*, 86–99. [CrossRef]

48. Alp, D.; Bulantekin, Ö. The microbiological quality of various foods dried by applying different drying methods: A review. *Eur. Food Res. Technol.* **2021**, *247*, 1333–1343. [CrossRef]
49. Lewicki, P.P.; Pawlak, G. Effect of Drying on Microstructure of Plant Tissue. *Dry. Technol.* **2003**, *21*, 657–683. [CrossRef]
50. Djantou, E.B.; Mbofung, C.M.F.; Scher, J.; Phambu, N.; Morael, J.D. Alternation drying and grinding (ADG) technique: A novel approach for producing ripe mango powder. *LWT Food Sci. Technol.* **2011**, *44*, 1585–1590. [CrossRef]
51. Santos, P.H.S.; Silva, M.A. Retention of Vitamin C in Drying Processes of Fruits and Vegetables—A Review. *Dry. Technol.* **2008**, *26*, 1421–1437. [CrossRef]
52. Gulati, T.; Datta, A.K. Mechanistic understanding of case-hardening and texture development during drying of food materials. *J. Food Eng.* **2015**, *166*, 119–138. [CrossRef]
53. Kwaw, E.; Ma, Y.; Tchabo, W.; Apaliya, M.T.; Wu, M.; Sackey, A.S.; Xiao, L.; Tahir, H.E. Effect of lactobacillus strains on phenolic profile, color attributes and antioxidant activities of lactic-acid-fermented mulberry juice. *Food Chem.* **2018**, *250*, 148–154. [CrossRef]
54. Yang, X.; Hong, J.; Wang, L.; Cai, C.; Mo, H.; Wang, J.; Fang, X.; Liao, Z. Effect of Lactic Acid Bacteria Fermentation on Plant-Based Products. *Fermentation* **2024**, *10*, 48. [CrossRef]
55. Knez, E.; Kadac-Czapska, K.; Grembecka, M. Effect of Fermentation on the Nutritional Quality of the Selected Vegetables and Legumes and Their Health Effects. *Life* **2023**, *13*, 655. [CrossRef] [PubMed]
56. Buckenhueskes, H.J. Quality improvement and fermentation control in vegetables. In *Advances in Fermented Foods and Beverages: Improving Quality, Technologies and Health Benefits*; Woodhead Publishing: Sawston, UK, 2015; pp. 515–539.
57. Chin, S.; Siew, E.; Soon, W. Drying characteristics and quality evaluation of kiwi slices under hot air natural convective drying method. *Int. Food Res. J.* **2015**, *22*, 2188–2195.
58. Chen, M.L.; Yang, D.J.; Liu, S.C. Effects of drying temperature on the flavonoid, phenolic acid and antioxidative capacities of the methanol extract of citrus fruit (*Citrus sinensis* (L.) Osbeck) peels. *Int. J. Food Sci. Technol.* **2011**, *46*, 1179–1185. [CrossRef]
59. Bernaert, N.; De Clercq, H.; Van Bockstaele, E.; De Loose, M.; Van Droogenbroeck, B. Antioxidant changes during postharvest processing and storage of leek (*Allium ampeloprasum* var. porrum). *Postharvest Biol. Technol.* **2013**, *86*, 8–16. [CrossRef]
60. Papoutsis, K.; Pristijono, P.; Golding, J.B.; Stathopoulos, C.E.; Bowyer, M.C.; Scarlett, C.J.; Vuong, Q.V. Effect of vacuum-drying, hot air-drying and freeze-drying on polyphenols and antioxidant capacity of lemon (*Citrus limon*) pomace aqueous extracts. *Int. J. Food Sci. Technol.* **2017**, *52*, 880–887. [CrossRef]
61. Lutz, M.; Hernández, J.; Henríquez, C. Phenolic content and antioxidant capacity in fresh and dry fruits and vegetables grown in Chile. *CYTA J. Food* **2015**, *13*, 541–547.
62. Nooshkam, M.; Varidi, M.; Bashash, M. The Maillard reaction products as food-born antioxidant and antibrowning agents in model and real food systems. *Food Chem.* **2019**, *275*, 644–660. [CrossRef]
63. Honda, M.; Kageyama, H.; Hibino, T.; Ichihashi, K.; Takada, W.; Goto, M. Isomerization of Commercially Important Carotenoids (Lycopene, β-Carotene, and Astaxanthin) by Natural Catalysts: Isothiocyanates and Polysulfides. *J. Agric. Food Chem.* **2020**, *68*, 3228–3237. [CrossRef]
64. Chamberlain, M.C.; O'Flaherty, S.; Cobián, N.; Barrangou, R. Metabolomic Analysis of *Lactobacillus acidophilus*, L. *gasseri*, L. *crispatus*, and *Lacticaseibacillus rhamnosus* Strains in the Presence of Pomegranate Extract. *Front. Microbiol.* **2022**, *13*, 863228. [CrossRef]
65. Marques, L.G.; Silveira, A.M.; Freire, J.T. Freeze-Drying Characteristics of Tropical Fruits. *Dry. Technol.* **2006**, *24*, 457–463. [CrossRef]
66. Silva-Espinoza, M.A.; Ayed, C.; Foster, T.; Del Mar Camacho, M.; Martínez-Navarrete, N. The Impact of Freeze-Drying Conditions on the Physico-Chemical Properties and Bioactive Compounds of a Freeze-Dried Orange Puree. *Foods* **2019**, *9*, 32. [CrossRef] [PubMed]
67. Tylewicz, U.; Nowacka, M.; Rybak, K.; Drozdzal, K.; Dalla Rosa, M.; Mozzon, M. Design of Healthy Snack Based on Kiwifruit. *Molecules* **2020**, *25*, 3309. [CrossRef] [PubMed]
68. Dalmau, M.E.; Bornhorst, G.M.; Eim, V.; Rosselló, C.; Simal, S. Effects of freezing, freeze drying and convective drying on in vitro gastric digestion of apples. *Food Chem.* **2017**, *215*, 7–16. [CrossRef]
69. Rudy, S.; Dziki, D.; Krzykowski, A.; Gawlik-Dziki, U.; Polak, R.; Rózyło, R.; Kulig, R. Influence of pre-treatments and freeze-drying temperature on the process kinetics and selected physico-chemical properties of cranberries (*Vaccinium macrocarpon* Ait.). *LWT Food Sci. Technol.* **2015**, *63*, 497–503. [CrossRef]
70. Bas-Bellver, C.; Barrera, C.; Betoret, N.; Seguí, L. Effect of Processing and In Vitro Digestion on Bioactive Constituents of Powdered IV Range Carrot (*Daucus carota*, L.). *Wastes. Foods* **2023**, *12*, 731. [CrossRef] [PubMed]
71. Cui, L.; Niu, L.; Li, D.; Liu, C.; Liu, Y.; Liu, C.; Song, J. Effects of different drying methods on quality, bacterial viability and storage stability of probiotic enriched apple snacks. *J. Integr. Agric.* **2018**, *17*, 247–255. [CrossRef]
72. Champagne, C.P.; Ross, R.P.; Saarela, M.; Hansen, K.F.; Charalampopoulos, D. Recommendations for the viability assessment of probiotics as concentrated cultures and in food matrices. *Int. J. Food Microbiol.* **2011**, *149*, 185–193. [CrossRef]
73. Betoret, E.; Betoret, N.; Arilla, A.; Bennár, M.; Barrera, C.; Codoñer, P.; Fito, P. No invasive methodology to produce a probiotic low humid apple snack with potential effect against Helicobacter pylori. *J. Food Eng.* **2012**, *110*, 289–293. [CrossRef]
74. Shekh, S.L.; Boricha, A.A.; Chavda, J.G.; Vyas, B.R.M. Probiotic potential of lyophilized *Lactobacillus plantarum*, G.P. *Ann Microbiol.* **2020**, *70*, 1–12. [CrossRef]

75. Rishabh, D.; Athira, A.; Preetha, R.; Nagamaniammai, G. Freeze dried probiotic carrot juice powder for better storage stability of probiotic. *J. Food Sci. Technol.* **2023**, *60*, 916–924. [CrossRef]
76. Bekić Šarić, B. Processing of agricultural products by lyophilization. In Proceedings of the II International Scientific Conference "Sustainable Agriculture and Rural Development", Belgrade, Serbia, 16–17 December 2021.
77. Guiné, R.P.F. The Drying of Foods and Its Effect on the Physical-Chemical, Sensorial and Nutritional Properties. *ETP Int. J. Food Eng.* **2018**, *2*, 93–100. [CrossRef]
78. Scherzinger, M.; Kulbeik, T.; Kaltschmitt, M. Autoclave pre-treatment of green wastes—Effects of temperature, residence time and rotation speed on fuel properties. *Fuel* **2020**, *273*, 117796. [CrossRef]
79. Guo, Q.; Sun, D.W.; Cheng, J.H.; Han, Z. Microwave processing techniques and their recent applications in the food industry. *Trends Food Sci. Technol.* **2017**, *67*, 236–247. [CrossRef]
80. Ye, M.; Zhou, H.; Hao, J.; Chen, T.; He, Z.; Wu, F.; Liu, X. Microwave pretreatment on microstructure, characteristic compounds and oxidative stability of Camellia seeds. *Ind. Crops Prod.* **2021**, *161*, 113193. [CrossRef]
81. Kong, F.; Singh, R.P. Chemical Deterioration and Physical Instability of Foods and Beverages. In *The Stability and Shelf Life of Food*; Elsevier: Amsterdam, The Netherlands, 2016; pp. 43–76.
82. Rolfe, C.; Daryaei, H. Intrinsic and Extrinsic Factors Affecting Microbial Growth in Food Systems. In *Food Safety Engineering*; Food Engineering Series; Springer: Cham, Switzerland, 2020; pp. 3–24.
83. Nimkarde, A.D.; Gopnarayan, S.P.; Vaidya, K.S. Effect of Microwaves on the pH and °Brix value of Cranberry, Grape, Blackberry and Lemon. *J. Adv. Appl. Sci. Res.* **2022**, *4*, 74–79. [CrossRef]
84. Alvi, T.; Khan, M.K.I.; Maan, A.A.; Shahid, M.; Sablani, S. Microwaves as sustainable approach for artificial ripening of date fruit cv. Khupra Reduce Fruit. *Waste. Food Biosci.* **2023**, *54*, 102829. [CrossRef]
85. Malik, F.; Nadeem, M.; Ainee, A.; Kanwal, R.; Sultan, M.; Iqbal, A.; Mahmoud, S.F.; Alshehry, G.A.; Jumayi, H.A.A.L.; Algarni, E.H.A. Quality Evaluation of Lemon Cordial Stored at Different Times with Microwave Heating (Pasteurization). *Sustainability* **2022**, *14*, 1953. [CrossRef]
86. Conesa, C.; Seguí, L.; Laguarda-Miró, N.; Fito, P. Microwaves as a pretreatment for enhancing enzymatic hydrolysis of pineapple industrial waste for bioethanol production. *Food Bioprod. Process.* **2016**, *100*, 203–213. [CrossRef]
87. Shrotri, A.; Kobayashi, H.; Fukuoka, A. Cellulose Depolymerization over Heterogeneous Catalysts. *Acc. Chem. Res.* **2018**, *51*, 761–768. [CrossRef]
88. Liu, H.; Ni, Y.; Yu, Q.; Fan, L. Evaluation of co-fermentation of L. plantarum and P. kluyveri of a plant-based fermented beverage: Physicochemical, functional, and sensory properties. *Food Res. Int.* **2023**, *172*, 113060. [CrossRef]
89. Peng, W.; Meng, D.; Yue, T.; Wang, Z.; Gao, Z. Effect of the apple cultivar on cloudy apple juice fermented by a mixture of Lactobacillus acidophilus, Lactobacillus plantarum, and *Lactobacillus fermentum*. *Food Chem.* **2021**, *340*, 127922. [CrossRef] [PubMed]
90. Santhirasegaram, V.; Razali, Z.; Somasundram, C. Effects of thermal treatment and sonication on quality attributes of Chokanan mango (*Mangifera indica* L.) juice. *Ultrason. Sonochem* **2013**, *20*, 1276–1282. [CrossRef] [PubMed]
91. Aadil, R.M.; Zeng, X.A.; Wang, M.S.; Liu, Z.W.; Han, Z.; Zhang, Z.H.; Hong, J.; Jabbar, S. A potential of ultrasound on minerals, micro-organisms, phenolic compounds and colouring pigments of grapefruit juice. *Int. J. Food Sci. Technol.* **2015**, *50*, 1144–1150. [CrossRef]
92. Wang, H.; Zhao, Q.S.; Wang, X.D.; Hong Z dong Zhao, B. Pretreatment of ultrasound combined vacuum enhances the convective drying efficiency and physicochemical properties of okra (*Abelmoschus esculentus*). *LWT* **2019**, *112*, 108201. [CrossRef]
93. Santos, N.C.; Almeida, R.L.J.; Albuquerque, J.C.; de Andrade, E.W.V.; Gregório, M.G.; Santos, R.M.S.; Rodrigues, T.J.A.; Carvalho, R.d.O.; Gomes, M.M.d.A.; Moura, H.V.; et al. Optimization of ultrasound pre-treatment and the effect of different drying techniques on antioxidant capacity, bioaccessibility, structural and thermal properties of purple cabbage. *Chem. Eng. Process. Process Intensif.* **2024**, *201*, 109801. [CrossRef]
94. Urquieta-Herrero, M.; Cornejo-Mazón, M.; Gutiérrez-López, G.F.; García-Pinilla, S. Effect of two pasteurization methods on the content of bioactive compounds and antioxidant capacity of nance (*Byrsonima crassifolia*) pulp and their kinetics of loss during refrigerated storage. *Rev. Mex. Ing. Quim.* **2021**, *20*, 663–678. [CrossRef]
95. Alongi, M. *Sviluppo di Alimenti Funzionali Mediante Interventi di Proceso e Formulazione Innovativi e Sostenibili*; Università degli Studi di Udine: Udine, Italy, 2020.
96. Debelo, H.; Li, M.; Ferruzzi, M.G. Processing influences on food polyphenol profiles and biological activity. *Curr. Opin. Food Sci.* **2020**, *32*, 90–102. [CrossRef]
97. Cai, Y.X.; Wang, J.H.; McAuley, C.; Augustin, M.A.; Terefe, N.S. Fermentation for enhancing the bioconversion of glucoraphanin into sulforaphane and improve the functional attributes of broccoli puree. *J. Funct. Foods* **2019**, *61*, 103461. [CrossRef]
98. Meng, F.B.; Lei, Y.T.; Li, Q.Z.; Li, Y.C.; Deng, Y.; Liu, D.Y. Effect of *Lactobacillus plantarum* and *Lactobacillus acidophilus* fermentation on antioxidant activity and metabolomic profiles of loquat juice. *LWT* **2022**, *171*, 114104. [CrossRef]
99. Garcia, C.; Remize, F. Lactic acid fermentation of fruit and vegetable juices and smoothies: Innovation and health aspects. In *Lactic Acid Bacteria in Food Biotechnology—Innovations and Functional Aspects*; Elsevier: Amsterdam, The Netherlands, 2022; pp. 27–46.
100. Sharma, R.; Garg, P.; Kumar, P.; Bhatia, S.K.; Kulshrestha, S. Microbial Fermentation and Its Role in Quality Improvement of Fermented Foods. *Fermentation* **2020**, *6*, 106. [CrossRef]

101. Hou, F.; Cai, Y.; Wang, J. Antioxidant Capacity Changes and Untargeted Metabolite Profile of Broccoli during Lactic Acid Bacteria Fermentation. *Fermentation* **2023**, *9*, 474. [CrossRef]
102. Bei, Q.; Liu, Y.; Wang, L.; Chen, G.; Wu, Z. Improving free, conjugated, and bound phenolic fractions in fermented oats (*Avena sativa* L.) with Monascus anka and their antioxidant activity. *J. Funct. Foods* **2017**, *32*, 185–194. [CrossRef]
103. Wu, C.; Li, T.; Qi, J.; Jiang, T.; Xu, H.; Lei, H. Effects of lactic acid fermentation-based biotransformation on phenolic profiles, antioxidant capacity and flavor volatiles of apple juice. *LWT* **2020**, *122*, 109064. [CrossRef]
104. Sapci, Z. The effect of microwave pretreatment on biogas production from agricultural straws. *Bioresour. Technol.* **2013**, *128*, 487–494. [CrossRef] [PubMed]
105. Almaiman, S.A.; Albadr, N.A.; Alsulaim, S.; Alhuthayli, H.F.; Osman, M.A.; Hassan, A.B. Effects of microwave heat treatment on fungal growth, functional properties, total phenolic content, and antioxidant activity of sorghum (*Sorghum bicolor* L.) grain. *Food Chem.* **2021**, *348*, 128979. [CrossRef]
106. Álvarez, A.; Poejo, J.; Matias, A.A.; Duarte, C.M.M.; Cocero, M.J.; Mato, R.B. Microwave pretreatment to improve extraction efficiency and polyphenol extract richness from grape pomace. Effect on antioxidant bioactivity. *Food Bioprod. Process.* **2017**, *106*, 162–170. [CrossRef]
107. Gupta, Y.; Barrett, B.; Vlachos, D.G. Understanding microwave-assisted extraction of phenolic compounds from diverse food waste feedstocks. *Chem. Eng. Process. Process Intensif.* **2024**, *203*, 109870. [CrossRef]
108. Popoola, O.O. Phenolic compounds composition and in vitro antioxidant activity of Nigerian Amaranthus viridis seed as affected by autoclaving and germination. *Meas. Food* **2022**, *6*, 100028. [CrossRef]
109. Sanchiz, A.; Pedrosa, M.M.; Guillamón, E.; Arribas, C.; Cabellos, B.; Linacero, R.; Cuadrado, C. Influence of boiling and autoclave processing on the phenolic content, antioxidant activity and functional properties of pistachio, cashew and chestnut flours. *LWT* **2019**, *105*, 250–256. [CrossRef]
110. Rojas, M.L.; Kubo, M.T.K.; Caetano-Silva, M.E.; Augusto, P.E.D. Ultrasound processing of fruits and vegetables, structural modification and impact on nutrient and bioactive compounds: A review. *Int. J. Food Sci. Technol.* **2021**, *56*, 4376–4395. [CrossRef]
111. Kumar, K.; Srivastav, S.; Sharanagat, V.S. Ultrasound assisted extraction (UAE) of bioactive compounds from fruit and vegetable processing by-products: A review. *Ultrason. Sonochem* **2021**, *70*, 105325. [CrossRef] [PubMed]
112. Huang, D.; Men, K.; Li, D.; Wen, T.; Gong, Z.; Sunden, B.; Wu, Z. Application of ultrasound technology in the drying of food products. *Ultrason. Sonochem* **2020**, *63*, 104950. [CrossRef] [PubMed]
113. Rajewska, K.; Mierzwa, D. Influence of ultrasound on the microstructure of plant tissue. *Innov. Food Sci. Emerg. Technol.* **2017**, *43*, 117–129. [CrossRef]
114. Vera Zambrano, M.; Dutta, B.; Mercer, D.G.; MacLean, H.L.; Touchie, M.F. Assessment of moisture content measurement methods of dried food products in small-scale operations in developing countries: A review. *Trends Food Sci. Technol.* **2019**, *88*, 484–496. [CrossRef]
115. Prosapio, V.; Norton, I. Influence of osmotic dehydration pre-treatment on oven drying and freeze drying performance. *LWT* **2017**, *80*, 401–408. [CrossRef]
116. Maskan, M. Drying, shrinkage and rehydration characteristics of kiwifruits during hot air and microwave drying. *J. Food Eng.* **2001**, *48*, 177–182. [CrossRef]
117. Janiszewska-Turak, E.; Rybak, K.; Pobiega, K.; Nikodem, A.; Gramza-Michałowska, A. Sustainable Production and Characteristics of Dried Fermented Vegetables. *Fermentation* **2022**, *8*, 659. [CrossRef]
118. Oyinloye, T.M.; Yoon, W.B. Effect of Freeze-Drying on Quality and Grinding Process of Food Produce: A Review. *Processes* **2020**, *8*, 354. [CrossRef]
119. Ramírez-Pulido, B.; Bas-Bellver, C.; Betoret, N.; Barrera, C.; Seguí, L. Valorization of Vegetable Fresh-Processing Residues as Functional Powdered Ingredients. A Review on the Potential Impact of Pretreatments and Drying Methods on Bioactive Compounds and Their Bioaccessibility. *Front. Sustain. Food Syst.* **2021**, *5*, 82. [CrossRef]
120. Huang, Y.; Sun, Y.; Mehmood, A.; Lu, T.; Chen, X. Unraveling the temporal changes of Maillard reaction products and aroma profile in coffee leaves during hot-air drying. *J. Food Compos. Anal.* **2024**, *128*, 106055. [CrossRef]
121. Somjai, C.; Siriwoharn, T.; Kulprachakarn, K.; Chaipoot, S.; Phongphisutthinant, R.; Wiriyacharee, P. Utilization of Maillard reaction in moist-dry-heating system to enhance physicochemical and antioxidative properties of dried whole longan fruit. *Heliyon* **2021**, *7*, e07094. [CrossRef] [PubMed]
122. Réblová, Z. Effect of temperature on the antioxidant activity of phenolic acids. *Czech J. Food Sci.* **2012**, *30*, 171–175. [CrossRef]
123. Parchem, K.; Piekarska, A.; Bartoszek, A. Enzymatic activities behind degradation of glucosinolates. In *Glucosinolates: Properties, Recovery, and Applications*; Academic Press: Cambridge, MA, USA, 2020; pp. 79–106.
124. Antal, T. Comparative study of three drying methods: Freeze, hot air-assisted freeze and infrared-assisted freeze modes. *Agron. Res.* **2015**, *13*, 863–878.
125. Vargas, L.; Kapoor, R.; Nemzer, B.; Feng, H. Application of different drying methods for evaluation of phytochemical content and physical properties of broccoli, kale, and spinach. *LWT* **2022**, *155*, 112892. [CrossRef]
126. Zura-Bravo, L.; Rodriguez, A.; Stucken, K.; Vega-Gálvez, A. Drying kinetics of probiotic-impregnated murta (*Ugni molinae* T.) berries. *J. Food Sci. Technol.* **2019**, *56*, 103–113. [CrossRef] [PubMed]
127. Lee, Y.; Oh, J.; Jeong, Y.S. Lactobacillus plantarum-mediated conversion of flavonoid glycosides into flavonols, quercetin, and kaempferol in Cudrania tricuspidata leaves. *Food Sci. Biotechnol.* **2015**, *24*, 1817–1821. [CrossRef]

128. Filannino, P.; Bai, Y.; Di Cagno, R.; Gobbetti, M.; Gänzle, M.G. Metabolism of phenolic compounds by Lactobacillus spp. during fermentation of cherry juice and broccoli puree. *Food Microbiol.* **2015**, *46*, 272–279. [CrossRef]
129. Vega-Galvez, A.; Uribe, E.; Pasten, A.; Camus, J.; Rojas, M.; Garcia, V.; Araya, M.; Valenzuela-Barra, G.; Zambrano, A.; Goñi, M.G. Low-Temperature Vacuum Drying on Broccoli: Enhanced Anti-Inflammatory and Anti-Proliferative Properties Regarding Other Drying Methods. *Foods* **2023**, *12*, 3311. [CrossRef]
130. Chu, Q.; Li, L.; Duan, X.; Zhao, M.; Wang, Z.; Wang, Z.; Ren, X.; Li, C.; Ren, G. Effect mechanism of different drying methods on the quality and browning for daylily. *LWT* **2023**, *182*, 114862. [CrossRef]

foods

MDPI

Article

Valorizing Astringent 'Rojo Brillante' Persimmon Through the Development of Persimmon-Based Bars

Sepideh Hosseininejad, Gemma Moraga and Isabel Hernando *

Instituto Universitario de Ingeniería de Alimentos—Food UPV, Universitat Politècnica de València, Camino de Vera, s/n, 46022 Valencia, Spain; sehos@doctor.upv.es (S.H.); gemmoba1@tal.upv.es (G.M.)
* Correspondence: mihernan@tal.upv.es

Abstract: This study developed a new energy bar using the astringent 'Rojo Brillante' variety of persimmons to address postharvest losses. The bar was formulated with dehydrated persimmons, walnuts, hazelnuts, and chia seeds to enhance their nutritional profile. The proximate composition was evaluated and the mechanical and optical properties, soluble tannins, carotenoids, and antioxidant activities were monitored during storage. In addition, in vitro gastrointestinal digestion was performed to determine the recovery index of the bioactive compounds. The results showed that the formulated energy bar contained higher levels of healthy fats, proteins, and fibers than other fruit energy bars. The mechanical properties of dehydrated persimmon effectively supported the consistency of the bar, eliminating the need for hydrocolloids or syrups. During storage, soluble tannin content decreased, mitigating astringency issues commonly found in persimmon products, whereas carotenoid levels and antioxidant activity remained stable. In vitro digestion analysis revealed a higher recovery index for soluble tannins (180.08%) than carotenoids (9.87%). This persimmon-based energy bar offers a sustainable and nutritious option for the snack industry, catering to consumer preferences for natural products while contributing to the reduction of agricultural waste.

Keywords: *Diospyros kaki*; energy bars; astringency; tannins; carotenoids; antioxidant activity; in vitro digestion; sustainability

1. Introduction

Interest in enhancing the nutritional properties of many food products is increasing because of evolving eating patterns and increasing customer demand for nutritious foods with functional ingredients [1]. In the contemporary era, a considerable proportion of consumers prioritize the purchase of food items that align with their personal health and wellness goals. Consequently, the development of novel food products with distinctive textural, sensory, or functional attributes represents a significant area of interest [1].

Fruit bars are highly nutritious and calorically concentrated food products. Furthermore, they typically have a longer shelf life than raw fruits and can be a healthy, convenient food choice that provides dietary fiber and other bioactive substances needed to meet a person's daily nutritional requirements [2].

Persimmon (*Diospyros kaki* Thunb.) is an important fruit crop in Spain, primarily cultivated on approximately 17,600 hectares, with the focus largely on the 'Rojo Brillante' variety, which represents more than 90% of production [3]. Persimmons are a rich source of bioactive compounds, including dietary fiber, minerals, vitamins, phenolic compounds, and carotenoids [4]. These compounds have been linked to several health benefits, including the prevention of certain diseases such as cancer, hypertension, diabetes, and atherosclerosis [5–7]. However, owing to their seasonal availability, perishable nature, and difficulties in storage and transportation, persimmons have experienced considerable postharvest losses in recent years [8]. In the case of astringent varieties, the losses were even more significant. The persimmon cultivar 'Rojo Brillante' is characterized by astringency due to its high soluble tannin content when harvested in a firm state [9]. As the fruit matures, the soluble tannin content declines [10]; however, the soft texture of the fruit renders it susceptible to rapid deterioration. Accordingly, postharvest deastringency treatment is required before commercialization, as fruits must be consumed when they still have a firm texture [10]. A variety of techniques have been used for deastringency treatment, including anaerobic treatments with CO2. This process precipitates tannins, rendering them undetectable when persimmon is consumed while simultaneously preserving the firmness of the product [11]. Nevertheless, the application of deastringency treatments represents an additional financial burden and may sometimes prove ineffective, engendering further losses.

To mitigate losses, persimmons can be processed (e.g., dried) to prolong their shelf life and serve as a promising functional ingredient for incorporation into novel food formulations. González et al. [12] demonstrated that hot-air drying has the effect of considerably reducing the amount of soluble tannins. The research team observed a significant reduction in soluble tannin content in dried persimmon slices after drying at 40 and 60 °C. These tannins undergo transformation into insoluble forms. It is thus possible to obtain non-astringent persimmon snacks from astringent fruits with no prior deastringency treatment. In addition, our previous study on the use of astringent persimmon flour as a functional ingredient in gluten-free muffins demonstrated the removal of astringency because of the insolubilization of tannins during baking [8]. Nevertheless, astringency issues have been documented in the context of various persimmon transformation processes. Castelló et al. [13] investigated the impact of thermal processing and storage on the astringency of a spreadable product derived from the 'Rojo Brillante' persimmon through osmotic processes. Despite undergoing a deastringency process and initial osmotic dehydration, which reduced the soluble tannin content, partial re-solubilization of insoluble tannins was observed when the product was subjected to elevated temperatures. Furthermore, during the storage period, an increase in the soluble tannin content was observed. It was determined that refrigeration was the optimal method to prevent astringency.

A review of the literature revealed no studies examining the production of persimmon-based bars. Several studies have been conducted on bars prepared from other dried fruits, including dates [14,15]. Ibrahim et al. [15] investigated the formulation of date-based bars with varying percentages of dates, ranging from 40 to 70%. The textural, sensory, and technological attributes of the bars were enhanced when 50% date paste was incorporated into the formulation, as reported by other researchers. In another study, Leguizamon-Delgado et al. [16] evaluated the physicochemical and sensory characteristics of mango-based bars prepared with dried mango as a viable alternative to agro-industrial waste. The results demonstrate the potential of this approach to use agro-industrial waste, with notable findings in terms of the total fiber content and functional compounds, including polyphenols and ascorbic acid, with antioxidant activity. Kumar et al. [17] used dried blends of papaya and guava for preparing fruit bars and reported that the optimal combination ratio was 50:50, as this combination yielded the highest sensory score, overall acceptability, and superior carotenoid and protein content.

Incorporating ingredients such as walnuts, hazelnuts, and chia seeds can help enhance the nutritional profile of fruit bars. These ingredients provide plant-based proteins and are rich in monounsaturated and polyunsaturated fats. Furthermore, they contain significant micronutrients, including minerals, vitamins, and other bioactive compounds, which augment the functional attributes of the final product. Walnuts are an exceptional source of omega-3 fatty acid content, particularly alpha-linolenic acid, and their consumption has been linked to health benefits and chronic disease prevention [18]. Hazelnuts are a notable source of vitamin E, a fat-soluble antioxidant, and are regarded as valuable ingredients in the food industry because of their distinctive flavor profile [19]. Chia seeds are a rich source of dietary fiber, particularly soluble fiber, which has been demonstrated to exert beneficial effects on the microbiota [20–22]. It is also noteworthy that the addition of dehydrated persimmons contributes to the cohesiveness of the fruit bar, resulting in a mix of textures and flavors when combined with nuts and seeds. Furthermore, dehydrated persimmon provides a natural sweetness that obviates the need for additional sweeteners.

The objective of this study was to obtain and characterize a new fruit bar formulated with persimmon of the astringent 'Rojo Brillante' variety as a strategy to reduce postharvest losses. The physicochemical properties and the stability of tannins (related to astringency), carotenoids, and their antioxidant activity were assessed during storage. Moreover, in vitro gastrointestinal digestion was conducted to evaluate the recovery index of bioactive compounds, including the oral, stomach, and small intestine stages.

2. Materials and Methods

2.1. Bars Preparation

Persimmon fruit (*Diospyros kaki* Thunb. cv. 'Rojo Brillante') was provided by the Instituto Valenciano de Investigaciones Agrarias (IVIA, Spain) and was not subjected to any deastringency treatment prior to analysis. The persimmons were harvested from a local grove in L'Alcudia, Valencia, Spain, in early December at the ripening stage V [9], which is characterized by an orange color. The persimmons were washed, peeled, and dried in an oven Binder model FD 260 standard (Binder GmbH, Tuttlingen, Germany) at 45 °C for 36 h until they reached a water content of 37 g water/100 g product. Walnuts, hazelnuts, and chia seeds (Hacendado trademark) were procured from a local supermarket.

To prepare the bars, 190 g of dried persimmons, 50 g of walnuts, 50 g of hazelnuts, and 10 g of chia seeds were ground together for 3 min using a Moulinex grinder (Moulinex, Barcelona, Spain) until a homogeneous mixture was formed. Subsequently, bars measuring 10 × 2 × 1.5 cm were prepared. The final formulation of the bars was selected following a series of laboratory tests in which different ingredient ratios were evaluated. The selected bar exhibited the greatest mechanical stability while incorporating the highest possible amount of dried persimmon. The bars were stored at a refrigerated temperature of 4 °C for 4 weeks.

2.2. Proximate Composition

Proximate analysis (moisture, ash, fat, protein, sugar, and fiber) was conducted in accordance with the methodology delineated by the AOAC [23]. Carbohydrates were calculated by the difference (Equation (1)), and the gross energy value was obtained by applying Equation (2) [24].

$$\text{Carbohydrate (\%)} = 100 - [\text{moisture (\%)} + \text{protein (\%)} + \text{fat (\%)} + \text{ash (\%)}] \quad (1)$$

$$\text{Energy (kJ/100 g)} = (\text{protein} \times 16.7) + (\text{fat} \times 37.7) + (\text{carbohydrate} \times 16.7) \quad (2)$$

2.3. Mechanical Properties

A TA-XT Plus Texture Analyzer (Stable Micro System, Ltd., Surrey, UK) with a Warner-Bratzler reversible blade (HDP/BS) was used to perform a shear test. The sample was positioned over a slotted blade insert and bisected in the transverse direction. The speed of the test was 1 mm/s and the maximum force (in Newtons) was recorded. Each sample was subjected to six replicates.

2.4. Optical Properties

The color was evaluated using a Minolta CM-3600d spectro-colorimeter (Minolta Co., Tokyo, Japan), with the illuminant D65 and the 10° observer serving as the reference to obtain the CIE L*a*b* color coordinates. Each sample was subjected to six replicates. The hue (h*) and chroma (C*) values were calculated using Equations (3) and (4), respectively. The color differences (ΔE*) were calculated using Equation (5), with the initial sample (t 0) designated as the reference point.

$$h^* = arctg\frac{b^*}{a^*} \quad (3)$$

$$C^* = \left[\left(a^{*2} + b^{*2}\right)^{1/2}\right] \quad (4)$$

$$\Delta E^* = \left[(\Delta L^*)^2 + (\Delta a^*)^2 + (\Delta b^*)^2\right]^{1/2} \quad (5)$$

2.5. Soluble Tannin Content (STC)

The STC was determined using the Folin–Ciocalteu colorimetric method, as described by Arnal et al. [25], in an ethanolic extract (extracted with 96% ethanol). The results are expressed in terms of mg of gallic acid equivalent (GAE) per 100 g of sample. Tannin extraction was performed in triplicates.

2.6. Total Carotenoid Content (TCC)

TCC was calculated in accordance with the protocol established by Nath et al. [26], using a lipidic extract extracted with acetone and diethyl ether. The results are expressed in mg of β-carotene per 100 g of sample. Carotenoid extraction was performed in triplicate.

2.7. Antioxidant Activity (FRAP and DPPH)

Antioxidant activity was measured using the FRAP and DPPH protocols described by Benzie et al. [27] and Matsumura et al. [28], respectively. FRAP results are reported as μmol of Trolox per gram of sample, whereas the DPPH values are expressed as inhibition %.

2.8. In Vitro Digestion

The methodology proposed by Brodkorb et al. [29] was used to simulate the biological fate of the ingested substances using an in vitro gastrointestinal tract model. All oral, gastric, and intestinal phases were replicated. The complete set of enzymes required for the analysis was obtained from Sigma-Aldrich (Madrid, Spain). In accordance with the methodology proposed by González et al. [30], the digestive process was conducted in a "Carousel 6 Plus" reaction station (Radleys, Saffron Walden, UK) under controlled conditions at a temperature of 37 °C, with agitation at 150 rpm, in the absence of light, and in an atmosphere of N_2. The recovery index was calculated to analyze the effect of in vitro digestion on the content of soluble tannins and carotenoids. The recovery index was used to quantify the amount of substance recovered after intestinal digestion. This was achieved by comparing the recovered amount with that of an undigested sample [31]. The recovery index (RI) was calculated using Equation (6):

$$Recovery\ index\ (\%) = \frac{DF}{UDF} \times 100 \quad (6)$$

where *DF* (digested fraction) is the amount of bioactive compounds in the digested fraction following the intestinal digestion phase, and *UDF* (undigested fraction) is the amount of bioactive compounds quantified in fresh samples.

2.9. Statistical Analysis

Statistical analysis of the data was conducted using the Statgraphics Centurion XVII program, using analysis of variance (ANOVA) and calculating the minimum Fisher's significant differences at a 95% significance level ($p < 0.05$).

3. Results and Discussion

3.1. Proximate Composition of the Bars

The proximate compositions of the persimmon-based bars are presented in Table 1. The moisture content was comparable to that reported for fruit bars formulated using date paste [14]. Although Parna et al. [14] introduced small amounts of water to aid the blending process of dates into the paste, our formulation did not require water addition, as the dehydrated persimmon served as a cohesive matrix for the bar.

Table 1. Proximate composition of the persimmon-based bar ($n = 2$).

Macronutrients	(g/100 g)
Moisture	28.83 ± 1.24
Fat	22.80 ± 0.83
Carbohydrate	39.37 ± 1.41
-Sugar	26.60 ± 1.27
-Fiber	5.80 ± 0.38
Protein	7.40 ± 0.63
Ash	1.60 ± 0.09

The bar is a rich source of healthy fats derived from walnuts, hazelnuts, and chia seeds. These unsaturated healthy fats contribute to optimal brain function, are crucial for cardiovascular health, and serve as sustainable energy sources [32]. In addition, the seeds and nuts assisted in maintaining equilibrium between the protein and fiber content of the persimmon-based bar, which exhibited higher levels than other fruit-based bars. Sun-Waterhouse et al. [2] reported protein and fiber contents of 2.74% and 2.54%, respectively, in snack bars containing apple polyphenol extracts. These values are more than twice as low as the protein and fiber values obtained for the persimmon bars in the present study. In date-based bars, the values were 3.70% and 5.51%, respectively, when the cultivar Nabtat Ali was used and 4.06% and 4.47% when bars were formulated with dates from the Sukkari cultivar [14]. All of these values were lower than those of the persimmon bars. In a separate study, Srivastava et al. [33] reported that the guava-orange fruit bars they prepared contained fat in the range of 1.40% to 4.80%, which is significantly lower than the fat content of the bars in this study (22.80%). The persimmon-based bar comprises sugars as the primary carbohydrate source (Table 1). However, because no additional sugars were incorporated during the manufacturing process, the values were lower than those of other fruit-based energy bars. In a separate study, Agahari et al. [33] reported a sugar content of 70.98% in apple bars formulated with added sugar. Parna et al. [14] reported higher carbohydrate values in date-based bars despite the absence of additional sugars in the formulation. In a study by Lucas-González et al. [34], the sugar content of fruit snacks prepared with fig, pomegranate, date, or apricot ranged from 30 to 55%.

For macronutrients, the energy value of the persimmon-based bar was 1640.62 kJ/100 g, which was higher than that reported for other fruit-based energy bars [15]. Therefore, this nutritionally dense and calorically concentrated product can be regarded as a convenient food option for athletes and for individuals with specific dietary requirements, such as individuals following a high-protein diet with no added sugar.

3.2. Mechanical Properties

A crucial aspect of developing a novel fruit-based bar is that the product retains its structural integrity and displays no indications of disaggregation. Thus, various hydrocolloids have been considered for the production of fruit-based bars. Dana-Lache et al. [35] evaluated the incorporation of low-acyl and high-acyl gellan into mango-based bars, whereas Ahmad et al. [36] studied the effect of pectin, starch, and ethyl cellulose addition in fruit bars made from papaya and tomato. Dehydrated persimmon proved to be an effective matrix for bar formulation owing to its cohesiveness, obviating the need for additional hydrocolloids.

Table 2 presents the mean values of the maximum force (N) recorded in the shear test conducted on the samples over a four-week storage period. The test enabled the identification of alterations in the mechanical characteristics of persimmon-based bars. The bars exhibited a notable increase ($p < 0.05$) in F(N) values during the initial two weeks of storage. During the third week, these values remained constant. In the final week, it decreased significantly ($p < 0.05$) until it reached a value that was not significantly different from that recorded in the first week ($p > 0.05$). Therefore, the mechanical properties remained consistent after one and four weeks of storage, and no evidence of disaggregation was observed.

Table 2. Maximum force (N), lightness (L*), chroma (C*), hue (h*), and color difference (ΔE*) of persimmon-based bars during storage ($n = 6$).

t (weeks)	F(N)	L*	C*	h*	ΔE*
0	7.82 ^{a} ± 0.72	37.24 ^{a} ± 1.26	24.63 ^{a} ± 1.68	75.29 ^{a} ± 0.66	-
1	10.27 ^{b} ± 0.74	32.27 ^{b} ± 0.97	14.84 ^{b} ± 0.66	68.30 ^{b} ± 1.41	11.23
2	12.32 ^{c} ± 1.05	28.46 ^{c} ± 1.09	12.02 ^{c} ± 1.24	64.93 ^{c} ± 1.63	15.79
3	12.98 ^{c} ± 0.70	30.4 ^{c} ± 1.57	12.9 ^{c} ± 0.93	67.07 ^{b} ± 3.19	13.83
4	10.55 ^{b} ± 0.65	29.85 ^{c} ± 1.03	11.68 ^{c} ± 0.52	63.55 ^{c} ± 2.74	15.31

Mean values in a column with different superscript letters differ significantly ($p < 0.05$) according to ANOVA (LSD multiple range test).

The evolution of mechanical properties during storage has been documented in multiple studies. Ibrahim et al. [15] observed a progressive hardening of the date-based bars after 12 days of storage, yielding similar results. Ahmad et al. [36] reported significant differences after 120 days of storage, depending on the hydrocolloid combination. The findings indicated that the incorporation of pectin and starch (at concentrations of 0.5, 1, and 1.5%) resulted in a notable ($p < 0.05$) increase in the hardness of the fruit-based bar prepared from a combination of papaya and tomato. However, the incorporation of starch and ethyl cellulose (at concentrations of 0.5% and 1%, respectively) resulted in a statistically significant reduction ($p < 0.05$).

3.3. Optical Properties

As illustrated in Table 2, there was a notable decline ($p < 0.05$) in the L* value during storage compared with week 0. Similarly, the h* value exhibited a significant ($p < 0.05$) reduction with prolonged storage, indicating a shift toward a redder hue. The chromatic purity or chroma, which is related to the value of C*, also demonstrated variation throughout the storage period, exhibiting a significant decrease ($p < 0.05$) with increasing storage time. The ΔE* value reached 15 at the conclusion of the fourth week. The primary color alterations occurred during the initial two weeks of storage; subsequently, most optical parameters exhibited minimal fluctuations ($p > 0.05$). These changes can be attributed to several biochemical reactions, including enzymatic browning due to polyphenol oxidases, non-enzymatic browning such as Maillard reactions, oxidative breakdown of ascorbic acid, and degradation of carotenoids [37].

3.4. Soluble Tannin Content (STC) and Antioxidant Activity

The STC and related antioxidant activity of persimmon-based bars stored for four weeks are presented in Table 3.

Table 3. Soluble tannin content (STC) and antioxidant activity of persimmon-based bars during storage ($n = 3$).

t (Weeks)	STC (mg GAE/100 g)	FRAP (μmol Trolox /g)	DPPH (Inhibition %)
0	78.41 ^{a} ± 1.67	11.12 ^{a} ± 0.28	90.44 ^{a} ± 1.43
1	70.55 ^{b} ± 3.22	8.46 ^{b} ± 0.55	64.79 ^{b} ± 2.56
2	57.01 ^{c} ± 6.89	6.84 ^{c} ± 0.47	44.85 ^{c} ± 1.33
3	50.61 ^{c} ± 6.05	6.24 ^{c} ± 0.53	43.91 ^{c} ± 4.24
4	37.25 ^{d} ± 2.46	6.54 ^{c} ± 0.45	35.02 ^{d} ± 2.46

Mean values in a column with different superscript letters differ significantly ($p < 0.05$) according to ANOVA (LSD multiple range test).

The STC of the persimmon-based bar was 78.41 ± 1.67 mg GAE/100 g at the outset of the storage period. Given the findings of previous studies examining the correlation between STC and astringency, it can be concluded that the STC values observed in this study fall within the range typically observed for non-astringent persimmon products [8,38]. As González et al. [12] have previously observed, hot-air drying represents an effective technique for reducing STC, enabling the production of non-astringent dehydrated persimmon products derived from astringent fruits that have undergone no preliminary deastringency treatment. The bars formulated in this study serve as a prime example of this phenomenon. During storage, a statistically significant decline ($p < 0.05$) was observed, reaching a final value of 37.25 ± 2.46 mg GAE/100 g after four weeks at 4 °C. As stated by Zhou et al. [39], the storage of dried persimmons results in a reduction of tannin levels, primarily due to the oxidation and polymerization processes, which transform tannins from soluble to insoluble forms. The enzymatic action of polyphenol oxidase (PPO) facilitates the oxidation of soluble tannins, resulting in their polymerization into larger, insoluble molecules. As these molecules precipitate, the astringency and bitterness associated with soluble tannins are reduced. Therefore, the persimmon-based bar did not exhibit astringency issues related to the solubilization of insoluble tannins during storage, in contrast to other 'Rojo Brillante' persimmon products, such as spreadables [13]. This represents a promising avenue for the valorization of the astringent 'Rojo Brillante' cultivar, which could have significant implications for the wider food industry.

Table 3 presents the findings pertaining to the antioxidant activity of the ethanolic extract as determined by the FRAP and DPPH methods. A notable decline ($p < 0.05$) in antioxidant activity was evident during the initial two weeks, exhibiting a trend similar to that of STC. The observed decline in the antioxidant activity of the persimmon-based bar may be attributed to the insolubilization and oxidation of tannins during storage [7,40].

The persimmon-based bar exhibited higher STC and antioxidant activity than the raw material and dehydrated persimmons. The STC in the dehydrated persimmon was 51.17 ± 11.11 mg GAE/100 g, with a FRAP value of 8.55 ± 0.96 (μmol Trolox/g) and a DPPH value of 69.77 ± 8.65 inhibition %. The increase is due to additional ingredients in the formulation, specifically walnuts and hazelnuts, which have high levels of phenolic compounds. The total phenolic content of walnuts ranged between 1558 and 1625 mg GAE/100 g, whereas hazelnuts have been found to have a lower total phenolic content, with values between 291 and 875 mg GAE/100 g [41]. The phenolic compounds present in chia seeds have been found to range from 53.5 to 71.2 mg GAE/100 g [42].

3.5. Total Carotenoid Content (TCC) and Antioxidant Activity

The results indicated that the TCC was higher in the dehydrated persimmon (129.16 ± 0.38 mg β-carotene/100 g) than in the persimmon-based bar (Table 4). However, the formulated bar exhibited higher antioxidant activity than the dehydrated persimmon (1.82 ± 0.39 inhibition %), which can be attributed to the presence of other fat-soluble components in the formulation, including vitamin E in hazelnuts [19], chia [43], and walnuts [44].

Table 4. Total carotenoid content (TCC) and antioxidant activity of the persimmon-based bar during storage ($n = 3$).

t (Weeks)	TCC (mg β-Carotene/100 g)	DPPH (Inhibition %)
0	77.87 ^{a} ± 2.67	32.60 ^{a} ± 5.68
1	75.78 ^{a} ± 11.26	33.61 ^{a} ± 5.54
2	73.72 ^{a} ± 3.34	51.81 ^{b} ± 7.21
3	72.79 ^{a} ± 7.43	40.38 ^{a} ± 3.44
4	69.33 ^{a} ± 4.31	33.61 ^{a} ± 5.54

Mean values in a column with different superscript letters differ significantly ($p < 0.05$) according to ANOVA (LSD multiple range test).

The primary cause of carotenoid loss in fruits during storage is oxidation, which is affected by several variables, including temperature, light, and oxygen exposure [45,46]. Song et al. [45] observed a reduction in the TCC of dehydrated pumpkins during storage at 4 °C; they reported that the main reason for significant losses of carotenoids is non-enzymatic oxidative degradation occurring in the presence of molecular oxygen. However, in the persimmon-based bar, TCC remained stable, with no statistically significant differences ($p > 0.05$) detected throughout the storage period. Walnuts and hazelnuts are especially rich in antioxidants, including tocopherols (vitamin E), which contribute to their high antioxidant efficacy [47], these antioxidants help protect carotenoids, such as beta-carotene, from oxidative degradation. Regarding antioxidant activity, the values remained constant, except for the second week, during which a significant increase ($p < 0.05$) was observed.

3.6. Effect of In Vitro Digestion on STC, TCC and Antioxidant Activity

Following in vitro digestion, a notable increase ($p < 0.05$) in STC was observed in comparison with the undigested bars, resulting in an RI% value of 180.08% (Table 5). These findings are consistent with those of previous studies on persimmon products [8,30]. The liberation of tannins linked to fiber and proteins in the food matrix and the higher solubility of tannins due to factors such as the pH environment in the gastrointestinal phase may be the reason for the observed increase in STC in the digesta [48]. The antioxidant activity, as determined by the FRAP method, exhibited a similar trend, with values increasing from 11.12 before in vitro digestion to 15.74 μmol Trolox/g after in vitro digestion. Conversely, DPPH assay demonstrated a decline in antioxidant activity, with a reduction in inhibition from 38.32% to 28.87%. Kamiloglu et al. [49] reported a significant negative impact on the antioxidant activity of black carrot, jam, and marmalade after in vitro digestion. Nevertheless, the addition of persimmon flour to muffin formulations resulted in a notable enhancement of antioxidant activity within the soluble fraction of the small intestinal digestate [8].

Table 5. Soluble tannin content (STC), antioxidant activity of the soluble fraction (FRAP-s and DPPH-s), total carotenoid content (TCC), antioxidant activity of the carotenoid fraction (DPPH-c), and RI% of the STC and TCC of the persimmon-based bar after in vitro digestion ($n = 2$).

After In Vitro Digestion	
STC (mg GAE/100 g)	141.21 ± 4.11
FRAP-s (μmol Trolox/g)	15.74 ± 1.03
DPPH-s (inhibition %)	38.32 ± 3.36
RI% STC	180.08
TCC (mg β-carotene/100 g)	7.69 ± 0.74
DPPH-c (inhibition%)	44.23 ± 13.95
RI% TCC	9.87

As previously documented [8], TCC was found to undergo a notable decline ($p < 0.05$) following the in vitro digestive process. This decline was observed in the persimmon-based bar, with the non-digested sample containing 77.87 mg β-carotene/100 g, and the digesta from the small intestine exhibiting a significantly reduced concentration of 7.69 mg β-carotene/100 g. Carotenoids that underwent intestinal micellization before absorption exhibited a decrease in RI below 10% (Table 5). Several factors may influence carotenoid absorption, including (i) food processing, (ii) meal composition, (iii) digestive enzyme activity, and (iv) cross-enterocyte transport efficiency, as described by Desmarchelier et al. [50]. The reduction in TCC content after in vitro digestion may be attributed to the lipophilic properties of the compounds and the matrix effect. The bioavailability of carotenoids in plant-based foods is typically low due to their entrapment within plant cell membranes and dietary fibers. [50]. However, regarding the antioxidant activity evaluated by DPPH in the lipid fraction, a notable increase was observed following in vitro digestion, as evidenced by a statistically significant difference ($p < 0.05$). A comparable trend was observed in muffins fortified with persimmon flour [8].

4. Conclusions

The development of persimmon-based energy bars using the astringent 'Rojo Brillante' variety represents a novel approach to mitigate substantial postharvest losses of this crop while introducing a nutrient-dense product with strong market potential. Dehydrated persimmon serves as a natural binder, obviating the necessity for added sugars or hydrocolloids, thus maintaining a clean-label appeal. The incorporation of walnuts, hazelnuts, and chia seeds significantly enhanced the nutritional value of the bars, particularly in terms of healthy fats, proteins, and dietary fiber. This makes the bars a balanced option for health-conscious consumers. The bars demonstrate minimal disaggregation and desirable firmness over a four-week storage period. The stability of carotenoids and the controlled reduction of soluble tannins during storage demonstrated the effectiveness of the formulation in controlling astringency and preserving bioactive compounds. The in vitro digestion results demonstrated that soluble tannins are more bioavailable, which lends further support to the health benefits of these compounds. Despite the reduction in carotenoid recovery, the product demonstrated considerable antioxidant capacity following digestion. Therefore, this persimmon-based energy bar represents a sustainable and functional alternative within the snack food industry, addressing both consumer demand for natural and nutritious products and the need to mitigate agricultural waste. Further research should concentrate on the prolongation of the bars' shelf life, which can be achieved by the assessment of diverse packaging and storage conditions. Furthermore, a sensory analysis should be conducted to identify potential enhancements that could facilitate a successful market launch.

Author Contributions: Conceptualization, S.H., G.M., and I.H.; methodology, S.H., and G.M.; formal analysis, S.H.; investigation, S.H.; data curation, S.H.; supervision, G.M., and I.H.; writing—original draft preparation, S.H.; writing—review and editing, G.M., and I.H.; project administration, G.M., and I.H. All authors have read and agreed to the published version of the manuscript.

Funding: This research was funded by MCIN/AEI/ 10.13039/501100011033, and by "ERDF A way of making Europe" grant number RTA2017-00045-C02-02.

Institutional Review Board Statement: Not applicable.

Informed Consent Statement: Not applicable.

Data Availability Statement: The original contributions presented in the study are included in the article, further inquiries can be directed to the corresponding author.

Acknowledgments: The authors thank Phillip Bentley for his assistance in correcting the language in the manuscript.

Conflicts of Interest: The authors declare that they have no conflicts of interest.

References

1. Aljaloud, S.; Colleran, H.L.; Ibrahim, S.A. Nutritional Value of Date Fruits and Potential Use in Nutritional Bars for Athletes. *Food Nutr. Sci.* **2020**, *11*, 463–480. [CrossRef]
2. Sun-Waterhouse, D.; Teoh, A.; Massarotto, C.; Wibisono, R.; Wadhwa, S. Comparative Analysis of Fruit-Based Functional Snack Bars. *Food Chem.* **2010**, *119*, 1369–1379. [CrossRef]
3. Ministerio de Agricultura, Pesca y Alimentación (MAPA). Available online: https://www.mapa.gob.es/es/estadistica/temas/estadistica-digital/powerbi-cultivos.aspx (accessed on 1 October 2024).
4. Son, E.J.; Hwang, M.K.; Lee, E.; Seo, S.G.; Kim, J.; Jung, S.K.; Kim, J.R.; Ahn, G.; Lee, K.W.; Joo, H. Persimmon Peel Extract Attenuates PDGF-BB-Induced Human Aortic Smooth Muscle Cell Migration and Invasion through Inhibition of c-Src Activity. *Food Chem.* **2013**, *141*, 3309–3316. [CrossRef] [PubMed]
5. Kamimoto, M.; Nakai, Y.; Tsuji, T.; Shimamoto, T.; Shimamoto, T. Antiviral Effects of Persimmon Extract on Human Norovirus and Its Surrogate, Bacteriophage MS2. *Food Sci.* **2014**, *79*, 941–946. [CrossRef] [PubMed]
6. Liu, F.; Wang, Y.; Li, R.; Bi, X.; Liao, X. Effects of High Hydrostatic Pressure and High Temperature Short Time on Antioxidant Activity, Antioxidant Compounds and Color of Mango Nectars. *Innov. Food Sci. Emerg. Technol.* **2014**, *21*, 35–43. [CrossRef]
7. González, C.M.; García, A.L.; Llorca, E.; Hernando, I.; Pedro, A.; Bermejo, A.; Moraga, G.; Quiles, A. Carotenoids in Dehydrated Persimmon: Antioxidant Activity, Structure, and Photoluminescence. *LWT* **2021**, *142*, 111007. [CrossRef]
8. Hosseininejad, S.; Larrea, V.; Moraga, G.; Hernando, I. Evaluation of the Bioactive Compounds, and Physicochemical and Sensory Properties of Gluten-Free Muffins Enriched with Persimmon 'Rojo Brillante' Flour. *Foods* **2022**, *11*, 3357. [CrossRef]

9. Tessmer, M.A.; Besada, C.; Hernando, I.; Appezzato-da-Glória, B.; Quiles, A.; Salvador, A. Microstructural Changes While Persimmon Fruits Mature and Ripen. Comparison between Astringent and Non-Astringent Cultivars. *Postharvest Biol. Technol.* **2016**, *120*, 52–60. [CrossRef]
10. Arnal, L.; Del Río, M.A. Removing Astringency by Carbon Dioxide and Nitrogen-Enriched Atmospheres in Persimmon Fruit Cv. "Rojo Brillante". *J. Food Sci.* **2003**, *68*, 1516–1518. [CrossRef]
11. Salvador, A.; Arnal, L.; Besada, C.; Larrea, V.; Hernando, I.; Pérez-Munuera, I. Reduced Effectiveness of the Treatment for Removing Astringency in Persimmon Fruit When Stored at 15 °C: Physiological and Microstructural Study. *Postharvest Biol. Technol.* **2008**, *49*, 340–347. [CrossRef]
12. González, M.; Hernando, I.; Moraga, G. Influence of Ripening Stage and De-Astringency Treatment on the Production of Dehydrated Persimmon Snacks. *J. Sci. Food Agric.* **2021**, *101*, 603–612. [CrossRef] [PubMed]
13. Castelló, M.; Heredia, A.; Domínguez, E.; Ortolá, M.; Tarrazó, J. Influence of Thermal Treatment and Storage on Astringency and Quality of a Spreadable Product from Persimmon Fruit. *Food Chem.* **2011**, *128*, 323–329. [CrossRef] [PubMed]
14. Parna, O.J.; Bhat, R.; Yeoh, T.K.; Al-Hassan, A.A. Development of Novel Fruit Bars by Utilizing Date Paste. *F ood Biosci.* **2015**, *9*, 20–22. [CrossRef]
15. Ibrahim, S.A.; Fidan, H.; Aljaloud, S.O.; Stankov, S. Application of Date (*Phoenix dactylifera* L.) Fruit in the Composition of a Novel Snack Bar. *Foods* **2021**, *10*, 918. [CrossRef] [PubMed]
16. Leguizamon-Delgado, M.A.; Duque-Cifuentes, A.L.; Quintero-Castaño, V.D. Physico-Chemical and Sensory Evaluation of a Mango-Based Fruit Bar. *Dyna* **2019**, *86*, 276–283. [CrossRef]
17. Kumar, A.L.; Madhumathi, C.; Sadarunnisa, S.; Latha, P. Quality Evaluation and Storage Study of Papaya Guava Fruit Bar. *J. Pharmacogn. Phytochem.* **2017**, *6*, 2082–2087.
18. Polmann, G.; Badia, V.; Danielski, R.; Salvador, S.R.; Block, J.M. Nuts and Nut-Based Products: A Meta-Analysis from Intake Health Benefits and Functional Characteristics from Recovered Constituents. *Food Rev. Int.* **2023**, *39*, 5021–5047. [CrossRef]
19. Zhao, J.; Wang, X.; Lin, H.; Lin, Z. Hazelnut and Its By-Products: A Comprehensive Review of Nutrition, Phytochemical Profile, Extraction, Bioactivities and Applications. *Food Chem.* **2023**, *413*, 135576. [CrossRef]
20. Olivos-Lugo, B.; Valdivia-Lo´pez, M.A.; Tecante, A. Thermal and Physicochemical Properties and Nutritional Value of the Protein Fraction of Mexican Chia Seed (*Salvia hispanica* L.). *Food Sci. Technol. Int.* **2010**, *16*, 89–96. [CrossRef]
21. Kosarsoy Agçeli, G. A New Approach to Nanocomposite Carbohydrate Polymer Films: Levan and Chia Seed Mucilage. *Biol. Macromol.* **2022**, *218*, 751–759. [CrossRef]
22. de Morais, V.N.; Gomes, M.J.C.; Grancieri, M.; de Paula Dias Moreira, L.; Toledo, R.C.L.; Costa, N.M.B.; da Silva, B.P.; Martino, H.S.D. Chia (*Salvia hispanica* L.) Flour Modulates the Intestinal Microbiota in Wistar Rats Fed a High-Fat and High-Fructose Diet. *Food Res. Int.* **2023**, *172*, 113095. [CrossRef] [PubMed]
23. AOAC. *Official Methods of Analytical of the Association of Official Analytical Chemists*, 15th ed.; Association of Official Analytical Chemists: Washington, DC, USA, 1990. Available online: https://www.aoac.org/ (accessed on 20 September 2023).
24. Bhat, R.; Sridhar, K.R. Nutritional Quality Evaluation of Electron Beam-Irradiated Lotus (*Nelumbo nucifera*) Seeds. *Food Chem.* **2008**, *107*, 174–184. [CrossRef]
25. Arnal, L.; Del Río, M.A. Effect of Cold Storage and Removal Astringency on Quality of Persimmon Fruit (*Diospyros kaki* L.) Cv. Rojo Brillante. *Food Sci. Technol. Int.* **2004**, *10*, 179–185. [CrossRef]
26. Nath, P.; Kale, S.J.; Kaur, C.; Chauhan, O.P. Phytonutrient Composition, Antioxidant Activity and Acceptability of Muffins Incorporated with Red Capsicum Pomace Powder. *J. Food Sci. Technol.* **2018**, *55*, 2208–2219. [CrossRef] [PubMed]
27. Benzie, I.F.F.; Strain, J.J. The Ferric Reducing Ability of Plasma (FRAP) as a Measure of Antioxidant Power: The FRAP Assay. *Anal. Biochem.* **1996**, *239*, 70–76. [CrossRef]
28. Matsumura, Y.; Ito, T.; Yano, H.; Kita, E.; Mikasa, K.; Okada, M.; Furutani, A.; Murono, Y.; Shibata, M.; Nishii, Y.; et al. Antioxidant Potential in Non-Extractable Fractions of Dried Persimmon (*Diospyros kaki* Thunb.). *Food Chem.* **2016**, *202*, 99–103. [CrossRef]
29. Brodkorb, A.; Egger, L.; Alminger, M.; Alvito, P.; Assunção, R.; Ballance, S.; Bohn, T.; Bourlieu-Lacanal, C.; Boutrou, R.; Carrière, F.; et al. INFOGEST Static In Vitro Simulation of Gastrointestinal Food Digestion. *Nat. Protoc.* **2019**, *14*, 991–1014. [CrossRef]
30. González, C.M.; Llorca, E.; Quiles, A.; Hernando, I.; Moraga, G. An In Vitro Digestion Study of Tannins and Antioxidant Activity Affected by Drying "Rojo Brillante" Persimmon. *LWT* **2022**, *155*, 112961. [CrossRef]
31. Lucas-González, R.; Viuda-Martos, M.; Pérez Álvarez, J.A.; Fernández-López, J. Changes in Bioaccessibility, Polyphenol Profile and Antioxidant Potential of Flours Obtained from Persimmon Fruit (*Diospyros kaki*) Co-Products during In Vitro Gastrointestinal Digestion. *Food Chem.* **2018**, *256*, 252–258. [CrossRef]
32. Djuricic, I.; Calder, P.C. Beneficial Outcomes of Omega-6 and Omega-3 Polyunsaturated Fatty Acids on Human Health: An Update for 2021. *Nutrients* **2021**, *13*, 2421. [CrossRef]
33. Agahari, R.P.; Khurdiya, D.S.; Lata; Kaur, C.; Kapoor, H. Antioxidant Activity and Quality of Soy Enriched Apple Bar. *Food Pro. Preserv.* **2004**, *28*, 145–159. [CrossRef]
34. Lucas-González, R.; Viuda-Martos, M.; Pérez-Alvarez, J.A.; Fernández-López, J. Antioxidant Potential and Quality Characteristics of Mediterranean Fruit-Based Extruded Snacks. *Int. J. Food Sci. Technol.* **2016**, *51*, 2674–2681. [CrossRef]
35. Danalache, F.; Carvalho, C.Y.; Brito, L.; Mata, P.; Moldao-Martin, M.; Alves, V.D. Effect of Thermal and High Hydrostatic Pressure Treatments on Mango Bars Shelf-Life under Refrigeration. *J. Food Eng.* **2017**, *212*, 113–120. [CrossRef]

36. Ahmad, S.; Vashney, A.; Srivasta, P. Quality Attributes of Fruit Bar Made from Papaya and Tomato by Incorporating Hydrocolloids S. *Int. J. Food Prop.* **2007**, *8*, 88–99. [CrossRef]
37. Verma, R.; Bisen, B.P. Studies on Sensory Evaluation of Guava and Papaya Mixed Fruit Bar during Storage. *Pharmacogn. Phytochem.* **2020**, *9*, 1052–1056.
38. González, C.M.; Gil, R.; Moraga, G.; Salvador, A. Natural Drying of Astringent and Non-Astringent Persimmon "Rojo Brillante". Drying Kinetics and Physico-Chemical Properties. *Foods* **2021**, *10*, 647. [CrossRef]
39. Zhou, M.; Chen, J.; Bi, J.; Li, X.; Xin, G. The Roles of Soluble Poly and Insoluble Tannin in the Enzymatic Browning during Storage of Dried Persimmon. *Food Chem.* **2022**, *366*, 130632. [CrossRef]
40. Safdar, M.N.; Kausar, T.; Nadeem, M.; Murtaza, M.; Sohail, S.; Mumtaz, A.; Siddiqui, N.; Jabbar, S.; Afzal, S. Extraction of Phenolic Compounds from (*Mangifera indica* L.) and Kinnow (*Citrus reticulate* L.) Peels for the Development of Functional Fruit Bars. *Food Sci. Technol.* **2021**, *42*, e09321. [CrossRef]
41. Persic, M.; Mikulic-petkovsek, M.; Slatnar, A.; Solar, A.; Veberic, R. Changes in Phenolic pro Fi Les of Red-Colored Pellicle Walnut and Hazelnut Kernel during Ripening. *Food Chem.* **2018**, *252*, 349–355. [CrossRef]
42. Porras-loaiza, P.; Jim, T.J.-M.; Sosa-morales, M.E.; Palou, E.; Lopez-Malo, A. Original Article Physical Properties, Chemical Characterization and Fatty Acid Composition of Mexican Chia (*Salvia hispanica* L.) Seeds. *Int. J. Food Sci. Technol.* **2014**, *49*, 571–577. [CrossRef]
43. Bodoira, R.M.; Penci, M.C.; Ribotta, P.D.; Martínez, M.L. Chia (*Salvia hispanica* L.) Oil Stability: Study of the Effect of Natural Antioxidants. *LWT* **2017**, *75*, 107–113. [CrossRef]
44. Pycia, K.; Kapusta, I.; Jaworska, G.; Jankowska, A. Antioxidant Properties, Profile of Polyphenolic Compounds and Tocopherol Content in Various Walnut (*Juglans regia* L.) Varieties. *Eur. Food Res. Technol.* **2019**, *245*, 607–616. [CrossRef]
45. Song, J.; Wei, Q.; Wang, X.; Li, D.; Liu, C.; Zhang, M. Degradation of Carotenoids in Dehydrated Pumpkins as a Ff Ected by Di Ff Erent Storage Conditions. *Food Res. Int.* **2018**, *107*, 130–136. [CrossRef] [PubMed]
46. Vondráková, Z.; Trávníčková, A.; Malbeck, J.; Haisel, D.; Černý, R.; Cvikrová, M. The Effect of Storage Conditions on the Carotenoid and Phenolic Acid Contents of Selected Apple Cultivars. *Eur. Food Res. Technol.* **2020**, *246*, 1783–1794. [CrossRef]
47. Pycia, K.; Kapusta, I.; Jaworska, G. Impact of the Degree of Maturity of Walnuts (*Juglans regia* L.) and Their Variety on the Antioxidant Potential and the Content of Tocopherols and Polyphenols. *Molecules* **2019**, *24*, 2936. [CrossRef]
48. Ortega, N.; Macià, A.; Romero, M.; Reguant, J.; Motilva, M. Matrix Composition Effect on the Digestibility of Carob Flour Phenols by an In-Vitro Digestion Model. *Food Chem.* **2011**, *124*, 65–71. [CrossRef]
49. Kamiloglu, S.; Pasli, A.A.; Ozcelik, B.; Van Camp, J.; Capanoglu, E. Colour Retention, Anthocyanin Stability and Antioxidant Capacity in Black Carrot (*Daucus carota*) Jams and Marmalades: Effect of Processing, Storage Conditions and in Vitro Gastrointestinal Digestion. *J. Funct. Foods* **2015**, *13*, 1–10. [CrossRef]
50. Desmarchelier, C.; Borel, P. Overview of Carotenoid Bioavailability Determinants: From Dietary Factors to Host Genetic Variations. *Trends Food Sci. Technol.* **2017**, *69*, 270–280. [CrossRef]

foods

MDPI

Article

The Possibility of Using Oil Pomace as a Substitute for Walnuts (*Juglans regia* L.) in Muesli Bar Technology

Patryk Siczek [1] and Justyna Libera [2,*]

[1] Department of Plant Food Technology and Gastronomy, University of Life Sciences in Lublin, Skromna 8 Street, 20-704 Lublin, Poland; patryk.siczek@up.lublin.pl

[2] Department Engineering and Cereal Technology, University of Life Sciences in Lublin, Skromna 8 Street, 20-704 Lublin, Poland

* Correspondence: justyna.libera@up.lublin.pl; Tel.: +48-81-4623316

Abstract: The utilization of food industry waste to create innovative, high-quality products with reduced environmental impact is a growing trend in food technology. Walnut oil pomace, a byproduct of walnut oil production, is rich in nutrients and bioactive compounds, making it an excellent candidate for reuse in muesli bars as a replacement for walnuts. The aim of the study was to evaluate the possibility of replacing walnuts with oil pomace in muesli bar recipes and to assess whether the resulting product meets quality standards. Ground expeller walnut oil pomaces and aqueous extract were tested, with a bar containing ground walnuts serving as the reference sample. The bars were evaluated for sensory, physicochemical, and antioxidant properties, and their nutritional values were assessed. Results showed that the pomace-enriched bars exhibited satisfactory physicochemical properties, with texture, color, and safety (as measured by water activity and pH) comparable to the control bars. Sensory evaluations classified all bars as acceptable, with no significant differences in nutritional value. The study concludes that walnut pomace holds promise as a sustainable ingredient in food technology, potentially expanding product diversity while reducing environmental impact. This concept aligns with promoting sustainable practices in food manufacturing.

Keywords: walnut; muesli bar; by-product; oil pomace; functional food

1. Introduction

The walnut (*Juglans regia* L.) is valued for its taste, nutritional properties, and medicinal applications [1]. The beneficial effect on the body is determined by the presence of biologically active compounds in nuts, i.e., unsaturated fatty acids, dietary fiber, polyphenols and numerous minerals (zinc, magnesium, calcium) [2,3]. Green fruits, peels, leaves and bark are characterized by a high content of phenolic compounds, while the seeds are rich in essential fatty acids [4,5]. The pith and shell of green walnuts and green leaves are used to make tinctures, teas and infusions, and the leaves can be used for pickling vegetables [6]. Edible walnut seeds are used in the food industry as a raw material for oil extraction. In walnuts, the lipid profile is mostly dominated by PUFAs, with linoleic and linolenic acids being the greatest contributors [4]. Walnut cultivation and consumption continues to increase around the world [5–7], with the production of approximately 2.7 million tons in 2023 [8]. At the industrial level, walnuts are widely used as a source of desirable fatty acids, as an ingredient that enhances flavor and aroma [6]. Therefore, walnuts can be used in a wide range of food products, e.g., in bars, snacks, breakfast cereals, and energy bars, in the form of peanut butter, and as a dairy alternative.

The global food industry generates a substantial quantity of by-products annually, many of which remain unused, presenting a significant disposal issue [9]. These include, among others, soybean okra, fruit pomace, bran or pomace, including walnuts [10,11]. The latter are waste after oil extraction from seeds and, due to their high nutritional value, are used as feed, most often for ruminants and fish [12]. The walnut oil extraction industry itself is considered a significant source of waste, the amount of which depends on the pretreatment technology and oil pressing method used [13]. Precise data on the quantity of pomace produced per unit of walnut oil extracted are not widely available. The conversion of food waste into valuable products is referred to as the upcycling process [14]. By-products of the food industry can be used as ingredients for functional foods, which expands the range of products and at the same time reduces the burden on the environment [15]. It is not always possible to reduce the number of by-products, so they should be treated as valuable additional resources for production [16].

Walnuts and their waste products contain a large number of polyphenols, polysaccharides, and other biologically active substances, which have strong development potential for their added value [17]. Nuts have proven beneficial effects on humans, and in combination with a balanced diet they can significantly improve health [2,17–19]. Modern lifestyles have led to changes in dietary habits and, consequently, to an increase in the consumption of foods rich in saturated fats, sugars and salt, which contribute to the development of obesity and cardiovascular diseases. Walnut pomace has the potential to be used as an additive in dietary products or as an ingredient in functional foods. A result of milling this pomace is so-called walnut flour, which is a good source of protein and carbohydrates [10]. Including products containing walnut pomace in your daily diet can help improve your body's condition due to the richness of antioxidants [20,21]. Polyphenols in walnuts have antioxidant, anti-inflammatory, anti-tumor, and cardiovascular-protective abilities [22]. The ellagic acid content (ca. 0.5–1.7 mg g^{-1}) in pomaces from walnut *Diaphragma Juglandis* fructus was considerable and comparable to or even higher than that in blackberries, cloudberries, strawberries, raspberries, fruit juices, and walnut kernels [23]. Most phenolic compounds are found in the fatty nut pulp, while these compounds remain in the pomace after cold pressing [24,25]. The use of walnut pomace in the technology of muesli bars can not only improve their nutritional quality, but also have a positive impact on the technological process, e.g., on the correct color or improved water absorption [10,26].

Muesli bars are a popular snack, making it worthwhile to expand the range with innovative variations on classic recipes, especially since consumption forecasts indicate that the global snack bar industry will grow by about USD 4 billion between 2019 and 2025 [27]. This study aimed to evaluate the possibility of substituting walnuts in muesli bars with oil pomace to promote waste reutilization in food technology. This approach seeks to balance the use of by-products from food processing with preserving the high quality and sensory appeal of the modified product. The use of an aqueous extract of oil pomace in muesli bar technology has not yet been described in the literature, making its inclusion in this study an innovative approach.

2. Materials and Methods

2.1. Materials

All ingredients used in the study were purchased in a retail supermarket in Poland, immediately before the study began. Organic oat flakes (Melvit), multi-flower nectar honey (Apis), and virgin coconut oil (goBIO) were used. Dried cranberries, sunflower seeds, coconut flakes, flax seeds and walnuts were purchased from the Bakalland brand. The walnut oil pomace was obtained from the oil mill of Farmy Roztocza, Księżpol, Poland. The reagents used for analyses were analytical grade.

2.2. Preparing of Walnut Oil Pomaces Extract

Aqueous extracts of oil pomace were prepared based on prior, unpublished research results. Exactly 30 g of dried pomace was ground precisely in a grinder Ronic Partner (Ronic Poland Sp. z o.o., Łódź, Poland). Then, 350 mL of hot distilled water was added, and the suspension was shaken and heated for 30 min on a laboratory shaker (Water Bath Shaker Type 357, Elpan, Wrocław, Poland) at 150 rpm and at a water temperature of 85 ± 5 °C in the bath. Then, the suspension was thoroughly separated in an MPW-350R centrifuge (MPW

Med. Instruments, Warszawa, Poland) at 5000× g for 10 min. The supernatant was filtered through Whatman no. 1 paper, and the obtained volume of the extract (approximately 70 mL) was transferred to a round-bottom flask and concentrated in a vacuum evaporator (Rotavapor® R-215, Büchi®, Flawil, Switzerland) in a water bath at 50–60 °C and 250–300 hPa pressure. After evaporating 50% of the solvent, the final extract was obtained and transferred to 12 plastic Eppendorf containers with a capacity of 2 mL each. The mass of the concentrated extract was 33 g and the volume was 24 mL. On this basis, the density was estimated at 1.375 g/dm^3. The containers were tightly closed and stored in the freezer until the muesli bars were formulated. Due to its thick consistency, the extract was diluted with water in a 1:1 ratio before application.

2.3. Formulation of Muesli Bars

Oat flakes and dried cranberries were ground in a grinder (Ronic Partner, Ronic Poland Sp. z o.o., Łódź, Poland) for 5 s at the device's maximum power. Next, the remaining dry ingredients (linseed and sunflower seeds) were added and gently mixed with a spatula. Three mixture variants were prepared: with the addition of Ground Walnut oil Pomace (GWP variant), an aqueous Extract of Walnut oil Pomace (EWP variant), or ground walnuts (control variant, CON). The added amounts were 9.5% ground walnut or ground oil walnut pomace or 0.3% extract. Based on prior calculations, it was determined that 3 g of extract corresponded to 9.5 g of ground pomace. Walnuts and pomace were ground in a grinder (Ronic Partner, Ronic Partner Sp. z o.o., Łódź, Poland) for 5 s at maximum power immediately before being added to the mixture. The remaining ingredients (coconut oil and honey) were liquefied in a water bath and then combined with the dry ingredients. Each mixture was kneaded by hand for 60 s until a homogeneous mass was achieved. Rectangular bars, each weighing 30 g, were then formed, packaged in string bags, and refrigerated at 4 °C for 5 h before undergoing analysis. The recipes are presented in Table 1.

Table 1. Formulation of muesli bars for research variants.

Ingredients	Variants of Muesli Bars		
	GWP	EWP	CON (Control)
oat flakes	355 g	355 g	355 g
honey	240 g	240 g	240 g
coconut oil	100 g	100 g	100 g
dried cranberries	95 g	95 g	95 g
sunflower seeds	60 g	60 g	60 g
shredded coconut	30 g	30 g	30 g
linseed	25 g	25 g	25 g
ground walnuts	—	—	95 g
ground walnut oil pomace	95 g	—	—
extract of walnut oil pomace	—	3 g	—
distilled water	—	3 g	—
Total mass of mixture	1000 g	908 g	1000 g

Explanatory notes: GWP—Ground Walnut oil Pomaces; EWP—aqueous Extract of Walnut oil Pomace; CON—ground walnuts (control variant).

The extract from the muesli bars was analyzed to determine pH, total phenolic content (TPC), and DPPH assays. It was prepared as follows: the mixture was obtained by homogenizing muesli bars with distilled water in a ratio of 1:9 for 60 s using an MQ5025 Multiquick 5 Vario hand blender (Braun Household GmbH, Neu-Isenburg, Germany). The suspension was then thoroughly separated in an MPW-350R centrifuge (MPW Med. Instruments, Warszawa, Poland) at 5000× g for 10 min. The supernatant was filtered through Whatman no. 1 paper into Falcon plastic containers and stored in the refrigerator until the muesli bars were analyzed.

2.4. Physicochemical Characteristic of Muesli Bars

The water activity (a_w) of muesli bars was determined in triplicate using an a_w-analyzer (LabMaster Novasina AG, Lachen, Switzerland) at 25 °C, according to the manufacturer's instructions.

The pH measurement was carried out using a CPC-501 digital pH-meter (Elmetron, Zabrze, Poland) equipped with a pH ERH-111 electrode (Hydromet, Bytom, Poland) at 25 °C, according to the manufacturer's instructions.

The color of the muesli bars were measured using a Konica Minolta CR-5 chroma meter (Konica Minolta Business Solution Europe GmbH, Langenhagen, Germany) applying the CIELAB scale (L^*, a^*, and b^*). Color measurements were performed at room temperature at six different locations on the bars. The conditions were 13 mm port size, illuminant D65 and 10° standard observer. The Konica Minolta's white and black standards were used to calibrate the spectrophotometer. The total color difference (ΔE^*) was calculated based on ΔL^*, Δa^* and Δb^* results for each sample according to the control variant, as follows: $\Delta E^* = [(\Delta L^*)^2 + (\Delta a^*)^2 + (\Delta b^*)^2]^{1/2}$.

Texture profile analysis (TPA) of the muesli bars was determined in 15 replications using a Texture Analyzer (TA.XT2, Stable Micro Systems, Godalming, UK) equipped with the P/75 attachment and the HDP/90 stage. A double compression cycle test was performed at up to 50% compression of the original portion height with a cylinder probe. A time of 5 s was allowed to elapse between the two compression cycles. Force-time deformation curves were obtained with a 50 kg load cell applied at a cross-head speed of 5 mm s^{-1}. The TPA test parameters were quantified as hardness, springiness, cohesiveness, gumminess and chewiness. Calculations of the texture parameters were performed using standard computational methodologies implemented in the device control software (Texture Exponent v32).

2.5. Antioxidant Properties of Muesli Bars

Total phenolic compounds (TPC) in the extract of muesli bars were determined using the Folin–Ciocalteau method according to Singleton et al. [28]. The results showed the value of gallic acid equivalent (mg GAE mL^{-1}) and were determined from the following regression equation based on the established calibration curve (y = 3.6373x + 0.0994; $R^2 = 0.9872$, where y is the absorbance and x is the gallic acid concentration).

The antioxidant activity as DPPH radical scavenging activity was determined spectrophotometrically (at 517 nm) using 2,2-diphenylpicrylhy-drazyl (DPPH) radicals. Refer to the method of Brand-Williams et al. [29] The DPPH radical scavenging activity was determined using the formula:

$$\text{DPPH radical scavenging (\%)} = 1 - [(\text{Abs}_{\text{control}} - \text{Abs}_{\text{sample}})/\text{Abs}_{\text{control}}] \times 100\%$$

2.6. Sensory Evaluation of Muesli Bars

For the sensory evaluation of muesli bars, the Quantitative Descriptive Profile (QDP) described by Szydłowska et al. [30] was used with a slight modification by the authors. The study was conducted in accordance with the Declaration of Helsinki, and the protocol was approved by the Ethics Committee for no. KE/38/2024 (Keep). The freshly manufactured samples were assessed by 23 panelists (students and employees) of the Faculty of Food Science and Biotechnology of the University of Life Sciences in Lublin. Before evaluation, the panelists were trained in the methodology. The authors prepared descriptors for the sensory test (Table 2). Bars were divided into 10 g portions and placed in plastic boxes (125 mL) covered with lids and kept at room temperature (22 °C) for 30 min before analysis. Two bars assessed (EWP and GWP) were coded, and the CON was designated as the reference sample. The authors' modification of the method consisted of the sensory panel assessing the quality of the coded muesli bars based on qualitative features, defining the evaluated feature as "less intense" ($-5 \div -1$) or "more intense" ($+1 \div +5$) compared to the control sample (0 value). The task of the panelist was to determine the intensity of each of the quality features and put their assessment on an unstructured, 100 mm graphic scale (from −5 to +5 contractual units, c.u.).

Table 2. Descriptors and the marks of anchors defined in the QDP.

Quality Features	Descriptor	The Marks of Anchors
Odor	intensity of nut odor intensity of fruity odor intensity of grain odor intensity of storage odor intensity of other odor	less intense (−5 c.u.) — the reference sample CON (0 c.u.)—more intense (+5 c.u.)
Flavor	intensity of nut flavor intensity of fruity flavor intensity of grain flavor intensity of sweet flavor intensity of floury flavor intensity of bitter flavor intensity of other flavor	less intense (−5 c.u.) — the reference sample CON (0 c.u.)—more intense (+5 c.u.)
Texture	hardness viscosity	less intense (−5 c.u.) — the reference sample CON (0 c.u.)—more intense (+5 c.u.)
Overall quality		worse—better (than the reference sample CON)

2.7. Proximate Nutritional Value of Muesli Bars

The nutritional value of the muesli bars was estimated based on the recipe composition (the exact amount of raw materials used), using the information included on the labels of the purchased raw materials. The energy value was computed using the Atwater general factor system: carbohydrate (4 kcal g^{-1}), lipid (9 kcal g^{-1}), and protein (4 kcal g^{-1}) [31].

2.8. Statistical Analysis

The collected data were analyzed using one-way analysis of variance (ANOVA) to assess the effects of muesli bar formulations on selected parameters. The experiment was conducted in two independent series, with each analysis performed in at least triplicate for accuracy. Statistical calculations were conducted using Microsoft Office Excel 2013 (Microsoft Corporation, Redmond, WA, USA) and Statistica 10 (StatSoft Polska Sp. z o.o., Kraków, Poland). All groups met the assumption of normality ($p > 0.05$). Post hoc analysis was performed using Tukey's test to identify significant patterns and relationships among subgroups. Statistical significance was set at $p \leq 0.05$. Results are presented as means ± standard deviation.

3. Results and Discussion

Water activity (aw) is an important indicator that characterizes the content of free water necessary for microbial growth, and therefore its values are useful for predicting the microbiological stability of the product. High aw-values (>0.8) pose a microbiological risk to the product because they are optimal for microorganism growth; for mycotoxigenic mold, the lower growth limit has been reported to be 0.78 aw [32]. Therefore, it is important that the product has lower water activity, providing a natural margin of safety. Water activity (Table 3) is an important parameter that determines the quality and safety of food. The aw-values of all the assessed muesli bars were similar (from 0.64 to 0.66). The EWP bar had the highest aw-value, but it was not a statistically significant difference ($p \leq 0.05$). Similar water activity was found in high-protein muesli bars (0.63–0.69) [30] and bars with date paste (0.65–0.72) [33], while higher aw-values (0.78–0.85) were also reported in articles of this parameter in high-energy protein bars [33].

Table 3. Physicochemical characteristic of muesli bars.

Parameter		Variants of Muesli Bars		
		GWP	EWP	CON (Control)
Water activity [−]		0.644 [a] ± 0.014	0.661 [a] ± 0.008	0.657 [a] ± 0.002
pH value [−]		5.20 [b] ± 0.06	4.57 [c] ± 0.05	5.43 [a] ± 0.03
Color parameter	*L** (lightness)	51.94 [a] ± 2.38	52.84 [a] ± 2.67	52.29 [a] ± 2.04
	*a** (redness)	4.99 [a] ± 0.74	5.46 [a] ± 0.82	4.77 [a] ± 0.52
	*b** (yellowness)	18.57 [a] ± 1.68	18.66 [a] ± 1.43	17.05 [b] ± 1.44
	Δ*E** (total color difference between tested bars and CON) [1]	1.58	1.84	-
Texture profile analysis (TPA)	Hardness [N]	232 [a] ± 122	189 [a] ± 59	252 [a] ± 50
	Springiness [−]	0.08 [b] ± 0.01	0.11 [a] ± 0.04	0.07 [b] ± 0.01
	Cohesiveness [−]	0.12 [b] ± 0.05	0.16 [a] ± 0.04	0.10 [b] ± 0.01
	Gumminess [g]	260 [a] ± 191	305 [a] ± 161	253 [a] ± 74
	Chewiness [g]	23.0 [ab] ± 20.1	35.5 [a] ± 29.1	17.3 [b] ± 6.7

Explanatory notes: [a–c] means followed by the same letter in the row do not differ statistically among themselves by the Tukey test ($p \leq 0.05$). [1] Δ*E** values below 2.5 are imperceptible.

The next quality parameter of muesli bars, which strictly determines their quality, is the pH value (Table 3). The bars were characterized by a low pH (ranging from 4.57 to 5.43), and the test samples had significantly ($p \leq 0.05$) lower values compared to the control sample, which proves their microbiological safety. Other researchers found higher values of the pH parameter (6.3–7.0) in high-protein muesli bars [30].

The color (Table 3) of muesli bars is an important parameter determining consumers' choices. In this study, the CON muesli bar was used as a reference sample, and the colors of the test bars (GWP and EWP) were compared to it. The calculated total color difference (Δ*E**) should not exceed 2.5 units, as values above this threshold indicate different coloration. The Δ*E** values for the GWP and EWP samples were 1.6 and 1.8, respectively, demonstrating their similarity in color to the reference sample. All muesli bars had identical lightness, expressed as the *L** index (average 52.4 ± 0.5 c.u.), and redness (5.1 ± 0.4 c.u.). The yellowness, expressed by the *b** index, turned out to be the only difference in colors of the muesli bars. The control sample (17.0 ± 1.4 c.u.) had a significantly lower ($p \leq 0.05$) value so it was less yellow than the samples: GWP (18.6 ± 1.7 c.u.) and EWP (18.7 ± 1.4 c.u.). The obtained results of instrumental color parameters indicate that the tested bars were brown, typical of muesli/granola bars.

A common issue in snack bar production is the stickiness caused by soluble sugars, which impacts the texture and sensory qualities, resulting in a softer consistency [33]. Texture evaluation with the TPA double compression test showed that all muesli bars had similar hardness, although there were significant standard deviations (25–50% of mean value) during testing (Table 3). This was due to the heterogeneous structure of muesli bars. Plant additives can influence the texture of bars, e.g., dates can be used as an ingredient that retains moisture in the storage of this type of food product [33].

The average hardness of all muesli bars was 224 N ± 85, and variants GWP, EWP and CON did not differ statistically, respectively: 232 N, 189 N and 252 N. Similarly with the chewiness, which was average 273 g ± 149. Differences were found between springiness and cohesiveness in the tested bars. The sample with the extract performed significantly ($p \leq 0.05$) higher for these values (S: 0.11 ± 0.04, C: 0.16 ± 0.04) compared to the other samples. Both samples of bars with the addition of walnut pomace had greater chewiness than the control sample, especially EWP-bars ($p \leq 0.05$). The obtained results for cohesiveness are similar to those of Dimopoulou et al. [34], who tested muesli bars for diabetics. On the other side, hardness tests for bars containing *Coprinus comatus* protein powder showed higher values, from 304 to 429 N, so these bars were harder than muesli bars prepared by the authors of this study [34].

Muesli bars contain walnuts in their formulation in various forms, making them a source of antioxidant compounds. The total polyphenol content and the antioxidant activity defined as % DPPH inhibition was assessed (Table 4).

Table 4. Antioxidant properties of muesli bars.

Parameter	Variants of Muesli Bars		
	GWP	EWP	CON (Control)
TPC [mg GAE mL^{-1}]	0.11 ^{b} ± 0.01	0.06 ^{c} ± 0.00	0.13 ^{a} ± 0.01
DPPH [% radical scavenging]	52.78 ^{b} ± 0.71	36.03 ^{c} ± 0.65	56.71 ^{a} ± 0.87

Explanatory notes: $^{a–c}$ means followed by the same letter in the row do not differ statistically among themselves by Tukey test ($p \leq 0.05$).

The highest DPPH inhibition value was achieved by the variant that contained walnuts, i.e., the control sample (56.7%), while a slightly lower value was achieved by the variant containing walnut oil pomace (52.8%). The EWP bar had the lowest antioxidant activity value, at 1/3 less than the GWP bar. It can be seen that the DPPH inhibitory capacity decreases with the change in walnut form. The differences obtained between the variants were statistically significant at ($p \leq 0.05$). Similarly, the results for total polyphenol content (TPC) in bars were ranked. The CON sample with the addition of ground walnuts had the highest value (0.13 mg GAE mL^{-1}). The differences obtained between the variants were significant ($p \leq 0.05$). There are no published research results on muesli bars with walnut pomace to which our antioxidant activity results could be compared. Slightly higher antioxidant activity (61–69% DPPH inhibition) was reported by Szydłowska et al. [30], though it should be noted that their muesli bars were high in protein. Similar DPPH inhibition results (32.3–50.5%) were obtained in another study [35], where the authors attributed the antioxidant activity to the increased phenolic content. Moreover, the TPC activity was higher (0.46 mg GAE g^{-1}) in reformulation bars as compared to the control (0.13 mg GAE g^{-1}) due to the high content of functional food present in the protein bars [35]. Muesli bars seem to be a good source of antioxidants, especially since the manufacturing process can often be performed without heat treatment.

The muesli bars were subjected to sensory evaluation (Figure 1) immediately after formulation and cooling. All three variants were similar to each other in terms of the intensity of the nut and cereal odor and flavor. The panelists assessed both bars and found that GWP and EWP had a slightly sweeter (2.0–2.8 c.u.) taste, and the storage odor was less noticeable (−3.0 c.u.). This is probably related to the lower fat content in these samples. According to the panelists' evaluation, both variants (GWP and EWP) differed in their bitter flavor and the presence of the other flavor. The GWP muesli bar had a more intense (2.3 c.u.) bitter taste than the others. The GWP and EWP variants had similar levels of fruity odor and flavor, with these features being more intense than in the CON bar. The texture of the bars tested was assessed as harder than in the control sample, and the viscosity as lower. The GWP and EWP variants were highly (above 4 c.u.) assessed in the sensory evaluation, and the EWP bar achieved the highest score (+4.89 c.u.) in the overall quality category. A similar sensory evaluation of muesli bars was conducted by Szydłowska et al. [30], but they used a different scale from 1 to 10. In their study, muesli bars were characterized by the intensity of the cereal and nutty aroma impression at the level of about 5 c.u. The intensity of the taste was noted as follows: cereal; nutty and salty, and the overall quality was noted at the level of 6.8–7.2 c.u. [30]. Ibrahim et al. conducted a sensory evaluation of the bars to determine the optimal sensory characteristics and found that the overall acceptability, appearance, taste, sweetness and texture were perceived without significant differences between the control sample and those supplemented with different concentrations of the date snack bar [33]. Therefore, it seems that different recipes of muesli bars do not negatively affect their quality and sensory characteristics.

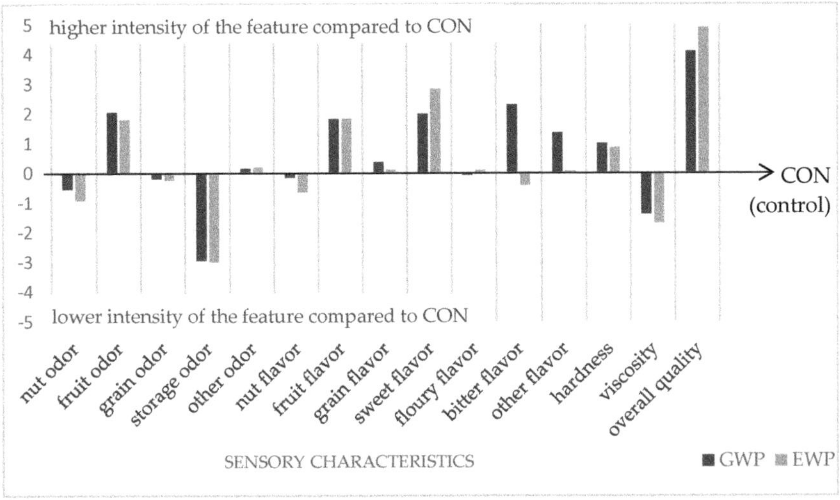

Figure 1. Results of sensory evaluation of fresh muesli bars (QDP method). Explanatory note: a value of zero on the vertical axis indicates the reference sample (CON). Values above mean higher intensity of the feature, and values below mean lower intensity of the feature.

The recipe composition and nutritional value of a muesli bar are important because it is often classified as a high-sugar food. If the bar recipe includes a source of protein, fiber, and polyunsaturated fats, and is therefore designed appropriately, it can be considered a functional food [36]. The nutritional value of a food product, defined by the type and quantity of nutrients it contains and their bioavailability, determines its overall usefulness. Nutritional value also serves as an indicator of nutrient density, reflecting the amount of energy released within the body during the biological oxidation of macronutrients [33]. One muesli bar provides 6–7% of the daily energy requirement and can be a valuable snack between meals, but due to the high sugar content (ca. 7 g), these portions should not be exceeded (Table 5). In the present study, muesli bars contained dietary fiber above 3 g per 100 g, and could therefore be considered a food is a source of this ingredient [37]. Such information could be included on the bar label in the form of a nutrition claim, as this would help consumers make informed food choices.

The energy density values of the muesli bars calculated in this study (374–435 kcal) are comparable to the average values of bars available on the Australian market (423 kcal). Additionally, they have similar average protein (8 g) and carbohydrate (57 g) content per 100 g of product [38]. Comparing the obtained results to other studies on protein-rich bars, it can also be concluded that the energy value of 100 g of the product may range from 315 [34] to 390 kcal [30] and this is due to the large share of protein in their recipe composition. However, due to the presence of walnuts and coconut oil in the recipe composition of our bars, we obtained a higher (more by 5–10 g) fat content. The sample muesli bar, containing oil pomace instead of walnuts, has lower energy (374 kcal) and a lower fat content (18.5 g).

Table 5. Nutritional value of muesli bars (per 30 and 100 g of product and % daily value).

Parameter	GWP			EWP			CON		
	100 g of Product	Serving Size (30 g)	% of the DV *	100 g of Product	Serving Size (30 g)	% of the DV *	100 g of Product	Serving Size (30 g)	% of the DV *
Energy (kcal)	374	112	6%	389	123	6%	435	131	7%
Protein (g)	6.7	2.0	2%	6.3	1.9	2%	7.8	2.3	3%
Total Fat (g)	18.5	5.6	10%	17.9	5.4	10%	24.2	7.3	13%
- saturated (g)	10.6	3.2	n/a	10.6	3.2	n/a	11.3	3.4	n/a
Carbohydrates (g)	47.9	14.4	5%	53.6	16.1	6%	49.3	14.8	5%
- sugars (g)	21.3	6.4	n/a	22.1	6.7	n/a	22.5	6.8	n/a
- dietary fiber (g)	5.1	1.5	5%	4.4	1.3	4%	5.2	1.6	5%

* Percent Daily Values (DV) are based on a 2000 calorie diet. Explanatory note: n/a—not available.

4. Conclusions

Developing sustainable food sources to meet global nutritional needs is a priority for the food industry, and may involve the industrial and technological utilization of novel food sources. The global food industry produces substantial quantities of by-products annually, many of which remain unutilized, posing significant disposal challenges. The pomace generated as a by-product in the processing of edible oils can be repurposed to create products with properties comparable to conventional foods. The present study demonstrated that pomace derived from walnut oil and its extract can be incorporated into muesli bar technology as a substitute for walnuts. The addition of pomace did not adversely affect the acceptability of the product's appearance. Physicochemical analyses indicated that substituting walnuts with pomace did not alter the water activity, pH values, or color, or significantly modify the texture of the muesli bars. Sensory evaluation revealed that bars containing pomace received favorable ratings, comparable to those of control bars made with walnuts. These findings underscore the potential of utilizing waste materials in the development of innovative muesli bar formulations. Further investigations are warranted to explore the full potential of pomace in food technology.

Author Contributions: Conceptualization, J.L.; methodology, J.L. and P.S.; validation, J.L.; formal analysis, J.L. and P.S.; investigation, J.L. and P.S.; resources, P.S. and J.L.; data curation, J.L.; writing—original draft preparation, J.L. and P.S.; writing—review and editing, J.L.; visualization, J.L.; supervision, J.L.; project administration, J.L. All authors have read and agreed to the published version of the manuscript.

Funding: This research received no external funding.

Institutional Review Board Statement: This study was conducted in accordance with the Declaration of Helsinki, and approved by the Ethics Committee of Scientific Research Involving Humans at University of Life Sciences in Lublin (Resolution No. KE/38/2024 on 12 June 2024).

Informed Consent Statement: Informed consent was obtained from all subjects involved in the study.

Data Availability Statement: The original contributions presented in the study are included in the article, further inquiries can be directed to the corresponding author.

Conflicts of Interest: The authors declare no conflicts of interest.

References

1. Gruszecki, R.; Chmist, A.; Walasek, M. Surowce zielarskie orzecha włoskiego (Walnut raw materials in Polish). *Ann. Hortic.* **2019**, *29*, 5–14. [CrossRef]
2. de Souza, R.G.M.; Schincaglia, R.M.; Pimentel, G.D.; Mota, J.F. Nuts and human health outcomes: A systematic review. *Nutrients* **2017**, *9*, 1311. [CrossRef] [PubMed]
3. Moskwa, J.; Naliwajko, S.K.; Puścion-Jakubik, A.; Soroczyńska, J.; Socha, K.; Koch, W.; Markiewicz-Żukowska, R. In vitro assessment of the bioaccessibility of Zn, Ca, Mg, and Se from various types of nuts. *Foods* **2023**, *12*, 4453. [CrossRef] [PubMed]
4. Torabian, S.; Haddad, E.; Rajaram, S.; Banta, J.; Sabaté, J. Acute effect of nut consumption on plasma total polyphenols, antioxidant capacity and lipid peroxidation. *J. Hum. Nutr. Diet.* **2009**, *22*, 64–71. [CrossRef]

5. Stanicka, K.; Woźniak, M.; Sosnowska, K.; Mrówczyńska, L.; Sip, A.; Waśkiewicz, A.; Ratajczak, I. Aktywność biologiczna i profil fenolowy ekstraktuz łupiny orzecha włoskiego (Biological activity and phenolic profile of walnut shell extract in Polish). *Borgis-Postępy Fitoter.* **2022**, *1*, 17–22. [CrossRef]
6. Grosu, C.; Boaghi, E.; Paladi, D.; Deseatnicova, O.; Reşitca, V. Prospects of using walnut oil cake in food industry. In *Modern Technologies in the Food Industry*, 1st ed.; IBN: Tangier, Morocco, 2012; Volume 1, pp. 362–365, ISBN 978-9975-87-428-1.
7. Bakkalbaşı, E. Oxidative stability of enriched walnut oil with phenolic extracts from walnut press-cake under accelerated oxidation conditions and the effect of ultrasound treatment. *J. Food Meas. Charact.* **2019**, *13*, 43–50. [CrossRef]
8. Kumar, P.; Szalóki-Dorkó, L.; Szabó-Nótin, B.; Kereszturi, J.; Tormási, J.; Abrankó, L.; Dalmadi, I.; Székely, D.; Máté, M. Effect of extracted walnut pomace cake edible coating and packaging type on quality parameters of walnut (*Juglans regia* L.) kernels during long storage periods. *Appl. Sci.* **2024**, *14*, 10120. [CrossRef]
9. Food and Agriculture Organization of United Nations (FAO). *The Future of Food and Agriculture—Trends and Challenges*; FAO: Rome, Italy, 2017; ISBN 978-92-5-109551-5. Available online: https://www.fao.org/3/i6583e/i6583e.pdf (accessed on 12 November 2024).
10. Burbano, J.J.; Correa, M.J. Composition and physicochemical characterization of walnut flour, a by-product of oil extraction. *Plant Foods Hum. Nutr.* **2021**, *76*, 233–239. [CrossRef]
11. Gumul, D.; Berski, W.; Zięba, T. The influence of fruit pomaces on nutritional, pro-health value and quality of extruded gluten-free snacks. *Appl. Sci.* **2023**, *13*, 4818. [CrossRef]
12. Pycia, K.; Kapusta, I.; Jaworska, G. Walnut oil and oilcake affect selected the physicochemical and antioxidant properties of wheat bread enriched with them. *J. Food Process. Preserv.* **2020**, *44*, e14573. [CrossRef]
13. Masoodi, L.; Gull, A.; Masoodi, F.A.; Gani, A.; Nissar, J.; Ahad, T.; Nayik, G.A.; Mukarram, S.A.; Kovács, B.; Prokisch, J.; et al. An overview on traditional vs. green technology of extraction methods for producing high quality walnut oil. *Agronomy* **2022**, *12*, 2258. [CrossRef]
14. Bangar, S.P.; Chaudhary, V.; Kajla, P.; Balakrishnan, G.; Phimolsiripol, Y. Strategies for upcycling food waste in the food production and supply chain. *Trends Food Sci. Technol.* **2024**, *143*, 104314. [CrossRef]
15. Helkar, P.B.; Sahoo, A.K.; Patil, N.J. Review: Food Industry by Products used as a Functional Food Ingredients. *Int. J. Waste Resour.* **2016**, *6*, 248. [CrossRef]
16. Betlej, I.; Borysiuk, P.; Borysiak, S.; Rybak, K.; Nowacka, M.; Barlak, M.; Andres, B.; Krajewski, K.; Lipska, K.; Cebulak, T.; et al. Pomace from oil plants as a new type of raw material for the production of environmentally friendly biocomposites. *Coatings* **2023**, *13*, 1722. [CrossRef]
17. Martínez, M.L.; Labuckas, D.O.; Lamarque, A.L.; Maestri, D.M. Walnut (*Juglans regia* L.): Genetic resources, chemistry, by-products. *J. Sci. Food Agric.* **2010**, *90*, 1959–1967. [CrossRef] [PubMed]
18. Grosso, G.; Yang, J.; Marventano, S.; Micek, A.; Galvano, F.; Kales, S.N. Nut consumption on all-cause, cardiovascular, and cancer mortality risk: A systematic review and meta-analysis of epidemiologic studies. *Am. J. Clin. Nutr.* **2015**, *101*, 783–793. [CrossRef]
19. Shin, H.R.; Kim, J.; Song, S. Association between nut consumption and mortality risk: A 20-year cohort study in Korea with a stratified analysis by health-related variables. *Nutr. J.* **2024**, *23*, 113. [CrossRef]
20. Derebecka, N.; Kania, M.; Baraniak, J. Liść orzecha włoskiego (*Juglandis folium*)—Działanie przeciwdrobnoustrojowe oraz bezpieczeństwo stosowania w chorobach skórnych (Antimicrobial activities of common walnut leaf (*Juglandis folium*) and its safety in skin diseases in Polish). *Borgis-Postępy Fitoter.* **2012**, *3*, 197–202.
21. Simsek, M.; Süfer, Ö. Infusion of walnut (*Juglans regia* L.) shell tea: Multi response optimization and antioxidant potential. *J. Appl. Res. Med. Aromat. Plants* **2021**, *20*, 100278. [CrossRef]
22. Song, H.; Cong, Z.; Wang, C.; He, M.; Liu, C.; Gao, P. Research progress on walnut oil: Bioactive compounds, health benefits, extraction methods, and medicinal uses. *J. Food Biochem.* **2022**, *46*, e14504. [CrossRef]
23. Liu, R.X.; Zhao, Z.Y.; Dai, S.G.; Che, X.; Liu, W.H. Identification and quantification of bioactive compounds in *Diaphragma Juglandis* Fructus by UHPLC-Q-Orbitrap HRMS and UHPLC-MS/MS. *J. Agric. Food Chem.* **2019**, *67*, 3811–3825. [CrossRef] [PubMed]
24. Pop, A.; Păucean, A.; Socaci, S.A.; Alexa, E.; Man, S.M.; Mureșan, V.; Muste, S. Quality characteristics and volatile profile of macarons modified with walnut oilcake by-product. *Molecules* **2020**, *25*, 2214. [CrossRef] [PubMed]
25. Martínez, M.L.; Penci, M.C.; Ixtaina, V.; Ribotta, P.D.; Maestri, D. Effect of natural and synthetic antioxidants on the oxidative stability of walnut oil under different storage conditions. *LWT-Food Sci. Technol.* **2013**, *51*, 44–50. [CrossRef]
26. Mihai, A.L.; Negoiță, M.; Horneț, G.A.; Belc, N. Valorization potential of oil industry by-products as sources of essential fatty acids. *Processes* **2022**, *10*, 2373. [CrossRef]
27. Research and Markets. Snack Bar Market—Forecasts from 2020 to 2025. Available online: https://www.researchandmarkets.com/tag/energy-bar (accessed on 1 October 2024).
28. Singleton, V.L.; Orthofer, R.; Lamuela-Raventos, R.M. Analysis of total phenols and other oxidation substrates and antioxidants by means of Folin-Ciocalteu reagent. *Oxid. Antioxid. Part A Methods Enzymol.* **1999**, *299*, 152–178. [CrossRef]
29. Brand-Williams, W.; Cuvelier, M.E.; Berset, C. Use of a free radical method to evaluate antioxidant activity. *LWT-Food Sci. Technol.* **1995**, *28*, 25–30. [CrossRef]
30. Szydłowska, A.; Zielińska, D.; Łepecka, A.; Trząskowska, M.; Neffe-Skocińska, K.; Kołożyn-Krajewska, D. Development of functional high-protein organic bars with the addition of whey protein concentrate and bioactive ingredients. *Agriculture* **2020**, *10*, 390. [CrossRef]

31. Provision of Food Information to Consumers. *Commission Regulation (EU) No. 1169/2011 of the European Parliament and of the Council of 25 October 2011*; European Union: Brussels, Belgium, 2011.
32. Tapia, M.S.; Alzamora, S.M.; Chirife, J. Effects of water activity (aw) on microbial stability: As a hurdle in food preservation. In *Water Activity in Foods: Fundamentals and Applications*; John Wiley & Sons, Inc.: Hoboken, NJ, USA, 2020; pp. 239–271. [CrossRef]
33. Ibrahim, S.A.; Fidan, H.; Aljaloud, S.O.; Stankov, S.; Ivanov, G. Application of Date (*Phoenix dactylifera* L.) Fruit in the composition of a novel snack bar. *Foods* **2021**, *10*, 918. [CrossRef]
34. Dimopoulou, M.; Vareltzis, P.; Floros, S.; Androutsos, O.; Bargiota, A.; Gortzi, O. Development of a functional acceptable diabetic and plant-based snack bar using mushroom (*Coprinus comatus*) powder. *Foods* **2023**, *12*, 2702. [CrossRef]
35. AlJaloudi, R.; Al-Dabbas, M.M.; Hamad, H.J.; Amara, R.A.; Al-Bashabsheh, Z.; Abughoush, M.; Choudhury, I.H.; Al-Nawasrah, B.A.; Iqbal, S. Development and characterization of high-energy protein bars with enhanced antioxidant, chemical, nutritional, physical, and sensory properties. *Foods* **2024**, *13*, 259. [CrossRef]
36. Martirosyan, D.; Kanya, H.; Nadalet, C. Can functional foods reduce the risk of disease? Advancement of functional food definition and steps to create functional food products. *Funct. Foods Health Dis.* **2021**, *11*, 213–221. [CrossRef]
37. R European Commission. *Regulation (EC) No 1924/2006 of the European Parliament and of the Council of 20 December 2006 on Nutrition and Health Claims Made on Foods*; European Union: Brussels, Belgium, 2006.
38. Curtain, F.; Grafenauer, S. Comprehensive nutrition review of grain-based muesli bars in Australia: An audit of supermarket products. *Foods* **2019**, *8*, 370. [CrossRef] [PubMed]

foods

MDPI

Article

The Prebiotic Activity of a Novel Polysaccharide Extracted from *Huangshui* by Fecal Fermentation *In Vitro*

Mei Li [1], Jian Su [2], Jihong Wu [1,*], Dong Zhao [2], Mingquan Huang [1], Yanping Lu [2], Jia Zheng [2] and Hehe Li [1]

1 Key Laboratory of Brewing Molecular Engineering of China Light Industry, Beijing Technology and Business University, Beijing 100048, China; lm9810172023@163.com (M.L.); huangmq@th.btbu.edu.cn (M.H.)
2 Key Laboratory of Soild-State Fermentation and Resource Utilization of Sichuan Province/Key Laboratory of Strong Flavor Baijiu Soild-State Fermentation of China Light Industry/Engineering Technology Research Center of Baijiu Brewing Special Grain of China, Wuliangye Yibin Co. Ltd., Yibin 644007, China; zhengwanqi86@163.com
* Correspondence: wujihong@btbu.edu.cn; Tel.: +86-01068984929

Abstract: A novel polysaccharide, HSP80-2, with an average molecular weight of 13.8 kDa, was successfully isolated by the gradient ethanol precipitation (GEP) method from *Huangshui* (HS), the by-product of Chinese Baijiu. It was mainly composed of arabinose, xylose, and glucose with a molar ratio of 4.0: 3.1: 2.4, which was completely different from the previous reported HS polysaccharides (HSPs). Morphological observations indicated that HSP80-2 exhibited a smooth but uneven fragmented structure. Moreover, HSP80-2 exerted prebiotic activity evaluated by *in vitro* fermentation. Specifically, HSP80-2 was utilized by gut microbiota, and significantly regulated the composition and abundance of beneficial microbiota such as *Phascolarctobacterium*, *Parabacteroides*, and *Bacteroides*. Notably, KEGG pathway enrichment analysis illustrated that HSP80-2 enriched the pathways of amino sugar and nucleotide sugar metabolism (Ko00520), galactose metabolism (ko00052), and the citrate cycle (TCA cycle) (ko00020). Meanwhile, the contents of short-chain fatty acids (SCFAs) mainly including acetic acid, propionic acid, and butyric acid in the HSP80-2 group were remarkably increased, which was closely associated with the growth of *Lachnoclostridium* and *Parabacteroides*. These results showed that HSP80-2 might be used as a potential functional factor to promote human gut health, which further extended the high value utilization of HS.

Keywords: polysaccharide; *Huangshui*; gut microbiota; prebiotic activity; short-chain fatty acids

1. Introduction

It is widely acknowledged that various chronic diseases, such as diabetes, obesity and colon cancer, are closely related to human gut microbiota [1]. Individual health is heavily dependent on the gut bacteria, since these symbiotic bacteria provide humans with certain useful metabolites. Therefore, improving gut microbial diversity and richness is a significant focus in the development and implementation of specialized medicinal and functional diets [2]. Diets exert a significant impact on the composition of gut microbiota. Polysaccharides are an integral component of a healthy diet which can regulate the intestinal microflora [3]. In order to study the probiotic properties of polysaccharides, many researchers reproduced the probiotic properties of polysaccharides under anaerobic conditions through *in vitro* fermentation. Numerous studies have demonstrated that the process of polysaccharide fermentation could produce advantageous metabolites, especially short-chain fatty acids (SCFAs) [3,4]. SCFAs are chemical compounds consisting primarily of acetate, propionate, and butyrate, which possess the ability to not only modulate the makeup of the gut microbiota, but also regulate the host's metabolism [5]. In addition, research has shown that polysaccharides could adjust the structure of the gut microbiota by promoting the proliferation of beneficial bacteria, such as *Lactobacillus* [6] and *Bacteroidetes* [7].

Huangshui (HS), as one of the by-products of traditional Chinese Baijiu production, has garnered significant interest in recent years due to its abundance of organic substances and microorganisms [8]. HS polysaccharides (HSPs), which were one of the most important ingredients of HS, were usually obtained through water extraction and the ethanol precipitation method. The ethanol was added to an aqueous solution containing polysaccharides, causing the dehydration of the polysaccharide molecules, and the dehydration process was followed by conformational changes and aggregation, which were induced by the increased strength of intra-molecular hydrogen bonding. It was observed that different concentrations of ethanol could cause the precipitation of polysaccharide fractions with varying average molecular weights. Therefore, the gradient ethanol precipitation (GEP) method was identified as an efficient and expeditious technique to prepare polysaccharide fractions with a high homogeneity [9]. Yet, up to now, there were no reports on using the GEP method to deal with HS and obtained polysaccharides. In terms of biological activity research, previous studies showed that HSPs displayed several significant biological activities, such as antiinflammation, antioxidation immunoregulatory, and protective effect on intestinal barrier damage [9–12]. However, the impact of HSPs on microbiota and microbial metabolite patterns in the gut, or in other words, the effect on the composition of the intestinal microbiota composition and the ability to metabolize SCFAs, was still unknown.

Consequently, the aim of the present study was to characterize a new polysaccharide and evaluate its regulatory effects on the makeup of the gut microflora and the contents of SCFAs *in vitro*. It would provide a foundation to explore HSPs as potential prebiotics. Main contents were as follows. Firstly, the new polysaccharide, named HSP80-2, was obtained from HS by the GEP method and its structure and morphology were further analyzed. Then, an *in vitro* fecal fermentation model from four healthy individuals was designed to evaluate the role of HSP80-2 on the diversity and richness of intestinal microflora. Finally, changes in the bacterial community structure and SCFAs were estimated using 16S rRNA and gas chromatography, respectively, to evaluate the prebiotic activity of HSP80-2.

2. Materials and Methods

2.1. Materials and Reagents

The sample of HS was collected from the Sichuan Yibin Wuliangye Group Limited (Yibin, China). T-series dextran standards were purchased from shodex (Tokyo, Japan), and the monosaccharide standards were purchased from Aladdin Holdings Group Co, Ltd. (Beijing, China). Acetic acid, propionic acid, butyric acid, and valeric acid were purchased from Sigma-Aldrich Chemical Co. (St. Louis, MO, USA). All the other chemicals and reagents used were of analytical grade, and were purchased from Sinopharm Chemical Reagent Co., Ltd. (Shanghai, China).

2.2. Extraction and Purification of Crude HSP80 (cHSP80)

After 15 min of centrifugation at 8000 rpm, the solid residue in HS was removed. Subsequently, it was deproteinized by treating with the Sevage reagent ($CHCl_3/BuOH = 4:1$, v/v), and the supernatant was obtained by centrifugation (6000 rpm for 20 min). This process was repeated until all free protein was thoroughly removed. As shown in the schematic diagram of the processing process in Figure 1a, a given volume of ethanol with a purity of 100% (v/v) was added to the supernatant, with the final content of 40% to obtain the precipitate and crude HSP40 (cHSP40), and the supernatant underwent concentration using a vacuum rotary evaporator to remove the ethanol. Similarly, 100% (v/v) ethanol was added into the resultant supernatant, until the final concentration was 60%, to obtain another precipitate named crude HSP60 (cHSP60), and the supernatant was evaporated to remove the ethanol. Finally, the remaining supernatant was added with 100% (v/v) ethanol, until the final content was 80%, to obtain the precipitate which passed through dissolution, dialysis, and lyophilization to obtain the crude HSP80 (cHSP80).

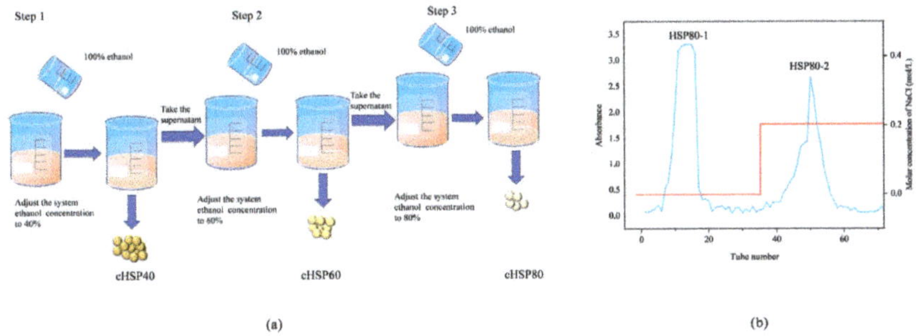

Figure 1. (**a**) The schematic diagram of the processing process of GEP method, and (**b**) the Stepwise elution curve of cHSP80 on a DEAE-FF column, in which HSP80-1 and HSP80-2 represented the fractions eluted with water and 0.20 mol/L of NaCl solution, respectively.

In order to obtain polysaccharides with a higher purity and lower molecular weight, which generally had biological activity, cHSP80 was chosen to be further purified in the present study. The specific operations were as follows. In 10 mL of deionized water, 150 mg of cHSP80 was dissolved. The solute was eluted by deionized water and NaCl (0.20 mol/L), respectively, at a rate of 0.2 mL/min using a DEAE-Sepharose FF (2.6 × 20 cm) ion-exchange column chromatography that had reached a state of equilibrium. The two fractions as shown Figure 1b, including HSP80-1 eluted by water and HSP80-2 eluted by 0.2 mol/L NaCl, were collected and subjected to dialysis, respectively, followed by freeze-drying to obtain pure polysaccharides. In the present study, HSP80-2 was selected for subsequent structural analysis and activity research.

2.3. *Determination of Molecular Weight of HSP80-2*

The average molecular weight of HSP80-2 was mensurated by high performance gel permeation chromatography (HPGPC, Shimadzu Corporation, Kyoto, Japan), combined with the TSK gel GMPWXL column (Tosoh Bioscience, Kyoto, Japan) [13]. An LC 20 high performance liquid chromatography pump (Shimadzu Corporation, Kyoto, Japan) equipped with a 7725i manual injector (Rheodyne L.P, Atlanta, GA, USA) was used. The elution solvent used to extract the samples consisted of 0.1% $NaNO_3$ and 0.06% NaN_3 at a rate of 0.6 mL/min. T-series dextrans standards with varying molecular weights of (6.30, 22.0, 49.4, 334.0, and 642.0 kDa) were employed for constructing the standard calibration curve to calculate the average molecular weight of HSP80-2.

2.4. *Monosaccharide Composition Analysis of HSP80-2*

The analysis of HSP80-2 monosaccharide composition had been modified based on previous methods [14]. 1.0 mg of HSP80-2 was dissolved in 3.0 mL of 2.0 mol/L trifluoroacetic acid (TFA) and hydrolyzed for 4 h at 120 °C. After hydrolysis was completed, 200 μL of polysaccharide hydrolysate was blown dry with N_2 to remove the excess TFA. Then, the desiccated hydrolysate of HSP80-2 and 100 μL of standard monosaccharides (2.0 mmol/L) were mixed with 250 μL of 0.6 mol/L aqueous NaOH and 500 μL of 0.4 mol/L PM, respectively, in methanol solution, which were heated in a water bath at 70 °C for 2 h. After cooling to an ambient temperature, the mixture was neutralized with 0.5 mol/L HCl. The obtained solution was extracted with the same volume of chloroform and water and the sample was subjected to filtration using a membrane with a pore size of 0.45 μm to analysis. The analytical column used was an Agilent Xtimate C_{18} (4.6 × 200 mm × 5 μm). The UV detection wavelength utilized was 250 nm. The composition of elution was acetonitrile and 0.05 mol/L sodium phosphate at a flow rate of 1.0 mL/min at 30 °C and 20 μL was injected.

2.5. Atomic Force Microscopy (AFM) Observation of HSP80-2

The 0.1 mg of the polysaccharide sample was dissolved into 10 mg of distilled water and stirred overnight to obtain polysaccharide solution. 5.0 μL of the HSP80-2 solution was dropped on mica surfaces and dried in the air. The cantilever driving frequency was specified as 70 kHz, while the scanning frequency was specified as 1 Hz. The spring constant was stated as 0.4 N/m with AFM (Bruker Dension lcon, Bruker AXS, Billerica, MA, USA).

2.6. Scanning Electron Microscope (SEM) Analysis of HSP80-2

The field-emission scanning electron microscopy (SU8020, Hitachi Int. Tokyo, Japan) was utilized to observe the scanning electron micrograph of HSP80-2. The dried HSP80-2 was stuck onto the sample table with conductive tape and sputtered gold film under the vacuum condition, observed at 500 × and 1000× magnifications.

2.7. FT-IR Spectroscopy Analyses of HSP80-2

The sample was combined with potassium bromide (KBr) and subsequently compressed into KBr sample discs. The FT-IR spectra of HSP80-2 was obtained by employing a Nicolet iS10 FT-IR spectrometer (Thermo Nicolet Corporation, Waltham, MA, USA). The infrared scanning, in range, spanning from 4000 to 500 cm^{-1} and the resolution was 4 cm^{-1}.

2.8. In Vitro Fermentation Evaluation of HSP80-2

The *in vitro* fermentation was performed according to the previously described methods with minor modifications [15]. The basal nutrient was prepared as described in the literature, and autoclaved at 121 °C for 20 min. Then, the fresh human feces were provided by four healthy young volunteers, including two males and two females aged from 20 to 27, without digestive diseases or probiotic or antibiotic intake within the last three months. The feces samples were diluted using normal saline to prepare the solid–liquid mixture (10%, *w*/*v*). The suspension was homogenized and centrifugated (500 rpm) at 4 °C for 5 min to gain the human fecal inoculum. In total, 3.0 mL of feces homogenate was mixed with 27.0 mL of a basic culture medium, containing 300.0 mg HSP80-2 or INL or without any other carbon source (BLK). Then, all groups were anaerobically fermented at 37 °C. Finally, four different fermentation periods (4, 8, 12, and 24 h) of the test samples were collected and stored at a temperature of −80 °C in order to facilitate further analysis.

2.9. Determination of SCFAs in the Fecal Fermentation Model of HSP80-2

In brief, sample solutions were subjected to centrifugation to remove impurities at 8000 rpm for 15 min. Subsequently, 400 μL of supernatant was combined with 20 μL of 2-ethylbutyric acid. The mixture was added with 100 μL of 50% H_2SO_4 to acidify, and shaken thoroughly. Next, 1000 μL of 2-ethylbutyric acid was added for extraction, and centrifugated at 500 rpm for 10 min at 4 °C. 1.0 μL of the sample supernatants was injected for analysis with the Agilent 7890 series GC system (Agilent Technologies, Palo Alto, CA, USA) with flame ionization detection. The carrier gas was N_2, and the initial temperature was 100 °C, and rising to 180 °C at a rate of 5 °C/min. The temperatures of injection and the detector were 200 °C and 250 °C [16], respectively.

2.10. Analysis of Gut Microbiota during In Vitro Fermentation

After 24 h of fermentation, total genomic DNA was extracted from three groups, including the BLK group without the carbon source, the INL group with the addition of inulin, and the HSP80-2 group with the addition of HSP80-2 immediately using the E.Z.N.A.® Soil DNA Kit (Omega Bio-tek, Norcross, GA, USA) according to the instructions. The quality and concentration of DNA were determined by 1.0% agarose gel electrophoresis and a NanoDrop2000 spectrophotometer (Thermo Scientific, Waltham, MA, USA) and kept at −80 °C prior to further use. The hypervariable region, V3-V4, of the bacterial 16S rRNA gene were amplified with primer pairs 338F (5′-ACTCCTACGGGAGGCAGCAG-3′) and 806R (5′-GGACTACHVGGGTWTCTAAT-3′) by a T100 Thermal Cycler PCR thermocycler (BIO-RAD, California city, CA, USA). Purified amplicons were pooled in equimolar amounts and paired-end sequenced on an Illumina PE300/PE250 platform (Illumina, San Diego, CA, USA) according to the standard protocols by Majorbio Bio-Pharm Technology Co. Ltd. (Shanghai, China). Raw FASTQ 16S rRNA gene sequences were merged by FLASH version 1.2.7. The optimized sequences were clustered into operational taxonomic units (OTUs) using UPARSE 7.1 with 97% sequence similarity level. The metagenomic function was predicted by PICRUSt2 (Phylogenetic Investigation of Communities by Reconstruction of Unobserved States) based on OTU representative sequences. Based on the OTUs information, alpha diversity indices, including Chao1 richness, Shannon index, Ace and Simpson were calculated with Mothur v1.30.1. The similarity among the microbial communities in different samples was determined by principal coordinate analysis (PCoA) based on the Bray–Curtis dissimilarity using Vegan v2.5-3 package. The distance-based redundancy analysis (db-RDA) was performed using Vegan v2.5-3 package.

2.11. Statistical Analyses

The data were given as means ± SD (n = 3), and were analyzed using a one-way ANOVA followed by the Duncan's test. If $p < 0.05$, the difference was considered statistically significant. SPSS for Windows, Version 17.0 (SPSS Inc., Chicago, IL, USA) was used for all statistical analyses.

3. Results and Discussion

3.1. Determination of Mw and Monosaccharide Compositions of HSP80-2

Natural polysaccharides possess a diverse array of biological functions that are linked to their chemical composition and structure, particularly their average molecular weights [17]. As presented in Figure 2a, the HPGPC spectrum demonstrated that HSP80-2 had a solitary symmetrical peak, and the average molecular weight was calculated at 1.38×10^4 Da with the standard curve of $y = -0.66x + 14.38$ ($R^2 = 0.9920$). Compared with another HS polysaccharide, HSP-3, obtained by 80% ethanol precipitation and eluted with 0.20 mol/L NaCl solution by DEAE-FF column chromatography [18], the average molecular weight of HSP80-2 was significantly smaller than that of HSP-3 (2.64×10^4 Da), revealing the ethanol as the extraction agent had an impact on the distribution of the average molecular weight of polysaccharides, and HSP80-2 prepared by GEP method might exhibit stronger biological activity due to its lower molecular weight [12].

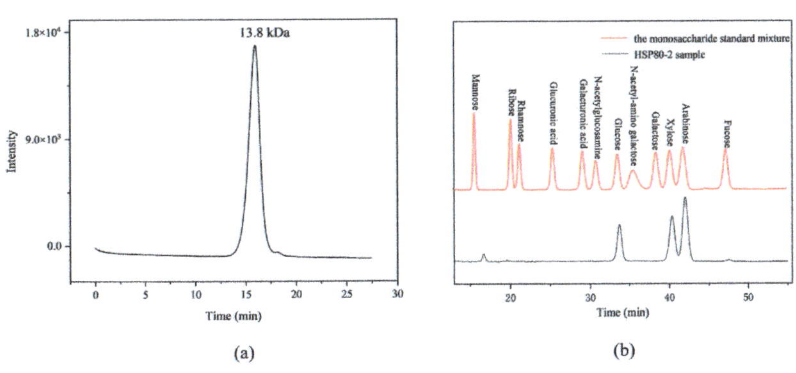

Figure 2. (**a**) HPGPC profile of HSP80-2 fractions on TSK gel GMPWXL column, and (**b**) HPLC analysis of the monosaccharide standards and HSP80-2 sample.

As shown in Figure 2b, the total ion chromatogram (TIC) of HPLC revealed that the composition of HSP80-2 primarily consisted of arabinose, xylose, and glucose with a molar ratio of 4.0:3.1:2.4., indicating that HSP80-2 was a heteropolysaccharide. According to the previous studies, HSP-3 contained rich mannose with a content of up to 46%. However, the monosaccharide composition of HSP80-2 was completely different from that of HSP-3. This evidence suggested that although the HSP80-2 and HSP-3 were both originated from the same HS sample, different HSPs had obviously distinct structures, which might be related to different extraction methods.

3.2. FT-IR Spectra of HSP80-2

FT-IR spectroscopy was commonly used to reveal typical characteristics of polysaccharide. The FT-IR spectrum of HSP80-2 was shown in Figure 3. The HSP80-2 exhibited a diverse array of stretching vibration characteristic peaks, including a hydroxyl group at around 3354 cm^{-1} and C-H at around 2923 cm^{-1} [19] which explicated that the presence of intermolecular hydrogen bonds within the molecular structure. Moreover, stretching peak at 1651 cm^{-1} and 1429 cm^{-1} belong to COO- [8], which indicated that HSP80-2 had the characteristic peak of a polysaccharide. The strong band distribution of 1100 cm^{-1} to 1000 cm^{-1} indicated the characteristic of C-O-C and C-O-H bonds, which evidenced the presence of a pyranose form of sugars [20]. The absorption peak at 812 cm^{-1} was attributed to the presence of α-pyran sugar in the HSP80-2 compound [21]. It was worth noting that HSP80-2 had a similar functional group structure to HSP-3 which both had characteristic absorption peaks of polysaccharides [20].

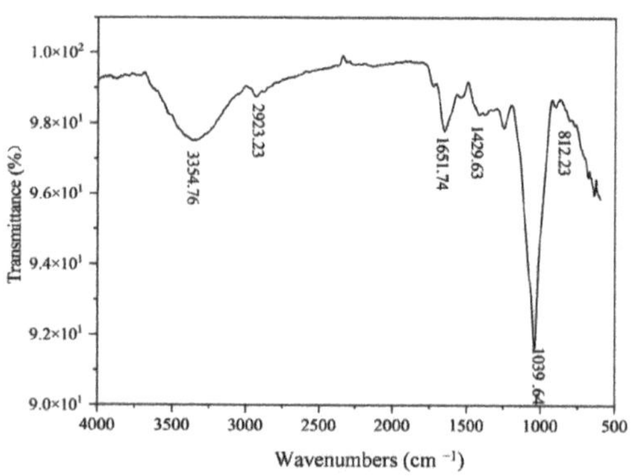

Figure 3. FT-IR spectrum analysis of HSP80-2.

3.3. Morphological Properties of HSP80-2

AFM technique can image single molecules and aggregates of polysaccharides, obtain quantitative information such as the diameter and length of single molecules, as well as morphological characteristics of molecular aggregates, and has been widely applied in the study of polysaccharides [22]. The planar images (Figure 4a,c) showed that HSP80-2 had a distinct morphology characterized by a short branch chain structure, and the 3D images (Figure 4b,d) showed the rough surfaces of HSP80-2. The molecular height of HSP80-2 was in the range of 1.09–2.98 nm, and the diameter of the dispersed particles was 66.12–107.15 nm. Interestingly, HSP-3 formed a nonlinear spherical structure due to the highly intertwined sugar chains which was totally different from the image structure characterization of HSP80-2. It suggested that the HSP80-2 had a more complex polysaccharide structure.

The molecular morphology of polysaccharides was investigated by SEM under 500× and 1000× magnifications. Figure 4e shows that the surface morphology of HSP80-2 was a flaky structure and uneven size. As shown in Figure 4f, HSP80-2 had a rather integrated and smooth surface at a high magnification. This could be owing to HSP80-2 having a low degree of branched and short strain structures [14].

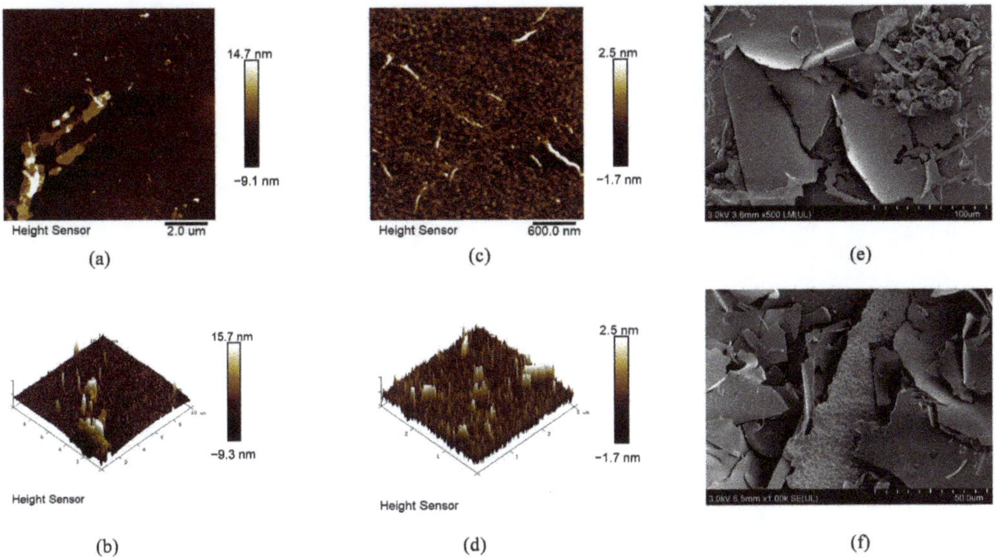

Figure 4. Molecular morphology of HSP80-2 was observed under AFM ((**a**,**c**): planar image, (**b**,**d**): cubic image; (**a**,**b**): 2.0 μm, (**c**,**d**): 600 nm), and SEM image ((**e**) 500× and (**f**) 1000×).

3.4. Changes in Gut Microbiota Population

It is widely recognized that the gut microbiota has a crucial role in both immune system functioning and energy metabolism [23]. Polysaccharides have been reported to elicit bioactive effects via the gut microbiota and its metabolites [24]. Consequently, the investigation of the correlation between polysaccharides and gut microbiota has promise for illness prevention and health promotion.

In the present study, a high-throughput sequencing study was conducted on the samples obtained from the INL, BLK, and HSP80-2 groups following 24 h fermentation to investigate the impact of HSP80-2 on the composition of gut microbiota. Table 1 displayed the alpha diversity metrics of the community, including the community richness (Chao1 and ACE) and the distributional diversity of the community (Simpson and Shannon). The study demonstrated that the BLK group had a higher functional role in preserving community abundance and diversity compared to the HSP80-2 and INL groups, the same results were obtained in the polysaccharides extracted from *Paecilomyces cicadae* TJJ1213 in the *in vitro* fermentation [25]. The reason might be that the intestinal microorganisms use inulin and polysaccharide to produce SCFAs, which could inhibit the growth of harmful bacterial communities and decrease the abundance of intestinal microorganisms [26].

Table 1. Alpha diversity of samples among different groups.

Groups	Index			
	Ace	Shannon	Chao 1	Simpson
BLK	313.65 ± 14.44 [a]	2.32 ± 0.13 [a]	312.23 ± 22.93 [a]	0.20 ± 0.01 [b]
INL	282.2 ± 44.49 [ab]	2.14 ± 0.06 [a]	270.03 ± 37.24 [ab]	0.27 ± 0.01 [a]
HSP80-2	234.76 ± 14.21 [b]	2.33 ± 0.13 [a]	236.93 ± 21.39 [b]	0.20 ± 0.03 [b]

[a, b] Significant differences ($p < 0.05$) are expressed with superscript letters within a row for each parameter ($n = 3$).

As shown in Figure 5a, principal component 1 (PC1) accounted for 52.97% of the variance observed among the three groups, and the principal component 2 (PC2) accounted for 45.12% of the variation, which suggested that the microbiota shifted in both HSP80-2 and INL groups comparing to the BLK group. The obtained results showed that the gut microbiota composition and structure of HSP80-2 and INL were completely different from that of BLK group, and further demonstrated that HSP80-2 and INL could change the composition of gut microbiota during a simulated colonic fermentation.

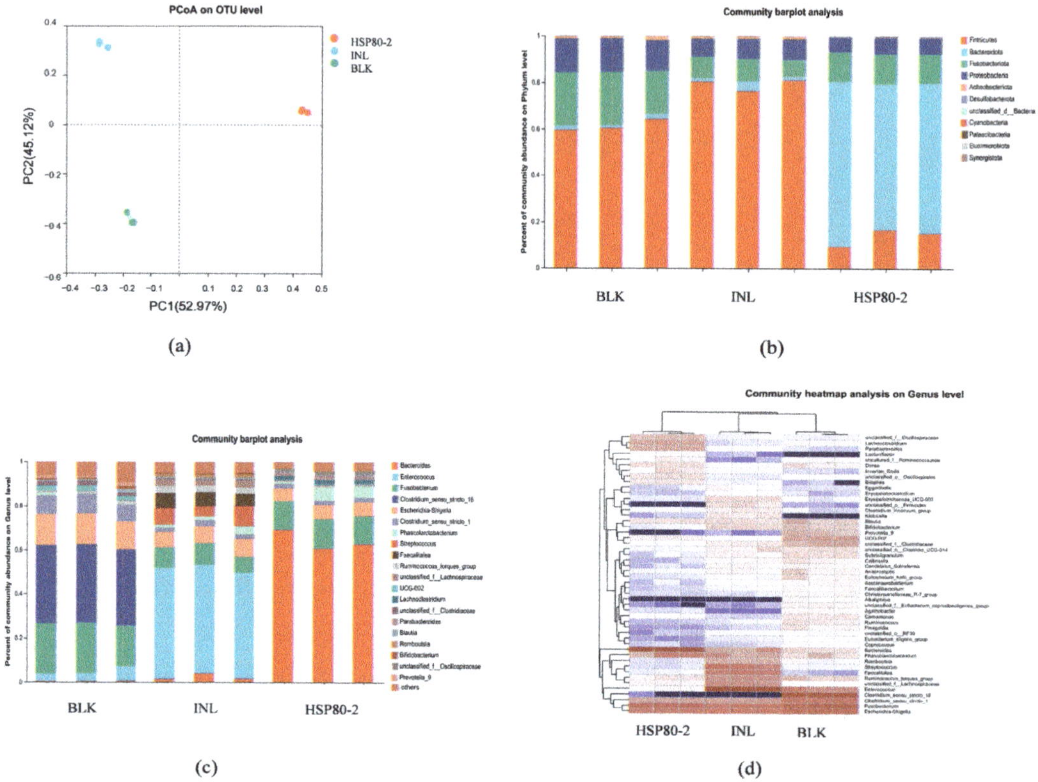

Figure 5. Principal component analysis (**a**) of gut microbiota. The relative abundance of bacterial community at the phylum level (**b**) and the genus level (**c**), and the heatmap analysis of the relative abundance of bacterial community at the genus level (**d**).

The relative bacterial community abundances of three groups at the phylum level were shown in Figure 5b, and the results indicated the bacterial community of INL, BLK, and HSP80-2 was mainly containing *Firmicutes*, *Bacteroidetes*, *Fusobacteria* and *Proteobacteria*. After 24 h of fermentation, the microbiota composition of HSP80-2 obviously differed from that in the BLK group and INL group. The HSP80-2 group exhibited a notable elevation in the abundance of *Bacteroidetes*, while the level of *Firmicutes* decreased compared with the BLK and INL groups. Previous study showed that *Bacteroidetes* could utilize more types of polysaccharides than *Firmicutes*, as *Bacteroidetes* can encode more polysaccharide degrading enzymes [27]. *Bacteroides* was known to inhabit the distal colon and possess the ability to enzymatically break down indigestible polysaccharides, resulting in the production of SCFAs. This capability was facilitated by the presence of many carbohydrate enzymes, including glycosidases and polysaccharide lyases [28]. In addition, the relative abundance of *Proteobacteria* in the HSP80-2 and INL groups were significantly lower compared with that of the BLK group. As reported, *Proteobacterias* was the largest phylum of bacteria and includes many pathogens, which were recognized as a significant factor in the development of intestinal microecological problems, leading to potential nutritional and metabolic disturbances in the host. Additionally, several bacterial genera within this phylum have been associated with immune-related disorders [29]. These results indicated that HSP80-2 could regulate the structure of bacterial communities to promote health of host.

In addition, the genus classification information was represented in Figure 5c. The BLK group was mainly composed of *Fusobacterium*, *Clostridium_sensu_stricto_18*, *Flavonifractor*, and *Escherichia-Shigella*, and clearly, the probiotics almost disappeared. In contrast to the BLK group and INL, the HSP80-2 group exhibited a higher abundance of *Bacteroides*, which was one of the most abundant human intestinal microbiotas and played an important role in the development and maintenance of healthy intestinal microbiota. *Bacteroides* were increasingly used as a model organism to study the physiology and function of intestinal symbiotic microorganisms, such as polysaccharide utilization, pathogenicity, bile acid metabolism, and the ecology and evolution of intestinal microbiota. Furthermore, the HSP80-2 group demonstrated elevated quantities of *Phascolarctobacterium* and *Lachnoclostridium* compared to the BLK group, similar results were also observed in polysaccharides isolated from *agaricus bisporus* [30]. The presence of a higher quantity of *Phascolarctobacterium* in the gastrointestinal microbiota of adult individuals had been found to promote weight reduction. Additionally, the presence of Dialister genes that encode enzymes involved in carbohydrate metabolism among gut microorganisms also play a role in facilitating weight loss [31]. The results were also observed in the INL group with increased *Phascolarctobacterium* when compared with the BLK group. Meanwhile, a reduction in the abundance of *Fusobacterium* was noted in both the INL and the HSP80-2 compared to the BLK group. Previous research had demonstrated a correlation between the gut microbiome and the development of colorectal cancer (CRC). Specifically, an excessive presence of *Fusobacterium* in the colon had consistently emerged as a reliable indicator in this context [32]. In summary, the HSP80-2 could regulate gut microbiota by increasing beneficial microbiota and reducing the abundance of harmful microbiota. Meanwhile, Figure 5d displayed a heat map that illustrated the relative abundance of the top 50 bacterial genera in the BLK, HSP80-2, and INL groups. In the HSP80-2 group, there was an observed rise in the abundance of profitable bacterial genera, including *Lachnoclostridium*, *Parabacteroides*, *Phascolarctobacterium*, and *Bacteroides* compared to the BLK groups. This result further confirmed that HSP80-2 could increase the abundance of beneficial microbiota, suggesting HSP80-2 displayed the potential to support and preserve intestinal health.

3.5. Prediction of HSP80-2 on the Metabolic Pathways of the Gut Microbiota by KEGG Analysis

The PICRUSt2 is used to predict the metagenomic pathways based on the Kyoto Encyclopedia of Genes and Genomes (KEGG) database. In the present study, a total of 44 KEGG pathways on level 2 were obtained, of which 41 exhibited significant differences compared to BLK (Figure S1). The level 2 KEGG pathways, with the top 14 relative abundance, were displayed at Figure 6a, which were mainly related to global and overview maps, carbohydrate metabolism, and glycan biosynthesis and metabolism. To further investigate metabolic pathways, KEGG pathway analysis on level 3 was carried out and 320 pathways were found, of which 256 pathways had significant differences between the BLK and HSP80-2 groups (Figure S2). The results indicate that carbohydrate metabolism-related pathways, including amino sugar and nucleotide sugar metabolism (Ko00520), galactose metabolism (ko00052), citrate cycle (TCA cycle) (ko00020), and other glycan degradation (ko00511) were enriched in the HSP80-2 group (Figure S2). In addition, the carbohydrates could regulate the production of amino acids by affecting the gut microbiota [33]. The metabolic pathways of most amino acids with HSP80-2 treatment were enriched (Figure S2), such as glycine, serine and threonine metabolism (ko00260), alanine, aspartate and glutamate metabolism (ko00250), lysine biosynthesis (ko00300), arginine and proline metabolism (ko00330), and valine, leucine and isoleucine biosynthesis (ko00290). These results suggested that HSP80-2 could be utilized by the gut microbiota and affect the production of amino acids. Additionally, the Level 3 KEGG pathways with the top 14 relative abundance were shown in Figure 6b, which were mainly related to metabolic pathways (ko01100), biosynthesis of secondary metabolites (ko01110), and microbial metabolism in diverse environments. Similar to previous research results [34], the pathways related to metabolic pathways (ko01100), biosynthesis of secondary metabolites (ko01110), biosynthesis of amino acids (ko01230), and amino sugar and nucleotide sugar metabolism (ko00520) were significantly increased in the HSP80-2 group compared with the BLK group, which further indicated that HSP80-2 could improve the metabolic pathway of amino acids.

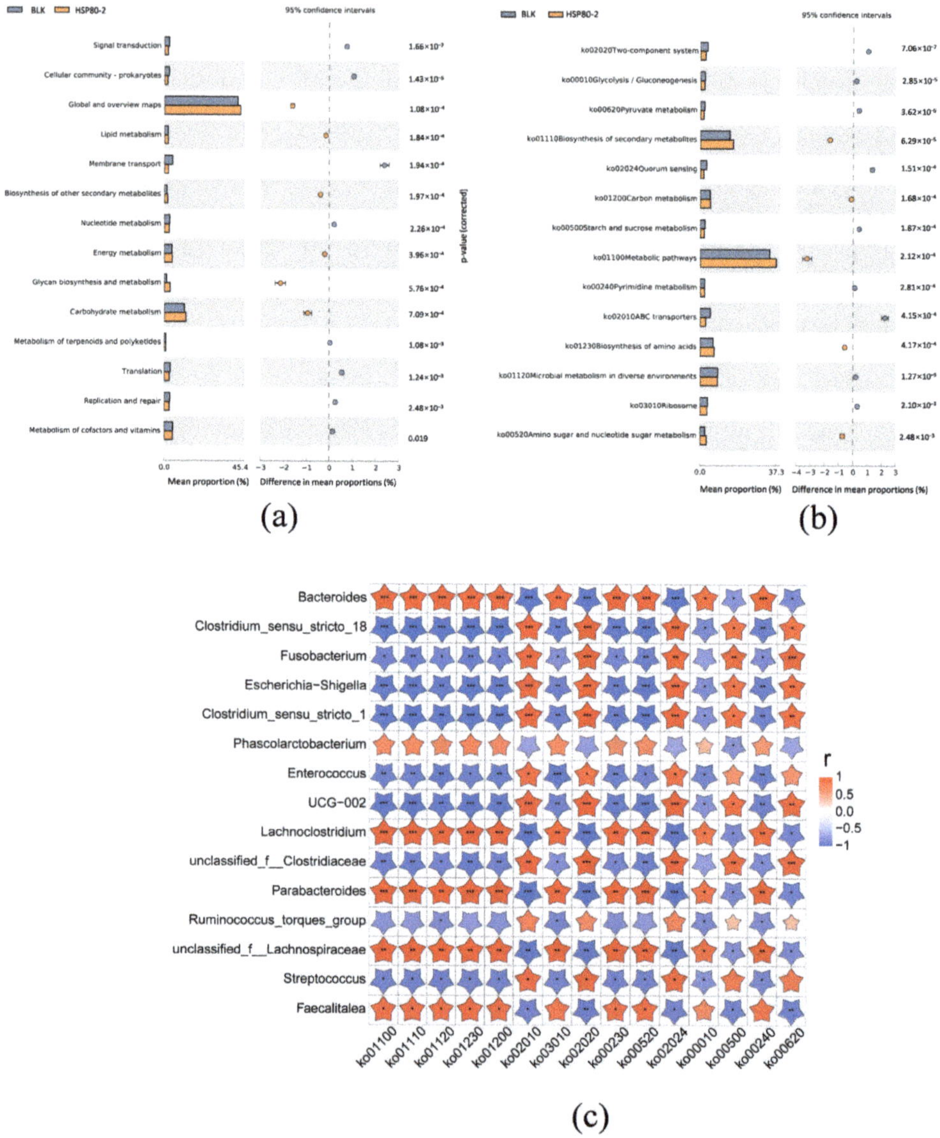

Figure 6. Difference of PICRUSt function prediction based on KEGG database after PHSP80-2 fermentation: (**a**) Relative abundance of the top 14 metabolic pathways base on KEGG categories at Level 2. (**b**) Relative abundances of the top 14 metabolic pathways base on KEGG categories at Level 3. (**c**) Heatmap showed Spearman correlations between 14 different metabolic pathways in Level 3 and 15 different genera, * $p < 0.05$, ** $p < 0.01$ and *** $p < 0.001$ denoted statistical significance between bacterial taxa and metabolic pathways.

In order to investigate the relationship between the main pathways and the gut microbiota, the Spearman correlation heatmap in Figure 6c showed the correlation between 14 metabolic pathways (the Level 3 KEGG pathways with the top 14 relative abundance) and 15 genera (the genera with top 15 relative abundance). The results indicate that *Bacteroides*, *Lachnoclostridium*, and *Parabacteroides* were significantly positively correlated with multiple metabolic pathways, such as Biosynthesis of amino acids (ko01230), Carbon metabolism (ko01200), Amino sugar and nucleotide sugar metabolism (ko00520), Purine metabolism (ko00230), and Pyrimidine metabolism (ko00240). It was worth noting that *Escherichia-Shigella*, *Clostridium_sensu_stricto_1*, and *Clostridium_sensu_stricto_18* was significantly negatively correlated with the aforementioned metabolic pathways. Based on the biological functions of these genera, we speculated that *Bacteroides*, *Lachnoclostridium*, and *Parabacteroides* play an active role in these metabolic processes.

3.6. SCFAs Production during In Vitro Fermentation

SCFAs are one of the most prominent by-products of polysaccharides fermentation by intestinal microbiota, which play a significant and beneficial function in intestinal epithelial cell regulation [24]. Inulin (INL) was a widely recognized type of prebiotic that was frequently employed as a positive control in *in vitro* fermentation investigations. The SCFAs levels during different fermentation periods of HSP80-2, INL, and BLK were detected by GC technique, respectively, and the results were shown in Table 2. With the increase of fermentation time in the BLK, INL, and HSP80-2 groups, the contents of total SCFAs gradually increased. Similar results were observed in the fermentation experiments of *aloe* polysaccharide [35], *Artemisia sphaerocephala* Krasch [36], and *Siraitia grosvenorii* polysaccharide [37]. It was worth noting that HSP80-2 produced substantially more SCFAs than BLK and INL at any time point. After 24 h of fermentation, total SCFAs concentration in the HSP80-2 group grew to 47.79 $\pm$ 2.71 mmol/L, significantly higher than in the BLK (14.86 $\pm$ 1.40 mmol/L) and INL (31.48 $\pm$ 1.40 mmol/L) groups. These results indicated that HSP80-2 and INL regulated the gut microenvironment might by producing SCFAs progressively, which was consistent with the fermentation results of extracting polysaccharides from *Ziziphus Jujuba* [38].

In fact, the acetic, propionic, and n-butyric were the primary SCFAs in HSP80-2. After fermentation for 24 h, the content of acetic acid and propionic acid in the HSP80-2 group were determined to be 25.63 $\pm$ 2.91 mmol/L and 13.12 $\pm$ 2.16, respectively, which was significantly higher than those of the INL (15.37 $\pm$ 0.69 mmol/L and 3.98 $\pm$ 0.25 mmol/L) group and BLK group (8.92 $\pm$ 1.01 mmol/L and 1.38 $\pm$ 0.21 mmol/L). Notably, the levels of acetic acid and propionic acid in the HSP80-2 group were significantly higher than those in BLK and INL at all fermentation periods. Acetate produced via acetyl-CoA and Wood-Ljungdahl pathways, which was the most abundant SCFAs of all substrates. At the same time, acetic acid was an important energy source for gut microbiota and liver cells, which could be utilized in brain, heart, and to inhibit enteropathogens [39]. The liver could absorb propionic acid to inhibit cholesterol synthesis and lower serum cholesterol levels, propionic acid was also a significant molecule in satiety signaling for interactions with FFAR2/3 [40]. Meanwhile, after fermentation for 24 h, the butyric acid in the HSP80-2 group was 8.74 $\pm$ 1.51 mmol/L, which was 2.31-fold of it in BLK (3.77 $\pm$ 0.33 mmol/L) but slightly lower than that in the INL group (11.96 $\pm$ 0.85 mmol/L). The phenomenon observed in this study was similar to the report by Wen et al. in the fermentation of *Sparassis crispa* polysaccharide [41]. Butyric acid had significant implications in the control of host genes, cellular development, and cellular death. Moreover, it was considered to be a primary energy source for epithelial cells [4]. These results collectively indicated that HSP80-2 had positive effects as inulin on promoting the production of SCFAs, thus, protecting intestinal health as a potential probiotic.

In order to link changes in microbiota composition with the different fermentation conditions and different SCFAs, the redundancy analysis at the phylum level and genus level were performed. At the level of phylum as represented in Figure 7a, the content of the

SCFAs were closely related to the relative abundance of *Bacteroides*. At the level of genera as shown in Figure 7b, the production of acetic acid and propionic acid was closely related to the growth of *Bacteroides*, *Lachnoclostridium*, and *Parabacteroides* which were positively regulated by HSP80-2.

Table 2. Changes in concentrations of SCFAs produced at different fermentation time.

Sample Group	Fermentation Time (h)	Short-Chain Fatty Acids Content (mmol/L)						
		Acetic Acid	Propionic Acid	i-Butyric Acid	N-Butyric Acid	i-Valeric Acid	n-Valeric Acid	Total
BLK	0	ND	ND	ND	ND	ND	ND	ND
	4	0.78 ± 0.07^{f}	0.18 ± 0.02^{d}	0.057 ± 0.01^{c}	0.33 ± 0.03^{f}	0.036 ± 0.008^{c}	0.016 ± 0.005^{e}	1.41 ± 0.10^{f}
	8	1.42 ± 0.11^{f}	0.23 ± 0.03^{d}	0.066 ± 0.02^{c}	0.52 ± 0.02^{f}	0.046 ± 0.009^{c}	0.017 ± 0.003^{e}	2.37 ± 0.14^{f}
	12	2.34 ± 0.10^{f}	0.48 ± 0.02^{d}	$0.075 \pm 0.01^{b,c}$	0.56 ± 0.03^{f}	$0.046 \pm 0.008^{b,c}$	$0.020 \pm 0.005^{c,d}$	3.51 ± 0.07^{f}
	24	8.92 ± 1.01^{d}	1.38 ± 0.21^{c}	$0.12 \pm 0.02^{a,b}$	3.77 ± 0.33^{d}	$0.049 \pm 0.008^{b,c}$	$0.019 \pm 0.004^{c,d}$	14.86 ± 1.40^{d}
HSP80-2	0	ND	ND	ND	ND	ND	ND	ND
	4	1.62 ± 0.13^{f}	0.24 ± 0.01^{d}	0.058 ± 0.02^{c}	0.36 ± 0.16^{f}	0.032 ± 0.003^{c}	0.015 ± 0.003^{e}	2.33 ± 0.34^{f}
	8	9.61 ± 1.01^{d}	1.36 ± 0.21^{c}	$0.091 \pm 0.01^{a,b,c}$	2.30 ± 0.39^{e}	$0.047 \pm 0.007^{b,c}$	$0.026 \pm 0.003^{c,d}$	13.16 ± 1.69^{d}
	12	13.21 ± 0.78^{c}	4.64 ± 0.07^{c}	$0.10 \pm 0.03^{a,b,c}$	3.50 ± 0.48^{d}	0.062 ± 0.007^{b}	0.029 ± 0.005^{c}	21.53 ± 1.05^{c}
	24	25.63 ± 2.91^{a}	13.12 ± 2.16^{a}	0.13 ± 0.07^{a}	8.74 ± 1.51^{d}	0.081 ± 0.007^{a}	0.075 ± 0.006^{a}	47.79 ± 2.71^{a}
INL	0	ND	ND	ND	ND	ND	ND	ND
	4	0.9 ± 0.43^{f}	0.21 ± 0.05^{d}	0.058 ± 0.003^{c}	0.35 ± 0.03^{f}	0.0408 ± 0.006^{c}	$0.019 \pm 0.005^{c,d}$	1.56 ± 0.07^{f}
	8	4.52 ± 0.08^{e}	0.89 ± 0.08^{d}	0.059 ± 0.01^{c}	1.05 ± 0.03^{c}	$0.046 \pm 0.009^{b,c}$	$0.019 \pm 0.001^{c,d}$	6.58 ± 0.10^{e}
	12	8.23 ± 0.31^{d}	1.01 ± 0.04^{c}	0.064 ± 0.005^{c}	4.65 ± 0.06^{b}	$0.050 \pm 0.004^{b,c}$	$0.025 \pm 0.001^{c,d}$	14.05 ± 0.40^{d}
	24	15.37 ± 0.69^{b}	3.98 ± 0.25^{b}	0.067 ± 0.01^{c}	11.96 ± 0.85^{a}	$0.048 \pm 0.011^{b,c}$	0.026 ± 0.02^{b}	31.48 ± 1.75^{b}

$^{a–f}$ Significant differences ($p < 0.05$) are expressed with superscript letters within a row for each parameter. ($n = 3$).
ND: Not detected.

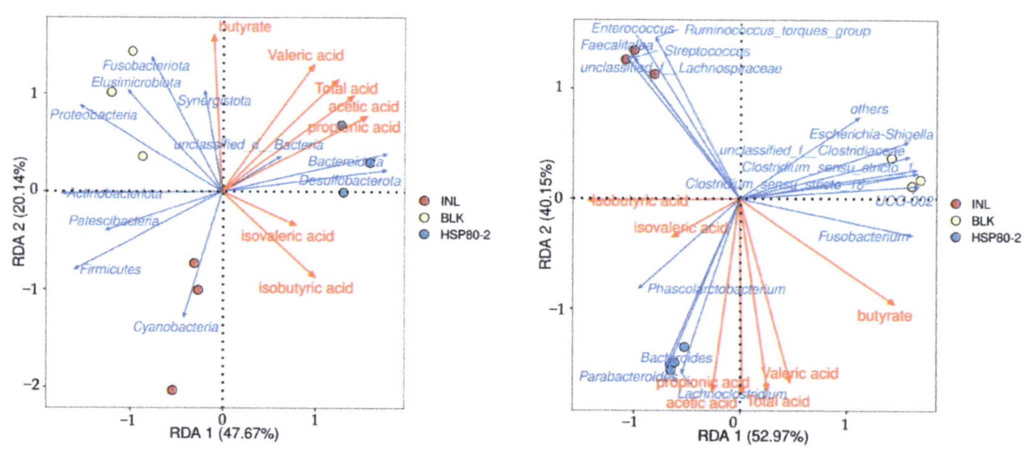

(a) (b)

Figure 7. The correlation analysis between SCFAs and microbial diversity at the phylum (**a**) and genus (**b**) level.

4. Conclusions

In conclusion, this study characterized a new polysaccharide from HS by the GEP method, and investigated its prebiotic activity by the *in vitro* fermentation model, which would provide a valuable reference for increasing the added value of HS, the by-product of Chinese Baijiu. Specifically, a novel heteropolysaccharide, HSP80-2, with an average molecular weight of 13.8 kDa was successfully extracted, which was mainly composed of arabinose, xylose, and glucose with a molar ratio of 4.0: 3.1: 2.4. In addition, HSP80-2 had a smooth surface fragmented structure and a short branch chain structure.

Furthermore, HSP80-2 exhibited prebiotic activity, which was attributed to the remarkable increasing contents of SCFAs, including acetic acid, butyric acid, and propionic acid, and the significant enhancement on the abundance of beneficial microbiota such as *Bacteroides* and *Phascolarctobacterium* after 24 h of fermentation in vitro. Also, it was worth noting that the production of acetic acid and propionic acid was closely related to the growth of *Lachnoclostridium* and *Parabacteroides*. Furthermore, the biosynthesis of amino acids was positively correlated with *Bacteroides*, *Lachnoclostridium*, and *Parabacteroides*, which were closely related to the treatment of HSP80-2. Although positive results had been achieved, the changes in HSP80-2 during *in vitro* fermentation and other products after fermentation were still worthy of further investigation. Meanwhile, as the result of the *in vitro* fermentation model cannot fully represent the interaction between HSP80-2 and gut microbiota, further research is needed to verify.

Supplementary Materials: The following supporting information can be downloaded at: https://www.mdpi.com/article/10.3390/foods12244406/s1, Figure S1: 41 significantly different KEEG pathways on level 2; Figure S2: 256 significantly different KEEG pathways on level 3.

Author Contributions: M.L.: Conceptualization, Writing—original draft. J.S.: Writing—review & editing, J.W.: Conceptualization, Writing—review & editing. D.Z.: Writing—review & editing, M.H.: Writing—review & edition, Y.L.: Writing—review & editing, J.Z.: Writing—review & editing and H.L.: Writing—review & editing. All authors have read and agreed to the published version of the manuscript.

Funding: This work was supported by the National Natural Science Foundation of China (31901815 and 31701567).

Data Availability Statement: The data presented in this study are available on request from the corresponding author.

Conflicts of Interest: The authors declare no conflict of interest.

References

1. Chen, C.; Huang, Q.; Fu, X.; Liu, R.H. *In vitro* fermentation of mulberry fruit polysaccharides by human fecal inocula and impact on microbiota. *Food Funct.* **2016**, *7*, 4637–4643. [CrossRef]
2. Wei, S.; Wang, C.; Zhang, Q.; Yang, H.; Deehan, E.C.; Zong, X.; Wang, Y.; Jin, M. Dynamics of microbial communities during inulin fermentation associated with the temporal response in SCFA production. *Carbohydr. Polym.* **2022**, *298*, 120057. [CrossRef] [PubMed]
3. Bai, Y.; Zhou, Y.; Li, X.; Zhang, R.; Huang, F.; Fan, B.; Tong, L.; Wang, F.; Zhang, M. Longan pulp polysaccharides regulate gut microbiota and metabolites to protect intestinal epithelial barrier. *Food Chem.* **2023**, *422*, 136225. [CrossRef] [PubMed]
4. Wu, D.; An, L.; Liu, W.; Hu, Y.; Wang, S.; Zou, L. *In vitro* fecal fermentation properties of polysaccharides from *Tremella fuciformis* and related modulation effects on gut microbiota. *Food Res. Int.* **2022**, *156*, 111185. [CrossRef] [PubMed]
5. Lee, B.; Huang, S.; Hou, C.; Chen, Y.; Chen, Y.; Hakkim Hazeena, S.; Hsu, W. Effect of polysaccharide derived from *dehulled adlay* on regulating gut microbiota and inhibiting Clostridioides difficile in an *in vitro* colonic fermentation model. *Food Chem.* **2023**, *410*, 135410. [CrossRef]
6. Wu, D.; Yuan, Q.; Guo, H.; Fu, Y.; Li, F.; Wang, S.; Gan, R. Dynamic changes of structural characteristics of snow chrysanthemum polysaccharides during *in vitro* digestion and fecal fermentation and related impacts on gut microbiota. *Food Res. Int.* **2021**, *141*, 109888. [CrossRef]
7. Li, Q.; Wu, W.; Chen, H.; Fang, X.; Han, Y.; Xie, M.; Gao, H. *In vitro* fecal fermentation characteristics of bamboo shoot (*Phyllostachys edulis*) polysaccharide. *Food Chem. X* **2021**, *11*, 100129. [CrossRef]
8. Huo, J.; Liao, Q.; Wu, J.; Zhao, D.; Sun, W.; An, M.; Li, Y.; Huang, M.; Sun, B. Structure elucidation and intestinal barrier protection of an alpha-D-glucan in *Huangshui*. *Int. J. Biol. Macromol.* **2022**, *223*, 595–605. [CrossRef]
9. Huo, J.; Li, M.; Wei, J.; Wang, Y.; Hao, W.; Sun, W.; Wu, J.; Huang, M. RNA-seq based elucidation of mechanism underlying the protective effect of Huangshui polysaccharide on intestinal barrier injury in Caco-2 cells. *Food Res. Int.* **2022**, *162*, 112175. [CrossRef]
10. Huo, J.; Pei, W.; Liu, G.; Sun, W.; Wu, J.; Huang, M.; Lu, W.; Sun, J.; Sun, B. *Huangshui* Polysaccharide Exerts Intestinal Barrier Protective Effects through the TLR4/MyD88/NF-κB and MAPK Signaling Pathways in Caco-2 Cells. *Foods* **2023**, *12*, 450. [CrossRef]

11. Huo, J.; Wu, J.; Huang, M.; Zhao, M.; Sun, W.; Sun, X.; Zheng, F. Structural characterization and immuno-stimulating activities of a novel polysaccharide from Huangshui, a byproduct of Chinese Baijiu. *Food Res. Int.* **2020**, *136*, 109493. [CrossRef] [PubMed]
12. Huo, J.; Wu, Z.; Zhao, H.; Sun, W.; Wu, J.; Huang, M.; Zhang, J.; Wang, Z.; Sun, B. Structure-activity relationship of antioxidant polysaccharides from *Huangshui* based on the HPLC fingerprint combined with chemometrics methods. *LWT* **2022**, *159*, 113201. [CrossRef]
13. Gao, J.; Lin, L.; Sun, B.; Zhao, M. Comparison Study on Polysaccharide Fractions from *Laminaria japonica*: Structural Characterization and Bile Acid Binding Capacity. *J. Agric. Food. Chem.* **2017**, *65*, 9790–9798. [CrossRef] [PubMed]
14. He, L.; Yan, B.; Yao, C.; Chen, X.; Li, L.; Wu, Y.; Song, Z.; Song, S.; Zhang, Z.; Luo, P. Oligosaccharides from *Polygonatum Cyrtonema* Hua: Structural characterization and treatment of LPS-induced peritonitis in mice. *Carbohydr. Polym.* **2021**, *255*, 117392. [CrossRef] [PubMed]
15. Gannasin, S.P.; Mustafa, S.; Adzahan, N.M.; Muhammad, K. *In vitro* prebiotic activities of tamarillo (*Solanum betaceum* Cav.) hydrocolloids. *J. Funct. Foods* **2015**, *19*, 10–19. [CrossRef]
16. Mao, Y.; Song, A.; Li, L.; Yang, Y.; Yao, Z.; Wu, J. A high-molecular weight exopolysaccharide from the *Cs-HK1 fungus*: Ultrasonic degradation, characterization and in vitro fecal fermentation. *Carbohydr. Polym.* **2020**, *246*, 116636. [CrossRef] [PubMed]
17. Long, X.; Hu, X.; Xiang, H.; Chen, S.; Li, L.; Qi, B.; Li, C.; Liu, S.; Yang, X. Structural characterization and hypolipidemic activity of *Gracilaria lemaneiformis* polysaccharide and its degradation products. *Food Chem. X* **2022**, *14*, 100314. [CrossRef]
18. Huo, J.; Wu, J.; Zhao, M.; Sun, W.; Sun, J.; Li, H.; Huang, M. Immunomodulatory activity of a novel polysaccharide extracted from *Huangshui on* THP-1 cells through NO production and increased IL-6 and TNF-α expression. *Food Chem.* **2020**, *330*, 127257. [CrossRef]
19. Yuan, Q.; Zhang, J.; Xiao, C.; Harqin, C.; Ma, M.; Long, T.; Li, Z.; Yang, Y.; Liu, J.; Zhao, L. Structural characterization of a low-molecular-weight polysaccharide from *Angelica pubescens Maxim.* f. *biserrata* Shan et Yuan root and evaluation of its antioxidant activity. *Carbohydr. Polym.* **2020**, *236*, 116047.
20. Wang, B.; Cao, J.; Zhang, B.; Chen, H. Structural characterization, physicochemical properties and α-glucosidase inhibitory activity of polysaccharide from the fruits of wax apple. *Carbohydr. Polym.* **2019**, *211*, 227–236. [CrossRef]
21. Cao, J.; Lv, Q.; Zhang, B.; Chen, H. Structural characterization and hepatoprotective activities of polysaccharides from the leaves of *Toona sinensis* (A. Juss) Roem. *Carbohydr. Polym.* **2019**, *212*, 89–101. [CrossRef] [PubMed]
22. Zhang, W.; Fan, X.; Gu, X.; Gong, S.; Wu, J.; Wang, Z.; Wang, Q.; Wang, S. Emulsifying properties of pectic polysaccharides obtained by sequential extraction from black tomato pomace. *Food Hydrocoll.* **2020**, *100*, 105454. [CrossRef]
23. Ai, J.; Yang, Z.; Liu, J.; Schols, H.A.; Battino, M.; Bao, B.; Tian, L.; Bai, W. Structural Characterization and *In Vitro* Fermentation Characteristics of Enzymatically Extracted Black Mulberry Polysaccharides. *J. Agric. Food. Chem.* **2022**, *70*, 3654–3665. [CrossRef] [PubMed]
24. Lu, X.; Xu, H.; Fang, F.; Liu, J.; Wu, K.; Zhang, Y.; Wu, J.; Gao, J. In vitro effects of two polysaccharide fractions from *Laminaria japonica* on gut microbiota and metabolome. *Food Funct.* **2023**, *14*, 3379–3390. [CrossRef] [PubMed]
25. Sun, Y.; Cui, X.; Duan, M.; Ai, C.; Song, S.; Chen, X. *In vitro* fermentation of κ-carrageenan oligosaccharides by human gut microbiota and its inflammatory effect on HT29 cells. *J. Funct. Foods* **2019**, *59*, 80–91. [CrossRef]
26. Tian, J.; Wang, X.; Zhang, X.; Chen, X.; Rui, X.; Zhang, Q.; Dong, M.; Li, W. Simulated digestion and fecal fermentation behaviors of exopolysaccharides from *Paecilomyces cicadae* TJJ1213 and its effects on human gut microbiota. *Int. J. Biol. Macromol.* **2021**, *188*, 833–843. [CrossRef]
27. Lapebie, P.; Lombard, V.; Drula, E.; Terrapon, N.; Henrissat, B. Bacteroidetes use thousands of enzyme combinations to break down glycans. *Nat. Commun.* **2019**, *10*, 2043. [CrossRef] [PubMed]
28. Liu, Y.; Li, Y.; Ke, Y.; Li, C.; Zhang, Z.; Wu, Y.; Hu, B.; Liu, A.; Luo, Q.; Wu, W. *In vitro* saliva-gastrointestinal digestion and fecal fermentation of *Oudemansiella radicata* polysaccharides reveal its digestion profile and effect on the modulation of the gut microbiota. *Carbohydr. Polym.* **2021**, *251*, 117041. [CrossRef]
29. Hrncirova, L.; Machova, V.; Trckova, E.; Krejsek, J.; Hrncir, T. Food Preservatives Induce *Proteobacteria* Dysbiosis in Human-Microbiota Associated Nod2-Deficient Mice. *Microorganisms* **2019**, *7*, 383. [CrossRef]
30. Fu, C.; Ye, K.; Ma, S.; Du, H.; Chen, S.; Liu, D.; Ma, G.; Xiao, H. Simulated gastrointestinal digestion and gut microbiota fermentation of polysaccharides from *Agaricus bisporus. Food Chem.* **2023**, *418*, 135849. [CrossRef]
31. Liu, C.; Guo, Y.; Cheng, Y.; Qian, H. A colon-targeted delivery system of torularhodin encapsulated in electrospinning microspheres, and its co-metabolic regulation mechanism of gut microbiota. *Food Hydrocoll.* **2023**, *135*, 108189. [CrossRef]
32. Tran, H.; Thu, T.; Nguyen, P.H.; Vo, C.N.; Doan, K.V.; Nguyen, N.M.C.; Nguyen, N.T.; Ta, V.; Vu, K.A.; Hua, T.D.; et al. Tumour microbiomes and *Fusobacterium* genomics in Vietnamese colorectal cancer patients. *NPJ Biofilms Microbomes* **2022**, *8*, 87. [CrossRef]
33. Zhang, D.; Liu, J.; Cheng, H.; Wang, H.; Tan, Y.; Feng, W.; Peng, C. Interactions between polysaccharides and gut microbiota: A metabolomic and microbial review. *Food Res. Int.* **2022**, *160*, 111653. [CrossRef]
34. Liu, T.; Zhao, M.; Zhang, Y.; Wang, Z.; Yuan, B.; Zhao, C.; Wang, M. Integrated microbiota and metabolite profiling analysis of prebiotic characteristics of *Phellinus linteus* polysaccharide *in vitro* fermentation. *Int. J. Biol. Macromol.* **2023**, *242*, 124854. [CrossRef]
35. Liu, C.; Du, P.; Cheng, Y.; Guo, Y.; Hu, B.; Yao, W.; Zhu, X.; Qian, H. Study on fecal fermentation characteristics of *aloe* polysaccharides *in vitro* and their predictive modeling. *Carbohydr. Polym.* **2021**, *256*, 117571. [CrossRef]

36. Li, J.; Pang, B.; Yan, X.; Shang, X.; Hu, X.; Shi, J. Prebiotic properties of different polysaccharide fractions from *Artemisia sphaerocephala* Krasch seeds evaluated by simulated digestion and *in vitro* fermentation by human fecal microbiota. *Int. J. Biol. Macromol.* **2020**, *162*, 414–424. [CrossRef] [PubMed]
37. Guo, Y.; Chen, X.; Gong, P.; Wang, M.; Yao, W.; Yang, W.; Chen, F. *In vitro* digestion and fecal fermentation of *Siraitia grosvenorii* polysaccharide and its impact on human gut microbiota. *Food Funct.* **2022**, *13*, 9443–9458. [CrossRef] [PubMed]
38. Han, X.; Zhou, Q.; Gao, Z.; Lin, X.; Zhou, K.; Cheng, X.; Chitrakar, B.; Chen, H.; Zhao, W. *In vitro* digestion and fecal fermentation behaviors of polysaccharides from *Ziziphus Jujuba* cv. *Pozao* and its interaction with human gut microbiota. *Food Res. Int.* **2022**, *162*, 112022.
39. Zhang, X.; Liu, Y.; Chen, X.; Aweya, J.J.; Cheong, K. Catabolism of *Saccharina japonica* polysaccharides and oligosaccharides by human fecal microbiota. *LWT* **2020**, *130*, 109635. [CrossRef]
40. Yu, C.; Ahmadi, S.; Shen, S.; Wu, D.; Xiao, H.; Ding, T.; Liu, D.; Ye, X.; Chen, S. Structure and fermentation characteristics of five polysaccharides sequentially extracted from sugar beet pulp by different methods. *Food Hydrocoll.* **2022**, *126*, 107462. [CrossRef]
41. Zhang, W.; Hu, B.; Liu, C.; Hua, H.; Guo, Y.; Cheng, Y.; Yao, W.; Qian, H. Comprehensive analysis of *Sparassis crispa* polysaccharide characteristics during the *in vitro* digestion and fermentation model. *Food Res. Int.* **2022**, *154*, 111005. [CrossRef] [PubMed]

MDPI AG
Grosspeteranlage 5
4052 Basel
Switzerland
Tel.: +41 61 683 77 34

Foods Editorial Office
E-mail: foods@mdpi.com
www.mdpi.com/journal/foods

www.ingramcontent.com/pod-product-compliance
Lightning Source LLC
Chambersburg PA
CBHW061521210926
51850CB00008B/1999
* 9 7 8 3 7 2 5 8 5 0 5 5 6 *